DIGITAL SIGNAL PROCESSING FOR RFID

Information and Communication Technology Series

The Wiley Series on Information and Communications Technologies (ICT) focuses on bridging the gap between advanced communication theories, practical designs, and end-user applications with respect to various next-generation wireless systems. Specifically, the ICT book series examines the practical difficulties of applying various advanced communication technologies to practical systems such as Wi-Fi, WiMax, B3G, etc. In addition to covering how things are designed in the standards, we look at why things are designed that way so as to link up theories and applications. The ICT book series also addresses application-oriented topics such as service management and creation, end-user devices, as well as coupling between end devices and infrastructures.

- Smart Data Pricing
  by Soumya Sen, Carlee Joe-Wong, Sangtae Ha, Mung Chiang

- WiFi, WiMAX and LTE Multi-hop Mesh Networks: Basic Communication Protocols and Application Areas
  by Hung-Yu Wei, Jarogniew Rykowski, Sudhir Dixit

- Networks and Services: Carrier Ethernet, PBT, MPLS-TP, and VPLS
  by Mehmet Toy

- Vehicle Safety Communications: Protocols, Security, and Privacy
  by Tao Zhang, Luca Delgrossi

- RF Circuit Design, 2nd Edition
  by Richard C. Li

- Equitable Resource Allocation: Models, Algorithms and Applications
  by Hanan Luss

- Fundamentals of Wireless Communication Engineering Technologies
  by K. Daniel Wong

- The Fabric of Mobile Services: Software Paradigms and Business Demands
  by Shoshana Loeb, Benjamin Falchuk, Thimios Panagos

- RF Circuit Design
  by Richard C. Li

- Digital Signal Processing Techniques and Applications in Radar Image Processing
  by Bu-Chin Wang

- Wireless Internet and Mobile Computing: Interoperability and Performance
  by Yu-Kwong Ricky Kwok, Vincent K.N. Lau

DIGITAL SIGNAL PROCESSING FOR RFID

Feng Zheng and Thomas Kaiser

University of Duisburg-Essen, Germany

Library of Congress Cataloging-in-Publication Data applied for.

ISBN: 9781118824313

A catalogue record for this book is available from the British Library.

Cover image: archibald1221, JacobH/Getty

Set in 10/12pt, TimesLTStd by SPi Global, Chennai, India.
Printed and bound in Singapore by Markono Print Media Pte Ltd

1   2016

to Zhiying and Anna Yuhan

Feng Zheng

to Petra and Hendrik

Thomas Kaiser

Contents

Preface

Identification is pervasive nowadays in daily life due to many complicated activities such as bank and library card reading, asset tracking, toll collecting, restricted access to sensitive data and procedures and target identification. This kind of task can be realized by passwords, biometric data such as fingerprints, barcode, optical character recognition, smart cards and radar. Radio frequency identification (RFID) is a technique to identify objects by using radio systems. It is a contactless, usually short distance, wireless data transmission and reception technique for identification of objects. An RFID system consists of two components: the tag (also called transponder) and the reader (also called interrogator).

Generally, signal processing is the core of a radio system. This claim also holds true for RFID. Several books are available now addressing other topics in RFID, such as the basics/fundamentals, smart antennas, security and privacy, but no book has appeared to address signal processing issues in RFID. We aim to complete this task in this book.

The book is organized as follows. Chapter 1 (Introduction) reviews some basic facts of RFID technology and gives an introduction about the scope of the book. In Chapter 2 (Fundamentals of RFID Systems), the operating principles and classification of RFID will be briefly introduced, some typical analogue circuits of RFID and their basic analysis will be addressed, channel models of RFID will be presented and RFID protocols will be briefly reviewed. In Chapter 3 (Basic Signal Processing for RFID), we will discuss some basic signal processing techniques and their applications in RFID. In Chapter 4 (RFID-oriented Modulation Schemes), we will address those modulation schemes that are suitable to RFID tags, which include binary amplitude shift keying and frequency/phase shift keying. The performance of these modulation schemes for RFID channels will be investigated. In Chapter 5 (MIMO for RFID), we examine the problems of transmit signal design and space-time coding at the tag for MIMO-RFID systems. In Chapter 6 (Blind Signal Processing for RFID), we will investigate the possibility of identifying multiple tags simultaneously from signal processing viewpoint in the PHY layer by using multiple antennas at readers and tags. In Chapter 7 (Anti-Collision of Multiple-Tag RFID Systems), we deal with the problem of identifying multiple tags from the viewpoint of networking. The basic tree-splitting and Aloha-based anti-collision algorithms for multi-tag RFID systems and their theoretical performance analysis will be examined. Some improvements for the corresponding algorithms will be discussed. Chapter 8 (Localization with RFID) is devoted to localization problems. Several localization algorithms/methods by using RFID systems will be described. In Chapter 9 (Some Future Perspectives for RFID), covert radio frequency identification by using ultra wideband and time reversal techniques, as an example of high-end RFID applications, and chipless tags, as an example of low-end RFID systems, will be presented.

This book is targeted at graduate students and high-level undergraduate students, researchers in academia and practicing engineers in the field of RFID. The book can be used as both a reference book for advanced research and a textbook for students. We try our best to make it self-contained, but some preliminary background on probability theory, matrix theory and wireless communications are helpful.

Acknowledgements

In July 2012, Professor T. Russell Hsing, a Co-Editor-in-Chief of the Wiley ICT Book Series, invited us to write a book proposal summarizing our recent research results. In the meantime, we were planning to deliver a lecture on RFID-related signal processing techniques. Therefore, the book idea for *Digital Signal Processing for RFID* came to us. Dr. Simone Taylor, Director of Editorial Development, and Diana Gialo, Senior Editorial Assistant at John Wiley, also supported this book idea. We received constant encouragement from Professor Hsing in writing and revising the detailed book proposal. Therefore, we wish to express our deep gratitude to Professor Hsing, Dr. Taylor, and Diana Gialo for their direct initiative of this book project.

We are grateful to the four anonymous reviewers for their constructive advice and comments on the initial book proposal. In particular, one reviewer suggested that we add a chapter addressing radar-embedded communications. This leads to the concept of coverting RFID, which forms the main part of Chapter 9. The reviewers also motivated us to add some sections on RFID protocols and MIMO principles. All these suggestions and comments helped improve the organization and quality of this book. In this regard, our thanks also go to Anna Smart, Acting Commissioning Editor at John Wiley & Sons, Ltd, for her coordinaton of the proposal reviewing.

We are particularly grateful to Liz Wingett, Clarissa Lim, Tiina Wigley, and Victoria Taylor, Project Editors at John Wiley & Sons, Ltd, for their superb support and coordination of the project.

The results in Chapter 5 were obtained with the support of German Research Foundation (DFG) via the project 'MIMO Backscatter-Übertragung auf Basis von Mehrantennen-Transpondern in RFID-basierten Funksystemen' (Project No. KA 1154/30-1). The support of DFG is greatly appreciated.

We are happy to acknowledge fruitful cooperation with Dr Bernd Geck and Mr Eckhard Denicke at the Leibniz University of Hannover and Dr. Kiattisak Maichalernnukul at Rangsit University in RFID-related projects. We are grateful to Professor Qing Zheng at Gannon University, Mr. Yuan Gao and Mr. Marc Hoffmann at the University of Duisburg-Essen for their carefully proofreading the book and helpful comments.

Finally, we want to thank our families Zhiying, Anna Yuhan, Petra and Hendrik for their unwavering love, support and patience. Without their spiritual support and tolerance in time, this book could not have been finished. Without their love, our expedition in this exciting field could never succeed. Therefore, we would like to dedicate this book to them.

Abbreviations

ACK	acknowledgement signal
ACMA	analytical constant modulus algorithm
AEE	average estimation error
AM	amplitude modulation
AME	average modulus error
AoA	angle of arrival
ASK	amplitude-shift keying
ASTC	Alamouti space-time coding
ATT	average transmission time
AWGN	additive white Gaussian noise
BER	bit error rate
AFSA1	adaptive frame size Aloha 1
AFSA2	adaptive frame size Aloha 2
BFSK	binary frequency-shift keying
BLF	backscatter link frequency
BPSK	binary phase-shift keying
BSP	blind signal processing
BSS	blind signal (or source) separation
C1G2	Class 1, Gen 2
CDMA	code-division multiple access
CIR	channel impulse response
CLT	Central Limit Theorem
CM	constant modulus
CMA	constant modulus algorithm
CPFSK	continuous-phase frequency-shift keying
CRB	Cramer–Rao bound
CRC	cyclic redundancy check
CROD	companion of real orthogonal design
CSI	channel state information
CSMA	carrier sense multiple accesses
CSMA/CA	carrier sense multiple access with collision avoidance
DAS	distributed antenna system
DC	direct current
DCF	distributed coordination function

DoA	direction of arrival
DR	divide ratio
DS	direct sequence
DSB	double sideband
DSTC	differential space-time coding
EIRP	equivalent isotropically radiated power
EPC	Electronic Product Code
FD	frequency-domain
FHSS	frequency hopping spread spectrum
FM	frequency modulation
FSK	frequency-shift keying
GPS	global positioning system
IC	integrated circuit
ID	identity
IDT	interdigital transducer
IoT	Internet of things
IR	impulse radio
ISI	inter-symbol interference
ISO	International Organization for Standardization
k-NN	k-nearest neighbours
LCD	least common denominator
LLS	linear least square
LMMSE	linear minimum mean square error
LNA	low-noise amplifier
LoS	line of sight
LPF	lowpass filter
LS	least square
LWLS	linear weighted least square
MAC	media access control
MIMO	multiple-transmit and multiple-receive antennas, or multiple-input multiple-output
MISO	multiple-transmit and single-receive antenna, or multiple-input single-output
ML	maximum likelihood
MLE	maximum likelihood estimation
MMSE	minimum mean square error
MUSIC	multiple signal characterization
NLoS	non line of sight
NRZ	non-return-to-zero
NSI	numbering system identifier
PA	power amplifier
PAM	pulse amplitude modulation
PPM	pulse position modulation
PC	protocol control
pdf	probability density function
PDoA	phase difference of arrival

PDP	power delay profile
PHY	physical or physical layer
PIE	pules-interval encoding
PLL	phase-locked loop
PM	phase modulation
PR	phase-reversal
PSD	power spectral density
PSK	phase-shift keying
QAM	quadrature amplitude modulation
QPSK	quadrature phase-shift keying
QT	query tree
RF	radio frequency
RFID	radio frequency identification
ROD	real orthogonal design
RSS	received signal strength
RSSE	received signal strength error
SAW	surface acoustic wave
SD	spatial-domain
SER	symbol error rate
SIMO	single-transmit and multiple-receive antennas, or single-input multiple-output
SISO	single-transmit and single-receive antenna, or single-input single-output
SNR	signal-to-noise (power) ratio
SS	spread spectrum
SSB	single sideband
STC	space-time coding
STT	signal travelling time
S-V	Saleh–Valenzuela
TDoA	time difference of arrival
TDSTT	time difference in signal travelling time
TH	time hopping
TID	tag's ID
ToA	time of arrival
TR	time reversal
TS	tree-splitting
UHF	ultrahigh frequency band
UMI	user-memory indicator
UWB	ultra wideband
VCO	voltage-controlled oscillator
WLAN	wireless local area network
WLS	weighted least square
XPC	extended protocol control

1

Introduction

1.1 What is RFID?

Identification is pervasive nowadays in daily life due to many complicated activities such as bank and library card reading, asset tracking, toll collecting, restricted accessing to sensitive data and procedures and target identification. This kind of task can be realized by passwords biometric data such as fingerprints, barcode, optical character recognition, smart card and radar. Radio frequency identification (RFID) is a technique to achieve object identification by using radio systems. It is a contactless, usually short distance, wireless data transmission and reception technique for identification of objects. An RFID system consists of two components:

- tag (also called transponder) – is a microchip that carries the identity (ID) information of the object to be identified and is located on/in the object;
- reader (also called interrogator) – is a radio frequency module containing a transmitter, receiver, magnetic coupling element (to the transponder) and control unit.

A passive RFID system works in the following way: the reader transmits radio waves to power up the tag; once the power of the tag reaches a threshold, the circuits in the tag start to work and the radio waves from the reader are modulated by the ID data inside the tag and backscattered to the reader and finally, the backscattered signals are demodulated at the reader and ID information of the tag is obtained.

RFID technology is quite similar to the well-known radar and optical barcode technologies, but an RFID system is different from radar in that backscattered signals from the tag are actively modulated in the tag (even for a passive tag or chipless tag), while backscattered signals in a radar system are often passively modulated by the scatterers of the object to be detected. An RFID system is different from an optical barcode system in that the information carrying tools are different: the RFID system uses radio waves as the tool, while the barcode system uses light or laser as the tool.

Many applications of RFID or barcode techniques are somewhat exchangeable, i.e., many ID identification tasks can be implemented by either RFID technique or barcode technique. However, optical barcode technology has the following critical drawbacks: (i) the barcode cannot be read across non-line-of-sight (NLoS) objects, (ii) each barcode needs care taken in order

Digital Signal Processing for RFID, First Edition. Feng Zheng and Thomas Kaiser.
© 2016 John Wiley & Sons, Ltd. Published 2016 by John Wiley & Sons, Ltd.

to be read and (iii) the information-carrying ability of the barcode is quite limited. RFID technology, using radio waves instead of optical waves to carry signals, naturally overcomes these drawbacks. It is believed that RFID can substitute, in the not-too-distant future, the widely used barcode technology, when the cost issue for RFID is resolved.

1.2 A Brief History of RFID

Many people date the origin of RFID back to the 1940s when radar systems became practical. In World War II, German airplanes used a specific manoeuvering pattern to establish a secret handshake between the pilot of the airplane and the radar operator in the base. Indeed, this principle is the same as that of modern RFID: to modulate the backscattering signal to inform the identity of an object. The true RFID, in the concept of modern RFID, appeared in the 1970s when Mario Cardullo patented the first transponder system and Charles Walton patented a number of inductively coupled identification schemes based on resonant frequencies. The first functional passive RFID systems with a reading range of several metres appeared in early 1970s [4]. Even though RFID has significantly advanced and experienced tremendous growth since then [1, 2], the road from concept to commercial reality has been long and difficult due to the cost of tags and readers. A major push that brought RFID technology into the mass market came from the retailer giant Wal-Mart, which announced in 2003 that it would require its top 100 suppliers to supply RFID-enabled shipments by the beginning of 2005[1]. This event triggered the inevitable movement of inventory tracking and supply chain management towards the use of RFID. Up to now, RFID applications have been numerous and far reaching. The most interesting and widely used applications include those for supply chain management, security and tracking of important objects and personnel [3, 5, 6].

Similar to other kinds of radio systems, the development of RFID has also been stimulated by necessity. Even though the progress in the design and manufacturing of antennas and microchips has smoothly driven performance improvement and cost decrease of RFID, booming development for it has not appeared until recently, since optical barcode technology has dominated the market for the last few decades. In recent years, many new technologies, such as smart antennas, ultra wideband radios, advanced signal processing, state-of-art anti-collision algorithms and so on, have been applied to RFID. In the meantime, some new requirements to object identification and new application scenarios of RFID have been emerging, such as simultaneous multiple object identification, NLoS object identification and increasing demand on data-carrying capacity of tag ID. It is this kind of application that calls for the deployment of RFID systems.

1.3 Motivation and Scope of this Book

Generally, signal processing is the core of a radio system. This claim also holds true for RFID. Several books are available now coping with other topics in RFID, such as basics, fundamentals, smart antennas, security and privacy, but no book has appeared to address signal processing issues in RFID. We aim to complete this task in this book.

The main purpose of this book is two-fold: first, it will be a textbook for both undergraduate and graduate students in electrical engineering; second, it can be used as a reference book

---

[1] see 'Wal-Mart Draws Line in the Sand' (www.rfidjournal.com/articles/view?462) and also 'Wal-Mart Expands RFID Mandate' (www.rfidjournal.com/articles/view?539).

for practice engineers and academic researchers in the RFID field. Therefore, the contents of this book include both fundamentals of RFID and the state-of-the-art research results in signal processing for RFID. For the former, we will discuss the operating principles, modulation schemes and channel models of RFID. For the latter, we will highlight the following research fields: space-time coding for RFID, blind signal processing for RFID, anti-collision of multiple RFID tags and localization with RFID. Also, due to the two-fold purpose of the book, some attention will be paid to pedagogical methods. For example, some concrete examples on the analysis of transmission efficiency of tree-splitting algorithms will be illustrated in detail before presenting general results in Chapter 7.

The book consists of the following chapters, after this one.

Chapter 2 – Fundamentals of RFID Systems. In this chapter, we will discuss the following issues: (i) operating principles of RFID, (ii) classification of RFID, (iii) analogue circuits for RFID and their basic analysis, (iv) channel models of RFID, (v) a brief review of RFID protocols and (vi) challenges in RFID. This chapter provides a basis for Chapters 3 to 9.

Chapter 3 – Basic Signal Processing for RFID. In this chapter, we will discuss some basic signal processing techniques and their applications in RFID, which include analogue/digital filtering and optimal estimation.

Chapter 4 – RFID-oriented Modulation Schemes. Since a passive RFID tag does not have an 'active' transmitter, some complicated signal modulation schemes in general communication systems cannot be applied to RFID. Instead, only very simple modulation schemes, namely, binary amplitude-shift keying and frequency/phase-shift keying, are suitable for an RFID tag. In this chapter, these modulation schemes, tailored to RFID channels, will be described. The performance of these modulation schemes for RFID channels will be investigated.

Chapter 5 – MIMO for RFID. In this chapter, we will discuss the following issues: (i) channel models of RFID systems with multiple antennas at both readers and tags (MIMO); (ii) signal design at the reader for RFID-MIMO systems (iii) space-time coding at the tag for RFID-MIMO systems and (iv) differential space-time coding at the tag for RFID-MIMO systems. Using multiple antennas in radio systems (especially in communication systems) is a general trend. Actually, employing multiple antennas has been incorporated into many existing communication standards. It is also believed that RFID systems equipped with multiple antennas will be deployed in the near future. Therefore, this chapter will be dedicated to the combination of RFID with MIMO. We will show that, by proper design, the bit-error-rate performance of the system can be greatly improved by using multiple antennas at the reader and tag.

Chapter 6 – Blind Signal Processing for RFID. In practice, one often meets the situation where several or many transponders are present in the reading zone of a single reader at the same time. Therefore, it is important to study the techniques to identify multiple tags simultaneously. In principle, two approaches can be used to do this job. The first one is to use collision avoidance techniques such as Aloha from a networking viewpoint. The second one is to use source separation techniques from a signal processing viewpoint. In this chapter, the second approach will be investigated, while Chapter 7 will be devoted to the first approach. It will be shown that, under a moderate SNR and when the number of measurements to the multiple tags in one snapshot is sufficiently high, the overlapped signals coming from the multiple tags can be separated at the reader receiver if the number of the tags is less than the number of receiving antennas at the reader.

Chapter 7 – Anti-Collision of Multiple-Tag RFID Systems. As already mentioned, there are two approaches to dealing with the multiple-tag identification problem. In this chapter, we will discuss this problem from the networking viewpoint. Basically, the traditional anti-collision algorithms in WLAN, such as tree splitting and slotted Aloha, can be applied to this problem. Since passive RFID systems are highly asymmetric, i.e., the reader is resource-rich, while tags have very limited storage and computing capabilities and are unable to hear the signal transmitted by other tags and to detect collisions, some advanced collision-avoidance algorithms in WLAN, such as carrier sense multiple access are difficult to implement in RFID tags. Therefore, basic tree-splitting and Aloha-based anti-collision algorithms for multi-tag RFID systems will be discussed in this chapter. The methods for the theoretical performance analysis of these algorithms will be addressed. It is found that the static Aloha yields very poor performance in both mean identification delay and transmission efficiency for multiple-tag RFID systems. Therefore, we propose two adaptive frame size Aloha algorithms, which have only a very light computational burden at the reader and no additional computational burden at the tag, but yield significant performance improvement.

Chapter 8 – Localization with RFID. In principle, the problem of localization with the help of RFID is similar to radar ranging problem. However, RFID ranging has its peculiar concerns. Since the distance between the reader and tag is usually short (typically of the order of less than 10 m), the round-trip signal delay is on the order of a few tens of nanoseconds. Because the available bandwidth of typical RFID signals is narrow, it is difficult to measure the time of arrival or time difference of arrival of the RFID signal. Thus baseband phase information is extremely useful for RFID localization problems. In this chapter, we will give an overview for RFID localization algorithms using various methods based on different kinds of information. To use the localization algorithms of the geometric approach, the range between readers and tags or angle of arrival (AoA) should be reliably measured or estimated from the measured information. Two approaches, namely frequency-domain phase difference of arrival (PDoA) approach and spatial-domain PDoA approach for measuring the range and AoA respectively, will be discussed. Finally, the challenging issue, that is, non-line-of-sight mitigation issue in RFID localization, will be addressed.

Chapter 9 – Some Future Perspectives for RFID. RFID systems discussed in preceding chapters belong to the middle class of RFID in the sense that IC chips are integrated inside the tags, but the power needed for signal transmission in the tags of this kind of RFID should be harvested from the reader's transmitted radio waves. This situation can be extended in two extreme ends: chipless tags and active tags. Using active tags, some advanced communication functionalities, such as covert radio frequency identification, can be realized. Using chipless tags, most tags can be printed by inkjet printers, thus greatly reducing the cost of manufacturing and packaging of tags. In this chapter, we will present a brief review for covert RFID and some chipless tags. For the first task, we need to use ultra wideband (UWB) technology and the time reversal (TR) technique. Therefore, some basics for UWB and TR will be also introduced. For the second task, two kinds of chipless tags, namely time-domain reflectometry-based chipless tags and frequency-domain spectral-signature-based chipless tags, will be discussed.

1.4 Notations

Throughout the book, we use $\mathbf{I}$ to denote an identity matrix, whose dimension is indicated by its subscript if necessary, $P_A(x)$ and $p_A(x)$ represent, respectively, the cumulative distribution function and probability density function (pdf) of a random variable A, $\mathbb{E}$ (or $\mathbb{E}_A$ if necessary) stands for the expectation of a random quantity with respect to the random variable A, $\mathbb{E}(\cdot|\cdot)$ denotes the conditional expectation, and Var(A) stands for the variance of A. The notation $\mathcal{N}(0, \sigma^2)$ stands for a Gaussian-distributed random variable with zero mean and variance σ^2. For a matrix or vector, the superscripts ^{T}, *, † denote the transpose, the element-wise conjugate (without transpose), and the Hermitian (conjugate) transpose, respectively, of the matrix or vector. The notations * and † also apply to a scalar. The symbol $\jmath$ is defined as $\jmath = \sqrt{-1}$. The function log is naturally based, if the base is not explicitly stated. We use `diag` to denote a diagonal matrix with the diagonal entries being specified by the corresponding arguments. The real part and imaginary part of a complex variable are denoted by `Re` and `Im`, respectively. We use $|\cdot|$ or $\det(\cdot)$ to denote the determinant of a matrix. Throughout the book, the symbols 0 or $\mathbf{0}$ denote scalar zero, vector zero or matrix zero with corresponding dimensions, depending on the context.

For other notations, we might use the same symbol to denote different things in different chapters or sections. If this case happens, we will explicitly explain what the symbol stands for.

References

[1] L. Boglione. RFID technology – are you ready for it? *IEEE Microwave Mag.*, 8(6):30–32, 2007.

[2] D. Dobkin and T. Wandinger. A radio-oriented introduction to RFID – protocols, tags and applications. *High Frequency Electronics*, 4(8):32–46, 2005.

[3] K. Finkenzeller. *RFID Handbook*, 3rd ed. John Wiley & Sons, Ltd, Chichester, 2010.

[4] A. R. Koelle, S. W. Depp, and R. W. Freyman. Short-range radiotelemetry for electronic identification, using modulated RF backscatter. *Proceedings of the IEEE*, 63:1260–1261, 1975.

[5] K. Michael, G. Roussos, G. Q. Huang, A. Chattopadhyay, R. Gadh, B. S. Prabhu, and P. Chu. Planetary-scale RFID services in an age of uberveillance. *Proc. IEEE*, 98:1663–1671, 2010.

[6] R. Weinstein. RFID: A technical overview and its application to the enterprise. *IT Professional*, 7(3):27–33, 2005.

2

Fundamentals of RFID Systems

2.1 Operating Principles

In this section, the basic operating principle of RFID will be discussed. Sending back the incident radio frequency (RF) power, which is modulated by the on-board information bits in a tag, is the communication principle used in passive RFID systems. The operation of an RFID involves four steps [26]: First, the reader emits electromagnetic power in the form of radio waves to the tag. Second, the antenna at the tag receives the electromagnetic power and thus charges the on-board capacitor. Third, once the energy built up in the capacitor reaches a threshold, it switches on RFID-tag circuit and then a modulated signal at the tag will be transmitted back to the reader. Finally, the returned signal is demodulated and the information bits are detected at the reader's receiver. The whole process is illustrated in Figure 2.1.

The conventional method for powering RFID tags wirelessly is to use a continuous-wave (which often has a constant envelope) power transmission from the reader, as specified by EPCglobal, Class 1, Generation 2 standard [7]. This provides a steady source of power for the tag to harvest, although very inefficiently.

There are several ways for the interaction between tag and reader for the tag to capture required energy: inductive coupling, backscattering coupling and capacitive coupling.

Inductive coupling is illustrated in Figure 2.2. In Figure 2.2, the capacitor C_r together with the coil of the reader's antenna forms a parallel resonant circuit, whose resonant frequency corresponds with the transmission frequency of the reader. The capacitor C_1 together with the coil of the tag's antenna forms another parallel resonant circuit, whose resonant frequency is tuned to the transmission frequency of the reader. Very high currents are generated at the antenna coil of the reader, which induce a voltage across the antenna coil of the tag. This voltage is rectified by the diode D_1, serving as the power supply for the data-carrying microchip of the tag.

The power captured by the coil of tag's antenna can be assessed by using transformer theory.

Both amplitude and phase of the returned signal are affected by the impendence of the tag, which can be again adjusted by a load connected to the microchip in the tag. Thus the tag's ID data information can be sent back to the reader. This kind of data modulation is called load modulation.

Digital Signal Processing for RFID, First Edition. Feng Zheng and Thomas Kaiser.
© 2016 John Wiley & Sons, Ltd. Published 2016 by John Wiley & Sons, Ltd.

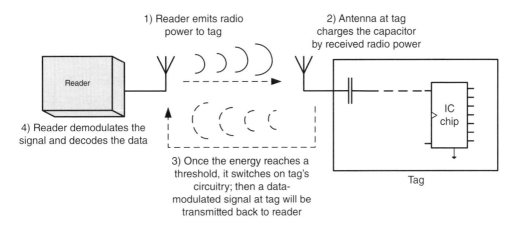

1) Reader emits radio
power to tag

2) Antenna at tag
charges the capacitor
by received radio power

Reader

IC
chip

4) Reader demodulates the
signal and decodes the data

3) Once the energy reaches a
threshold, it switches on tag's
circuitry; then a data-
modulated signal at tag will be
transmitted back to reader

Tag

Figure 2.1   An illustration of RFID principle.

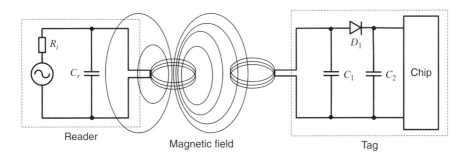

Reader

Magnetic field

Tag

Figure 2.2   An illustration of electromagnetic field coupling between reader and tag.

Inductive coupling is a near-field effect. Accordingly the distance between the coils of readers and tags must be kept within the range of the effect: normally this is about 0.15 wavelength of the frequency in use. Therefore, inductive coupling is often used on the lower RFID frequencies, that is, below 135 kHz or at 13.56 MHz.

When the frequency of the radio signal emitted by the reader is so high that the wavelength of the signal is much smaller than the gap between reader and tag, the radio signal will propagate away from the reader and the inductive coupling between the coils of reader and tag will become very weak. In this case, the power required to power up the tag's circuitry is intercepted by the tag's antenna and a part of the incident radio waves to the tag will be reflected back towards the RFID reader. This process is referred to as backscattering coupling, which is also called radiative coupling. The backscattering coupling is illustrated in Figure 2.3.

The power received by the tag's antenna via backscattering coupling can be assessed by free-space path loss law.

The way in which the signal is reflected back depends on the properties of the tag. Factors such as the cross sectional area and antenna properties and so on within the tag all can affect the strength of the reflected power. Therefore, changing factors such as adding or subtracting a load resistor across the antenna of the tag can change both amplitude and phase of the

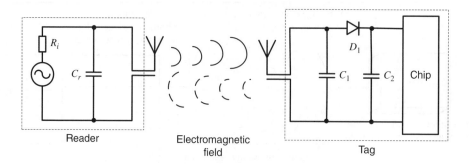

Figure 2.3   An illustration of backscattering coupling between reader and tag.

Table 2.1   Typical RFID operating frequencies and characteristics

	low frequency	high frequency	ultra high frequency	microwave
frequency	125~134 kHz	13.56 MHz	860–960 MHz	2.45/5.8 GHz
EM coupling	inductive	inductive	radiative	radiative
read range	≈ 1 m	≈ 1 m	3 m (typical)	4.5 m (typical)
			10 m (achievable)	
data rate	≈ 1 kbps	tens of kpbs	50–150 kbps	not specified

re-radiated signal. Using the tag's ID data to control the addition or subtraction of the load resistor can thus send the data information back to the reader. This kind of data modulation is called backscattering modulation.

Capacitive coupling is used for very short ranges (0.1 cm–1 cm) where a form of close coupling is needed. It uses electrodes–the plates of the capacitor, to provide the required coupling between tag and reader. Capacitive coupling operates best when items like smart cards are inserted into a reader: in this way the card is in very close proximity to the reader. Therefore, capacitive coupling is often used for smart cards. The data information in the tag is sent back to the reader by modulating the load in the tag.

In this book, our attention will be focused on the RFID with inductive coupling and backscattering coupling.

Typical operating frequencies of RFID and corresponding characteristics are summarized in Table 2.1.

2.2   Passive, Semi-Passive/Semi-Active and Active RFID

According to the power source from which the energy for RFID tags is obtained, RFID tags can be classified into passive tags (or passive RFID), semi-passive (or semi-active) tags (or semi-passive RFID) and active tags (or active RFID), as illustrated in Figure 2.4.

A passive tag has neither a battery nor a radio transmitter. The power for operating the tag chip is obtained by rectifying RF energy intercepted by the tag antenna. A semi-passive tag is equipped with a battery to provide power for the tag chip, but still uses the power captured

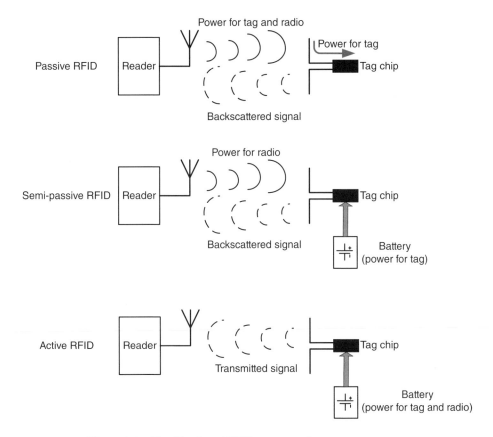

Figure 2.4 Classification of RFID tags according to power sources.

from the reader's emitted radio waves to communicate with the reader. Note that the battery of a semi-passive tag never provides the power for data transmission between tag and reader, but serves exclusively to supply the microchip and for the retention of the stored data. The power of the electromagnetic field received from the reader is the only power used for the data transmission between tag and reader [9].

An active tag is an architecturally conventional radio system. It has its own transmitter and receiver and uses a battery to power up its transmitter, receiver and chip [5]. From a pure technical perspective, active tags are not genuine 'RFID' tags, but short-range radio devices [9]. Almost all the advanced signal processing techniques can be applied to active tags without special consideration. Therefore, the contents of this book are confined on passive and semi-passive RFID.

Since passive tags have no internal power supply, they are much cheaper than active tags and have relatively long life span. The majority of RFID tags in the market are passive. Semi-passive tags are very similar to passive tags except for the addition of a small battery, which allows the tag chip to be immediately powered up once being woken up and removes the need for the antenna to be designed to collect power from the impinging signal. As a result, semi-passive tags can provide much longer read range than passive tags do.

As passive UHF RFID technology has matured, many new application scenarios have been proposed where a tag is expected to transmit ever-increasing amount of data. These scenarios include tags with expanded on-chip memory of 128 KB or more, tags including complex cryptographic security protocols, or tags that transfer stored sensor data in a semi-passive mode [24]. An extreme example of this trend is the Intel WiSP, a passive UHF RFID platform including a fully accessible, programmable 16-bit microcontroller with a variety of sensor peripherals [23]. The WiSP platform is currently being used for a variety of research applications. There are a variety of ways to increase the data rates of the communications between tag and reader.

2.3 Analogue Circuits for RFID

Figure 2.5 shows an example circuit for a load-modulation RFID tag with subcarrier communications [9]. This circuit consists of four functional parts: the tag antenna, power rectifier, timing clock and load modulator.

The power rectifier is mainly constructed by four diodes D_1–D_4, which form a typical bridge rectifier. The input voltage is provided by the induced high-frequency sinusoidal waves from the antenna coil. The output voltage is further smoothed by capacitor C_2. The zener diode D_5 (ZD 5V6) provides protection for possible surcharge of the bridge rectifier due to various reasons, for example, when the tag is very near to the reader. Due to the usage of D_5 and C_2, the voltage across the zener diode can be kept at exactly 5.6 V when the tag is well charged. The timing clock is mainly built up by the ripple counter IC 4024. The output Q_n is the nth stage of the counter, meaning that the frequency of the timing clock signal Q_n is the frequency of the input signal CLK divided by 2^n. The external timing signal CLK is produced by the induced sinusoidal signal from the antenna coil, where resistor R_2 provides protection for the ripple counter. Supposing that the operating frequency of the RFID is 13.56 MHz, the output Q_6 will provide an internal clocking signal of frequency 13.56 MHz$/2^6 = 212$ kHz.

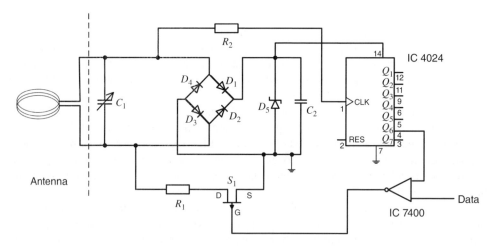

Figure 2.5 An example circuit for the load modulation RFID tag with subcarrier communications. (Reproduced with permission from Figure 3.18, K. Finkenzeller. RFID Handbook, 3rd ed. Wiley, Chichester, pp. 45, 2010.)

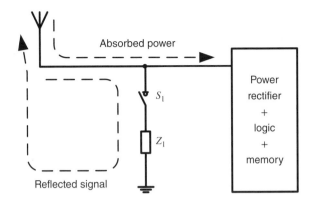

Figure 2.6   General equivalent circuit for RFID load modulation.

The load modulator consists of load modulation switch S_1 and resistor R_1. The switch is controlled by the internal clocking signal Q_n and data symbols via the logical 'and' operation. By controlling the state 'on' and 'off' of field-effect transistor S_1 via data symbols, the load of the tag antenna is changed. Therefore, the data symbols are modulated in the tag's returned signal. The equivalent circuit for the load modulation is shown in Figure 2.6. In practice, the load Z_1 can be either a resistive load or a capacitive load.

From Figure 2.5 it can be seen that a modulator with different subcarriers can be obtained by connecting one of the input terminals of the logic gate IC 7400 to different outputs of the counter IC 4024. The benefit of using subcarrier modulation is that the strong leakage interference signal from reader's transmitter to reader's receiver can be effectively isolated. The carrier interference caused by leakage is usually several magnitudes higher (up to 80 dB) than the tag modulation signal if the carrier frequency of tag's modulator is set to be the same as that of the reader's transmit signal [1].

The optional capacitor C_1 can be used to produce a resonant circuit with the antenna coil. Then the reading zone of this RFID can be significantly increased.

The same circuit can be also used for backscattering modulation. What we need to do is just to change the parameters of relevant capacitors and antenna coil so that the resonant frequency is equal to the operating frequency of the reader.

As can be seen from the circuit shown in Figure 2.5, to design an efficient power rectifier plays a key role for the successful operation of RFID. The power harvesting circuitry in common RFID tags is based on the Dickson charge pump topology [4] and operates with efficiencies of less than 60% [6] in voltage-limited tags. In [25], some power optimized waveforms are proposed to improve the power charging efficiency, where short, repeated and impacting bursts of power are used to charge the tag's capacitor.

2.4   Circuit Analysis for Signal Transfer in RFID

In this section, we will analyse quantitatively how the tag ID signal is transferred to the reader. The analysis is divided into two cases: load modulation and backscattering modulation.

2.4.1   Equivalent Circuit of Antennas in Generic Communication Links

First, let us discuss the equivalent circuit of antennas in generic communication links. Figure 2.7 shows a generic communication link with antennas and its equivalent circuit. It is assumed that the antennas comprise of linear reciprocal conducting or dielectric materials. A complete equivalent circuit for the antennas is shown in Figure 2.7(b) in the dashed box, where Z_{11} and Z_{22} are self-impedances of antennas at the transmitter side and receiver side,

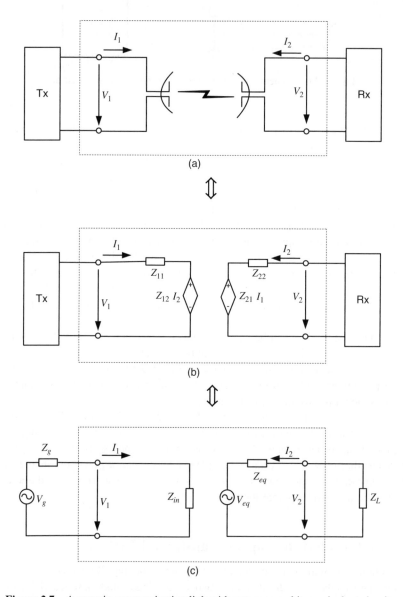

Figure 2.7   A generic communication link with antennas and its equivalent circuit.

respectively, and Z_{12} and Z_{21} are mutual impedances between the two antennas. The mutual impedances characterize the mutual electromagnetic coupling between two antennas. The Z-parameters of this two-antenna system depend only on the antenna construction, their separation and the propagation medium between them, independent of the transmitter and receiver configuration. To exactly calculate the Z-parameters, one should resort to electromagnetic wave propagation theory [3]. However, this is often difficult, if not impossible, for real systems. In practice, these parameters are often obtained through measurement.

The circuit in Figure 2.7(b) can be further simplified to Figure 2.7(c), where the transmitter is represented by the voltage generator V_g with an output impedance Z_g, the receiver is represented by a load with impedance Z_L, and the parameters Z_{in}, Z_{eq} and V_{eq} are given by:

$$Z_{in} = Z_{11} - \frac{Z_{12}Z_{21}}{Z_{22} + Z_L}, \tag{2.1}$$

$$Z_{eq} = Z_{22}, \tag{2.2}$$

$$V_{eq} = \frac{Z_{21}}{Z_{in} + Z_g} V_g. \tag{2.3}$$

Equations (2.1)–(2.3) can be obtained by equating the input-output relationship of the circuit in the dashed box of Figure 2.7(b) and that of Figure 2.7(c).

In the far field, the mutual impedances are small compared to the self impedances. Hence we have

$$Z_{in} \approx Z_{11},$$

$$V_{eq} \approx \frac{Z_{21}}{Z_{11} + Z_g} V_g.$$

Note that Figure 2.7(c) actually gives two equivalent circuits: one for transmitter equivalent circuit (the left-hand side of Figure 2.7c) and another for receiver equivalent circuit (the right-hand side of Figure 2.7c).

2.4.2   Load Modulation

For the case of load modulation, the communications between reader and tag are via inductive coupling, and hence the equivalent circuit for the antennas of the reader and tag can be approximated by an ideal transformer, which can be also obtained through Figure 2.7(b). Therefore, the equivalent circuit of the circuit in Figure 2.5 can be shown in Figure 2.8, where R_2 is the resistance shown in Figure 2.5, Z_2 represents the impedance of capacitor C_2 in Figure 2.5, Z_1 represents the impedance of resistor R_1 in Figure 2.5, and Z_0 denotes the lumped impedance of the reader's transmitter. Notice that R_1 in Figure 2.5 can also be substituted by a capacitor. Therefore, Z_1 can be either a resistance or a capacitance. The adjusting capacitor C_1 is neglected in this equivalent circuit. The signal $s(t)$ represents the tag's ID signal. It is a digital sequence consisting of zero and one. The switch S_1 in Figure 2.5 is controlled by $s(t)$. Mathematically this control is equivalent to *dividing* load Z_1 by signal $s(t)$. Therefore, the equivalent impedance of the parallel circuit Z_2 and Z_1 plus the regulatory signal $s(t)$ can be mathematically expressed as

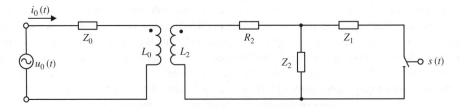

Figure 2.8   An equivalent circuit for load modulation analysis.

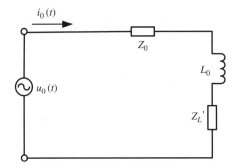

Figure 2.9   A further equivalent circuit for load modulation analysis.

$$Z_{\text{reg}} = \frac{Z_2 Z_1 / s(t)}{Z_2 + Z_1 / s(t)} = \frac{Z_1 Z_2}{Z_1 + Z_2 s(t)}.$$

The coupling between the coils of the reader and tag is approximated by an ideal transformer T_0, whose inductances at the primary winding (i.e. the reader's coil) and secondary winding (i.e. the tag's coil) are L_0 and L_2, respectively.

Now supposing that a voltage source $u_0(t)$ is applied to the reader's transmitter, we examine how the current at the reader side, denoted as $i_0(t)$, varies with the tag's ID signal $s(t)$. To find the current $i_0(t)$, a simple way is to eliminate transformer T_0 by referring the load at the tag side to the reader side, as shown in Figure 2.9.

The equivalent load impedance Z_L' reads

$$Z_L' = a^2 \left[j\omega L_2 + R_2 + \frac{Z_1 Z_2}{Z_1 + Z_2 s(t)} \right],$$

where a is the turn ratio between the turns of the reader's coil and the turns of the tag's coil. Therefore, current $i_0(t)$ reads

$$i_0(t) = \frac{u_0(t)}{Z_0 + j\omega L_0 + Z_L'} = \frac{u_0(t)}{Z_0 + j\omega L_0 + a^2 \left[j\omega L_2 + R_2 + \frac{Z_1 Z_2}{Z_1 + Z_2 s(t)} \right]}. \qquad (2.4)$$

2.4.3 Backscattering Modulation

As explained in Sections 2.1–2.2, once the reader starts transmitting read command, the tag's capacitor will capture power from the received radio waves in the first stage. When the captured power reaches a threshold, the tag starts to transmit its ID signal in the second stage. What we are concerned in this section is the relationship between the tag's transmitted ID signal and the received signal at the reader's receiver. In the second stage, the tag acts as a transmitter, and the reader acts as a receiver. Therefore, the equivalent circuit for the tag in this stage is shown in Figure 2.10.

In principle, the voltage V_{eq} in Figure 2.10 can be calculated based on Maxwell's equations and then the returned signal received by the reader's receiver can be computed based on the re-radiated V_{eq} from the tag. But this procedure is unnecessarily cumbersome in getting engineering insight for the key concern. In the case of far field, the power received at the tag's antenna and the power received at the reader's antenna resulted from the re-radiated wave at the tag's antenna are of primary concern. In the following, we show the power relationship among the aforementioned signals.

Let us consider a simplified equivalent circuit for the circuit as shown in Figure 2.10. This is shown in Figure 2.11, where Z_c represents a lumped impendence of the load.

Let $P_{rd,0}$ denote the transmitted power at the reader. The power density of the electromagnetic wave impinging to the tag antenna in free space is given by [3, 27]

$$S = \frac{P_{rd,0}G_{rd,t}}{4\pi r^2},$$ (2.5)

where $G_{rd,t}$ is the gain of the reader's transmit antenna, and r is the distance between the reader and tag. Let $A_{eff,tag}$ be the effective area of tag antenna, which is given by [3, 27]

$$A_{eff,tag} = \frac{\lambda^2}{4\pi}G_{tag},$$ (2.6)

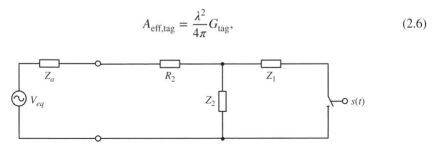

Figure 2.10   An equivalent circuit for the analysis of a tag's backscattering modulation.

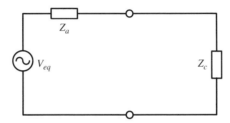

Figure 2.11   An illustration for calculating re-radiated power from the tag's antenna.

where G_{tag} is the gain of tag antenna, and λ is the wavelength of the transmitted radio signal. The power received by tag antenna depends on the load connected to the antenna. When a matched load (i.e. $Z_c = Z_a^*$) is connected to the antenna, the power received by the tag antenna is given by

$$P_{tag,0} = SA_{eff,tag}. \tag{2.7}$$

When an arbitrary load Z_c is connected to the tag antenna, a part of the received power will be re-radiated to the space. The re-radiated power at the tag is given by [18]

$$P_{tag,1} = KP_{tag,0}G_{tag}, \tag{2.8}$$

where

$$K = \frac{4R_a^2}{|Z_a + Z_c|^2} \tag{2.9}$$

with $Z_a = R_a + jX_a$. Suppose that a matched load is connected to the receive antenna at the reader. Then the received power at the reader is given by

$$P_{rd,1} = \frac{P_{tag,1}}{4\pi r^2}A_{eff,rd}, \tag{2.10}$$

where $\frac{P_{tag,1}}{4\pi r^2}$ is the power density of the re-radiated radio wave by the tag impinging to the reader's receive antenna, and $A_{eff,rd}$ is the effective area of reader antenna, which is given by

$$A_{eff,rd} = \frac{\lambda^2}{4\pi}G_{rd,r} \tag{2.11}$$

with $G_{rd,r}$ being the gain of reader's receive antenna. Substituting equations (2.5)–(2.9) and (2.11) into (2.10) yields

$$P_{rd,1} = \left(\frac{\lambda}{4\pi r}\right)^4 G_{tag}^2 G_{rd,t} G_{rd,r} P_{rd,0} \frac{4R_a^2}{|Z_a + Z_c|^2}. \tag{2.12}$$

Therefore, the amplitude of the voltage induced at the load of the reader's receive antenna, denoted by $|V_{rd,L}|$, is characterized by

$$|V_{rd,L}| = K_0 \sqrt{P_{rd,1}}$$

$$= K_0\left(\frac{\lambda}{4\pi r}\right)^2 G_{tag} \sqrt{G_{rd,t} G_{rd,r} P_{rd,0}} \frac{2R_a}{|Z_a + Z_c|}, \tag{2.13}$$

where K_0 is a constant that depends only on the impendence of the matched load at the reader's receive antenna.

Note that in the above derivation, we have assumed that an omnidirectional antenna is used in the tag. If this is not the case, the antenna gain G_{tag} in equations (2.12) and (2.13) may be different from receive mode and transmit mode. Then G_{tag} in equations (2.12) and (2.13) should be replaced with $\sqrt{G_{tag,r} G_{tag,t}}$, where $G_{tag,r}$ and $G_{tag,t}$ are the gain of tag's antenna in the receive mode and transmit mode, respectively, and also computed in the corresponding directions.

Now let us return to the circuit as shown in Figure 2.10. Substituting Z_c in equation (2.13) with

$$Z_c = R_2 + \frac{Z_1 Z_2}{Z_1 + Z_2 s(t)}$$

gives

$$|V_{rd,L}| = K_0 \left(\frac{\lambda}{4\pi r}\right)^2 G_{tag} \sqrt{G_{rd,t} G_{rd,r} P_{rd,0}} \frac{2R_a}{\left|Z_a + R_2 + \frac{Z_1 Z_2}{Z_1 + Z_2 s(t)}\right|}. \qquad (2.14)$$

This equation shows how the received signal $V_{rd,L}$ at the reader changes with the modulating signal $s(t)$ from the tag.

2.5 Signal Analysis of RFID Systems

The basic goal of an RFID system is to identify the tag ID from the received signal. To realize this goal, a basic way is to detect the difference in the received signal i_0 in equation (2.4) or $V_{rd,L}$ in equation (2.14), caused by different symbols in tag ID. The bigger the difference, the easier the identification. We call the signal analysis from this perspective *qualitative analysis*.

If the difference is not sufficiently big, more advanced techniques should be used. These techniques include the pre-processing at the input end (e.g. space-time coding) and post-processing at the output end (e.g. filtering in the time domain or in the frequency domain and maximum likelihood decoding from both the space domain and time domain). To implement such techniques in real systems, an exact or approximate relationship between the input signal, i.e. tag ID $s(t)$, and output signal, i.e. i_0 in equation (2.4) or $V_{rd,L}$ in equation (2.14), should be available.

Equations (2.4) and (2.14) show two nonlinear input-output relationships. Generally, this kind of the relationship makes both pre-processing and post-processing for the signals difficult. In most cases, an approximate linear relationship between the input and output will greatly facilitate problem solving. Often this is also desired from engineering point of view. Actually, in deriving the aforementioned equations, we have made several simplifying assumptions, which make the obtained input-output relationship characterize the truth only approximately. Therefore, to stick to the original nonlinear input-output relationship is unnecessary in many cases. We call the signal analysis from this perspective *quantitative analysis*.

In the following, we will discuss the signal analysis for RFID from the two aforementioned perspectives.

2.5.1 *Qualitative Analysis*

First let us define a nonlinear transform f:

$$f(x) = \frac{1}{x} \quad \forall x \in \mathbb{R}, \ x \neq 0. \qquad (2.15)$$

Then this nonlinear transform is applied to the current i_0 in equation (2.4) or voltage $V_{rd,L}$ in equation (2.14):

$$\check{i}_0(t) := f(i_0(t)) = \frac{1}{i_0(t)},$$

$$\check{V}_{rd,L} := f(V_{rd,L}) = \frac{1}{V_{rd,L}}.$$

The purpose using this transform is to facilitate the signal processing and analysis. Actually function f can be defined in a very general form. A general requirement for the generalized transform is that the function should be monotonically increasing or decreasing, so that one can uniquely solve the original variable from the transformed variable. It is clear that the function defined in (2.15) satisfies this requirement. The widely used linear amplifier in electronic systems is a special example of this generalized transform.

2.5.1.1 Load Modulation

Suppose that $u_0(t)$ is a constant and denote it by U_0. Let us consider the change in $\check{I}_0(t)$ by switching $s(t)$ from symbol '0' to '1' or vice versa. Define

$$
\begin{aligned}
D_1 &= \check{i}_0(t)|_{s(t)=1} - \check{i}_0(t)|_{s(t)=0} \\
&= \frac{Z_0 + j\omega L_0 + a^2 \left[j\omega L_2 + R_2 + \frac{Z_2 Z_1}{Z_2 + Z_1} \right]}{U_0} - \frac{Z_0 + j\omega L_0 + a^2 [j\omega L_2 + R_2 + Z_2]}{U_0} \\
&= -\frac{a^2}{U_0} \frac{Z_2^2}{Z_1 + Z_2}.
\end{aligned}
\tag{2.16}
$$

Equation (2.16) shows that the difference signal D_1 becomes very large if $Z_1 + Z_2$ approaches zero, which means that the parallel circuit consisting of Z_1 and Z_2 at the tag as shown in Figure 2.8 is in resonance when symbol '1' is transmitted (i.e. when the switch is closed).

2.5.1.2 Backscattering Modulation

For a given communicating scenario, the parameters K_0, λ, r, G_{tag}, $G_{rd,t}$, $G_{rd,r}$, $P_{rd,0}$ and R_a are constants. Therefore, we define a new constant

$$\bar{K}_0 := 2K_0 \left(\frac{\lambda}{4\pi r} \right)^2 G_{tag} \sqrt{G_{rd,t} G_{rd,r} P_{rd,0}} R_a.$$

Then, from equation (2.14) one gets

$$|\check{V}_{rd,L}| = \bar{K}_0^{-1} \left| Z_a + R_2 + \frac{Z_1 Z_2}{Z_1 + Z_2 s(t)} \right|.$$

Consider the change in $|\check{V}_{rd,L}|$ by switching $s(t)$ from symbol '0' to '1' or vice versa. Define

$$D_2 = |\check{V}_{rd,L}|\Big|_{s(t)=1} - |\check{V}_{rd,L}|\Big|_{s(t)=0}$$

$$= \bar{K}_0^{-1}\left(\left|Z_a + R_2 + \frac{Z_2 Z_1}{Z_2 + Z_1}\right| - |Z_a + R_2 + Z_2|\right). \tag{2.17}$$

Equation (2.17) again shows that the difference signal D_2 becomes very large if $Z_1 + Z_2$ approaches zero, which means that the parallel circuit consisting of Z_1 and Z_2 as shown in Figure 2.10 is in resonance when symbol '1' is transmitted (i.e. when the switch is closed).

From this analysis we can conclude that a guideline for the design of the tag's circuit is: a part of the circuit that is corresponding to a change of the circuit when different symbols are transmitted should be near to a state of resonance. By so doing, the different symbols can be easily differentiated by the reader.

2.5.2 Quantitative Analysis

In signal processing, it is highly desirable to obtain a linear input-output relationship. In this subsection, we will derive such an input-output relationship.

2.5.2.1 Load Modulation

Let us again suppose that $u_0(t)$ is a constant U_0. Define

$$Z_e = Z_0 + j\omega L_0 + a^2[j\omega L_2 + R_2].$$

From equation (2.4) we have

$$i_0(t) = \frac{U_0}{Z_e + a^2 \frac{Z_1 Z_2}{Z_1 + Z_2 s(t)}}$$

$$= \frac{U_0}{Z_e Z_1 + a^2 Z_2 Z_1}[Z_1 + Z_2 s(t)]\frac{1}{1 + \frac{Z_e Z_2}{Z_e Z_1 + a^2 Z_2 Z_1}s(t)}$$

$$\approx \frac{U_0}{Z_e Z_1 + a^2 Z_2 Z_1}[Z_1 + Z_2 s(t)]\left[1 - \frac{Z_e Z_2}{Z_e Z_1 + a^2 Z_2 Z_1}s(t)\right] \tag{2.18}$$

$$\approx \frac{U_0}{Z_e + a^2 Z_2} + \frac{a^2 U_0 Z_2^2}{Z_1(Z_e + a^2 Z_2)^2}s(t). \tag{2.19}$$

In deriving equation (2.18), we have used the Taylor series expansion of the complex function $\frac{1}{1+x}$, up to the first order of x. Therefore, equation (2.18) holds true under the condition

$$\left|\frac{Z_e Z_2}{(Z_e + a^2 Z_2)Z_1}s(t)\right| \ll 1. \tag{2.20}$$

In deriving equation (2.19), the higher-order term related with $s^2(t)$ is omitted. Therefore, equation (2.19) holds true under the condition

$$\left| \frac{Z_e}{a^2 Z_1} s(t) \right| \ll 1. \tag{2.21}$$

From equation (2.19) we can see that, under Conditions (2.20) and (2.21), the input and output of the load modulation RFID system exhibit an affine linear relationship. When the constant term $\frac{U_0}{Z_e + a^2 Z_2}$ in equation (2.19) is calibrated, a pure linear relationship for the input and output can be obtained.

2.5.2.2 Backscattering Modulation

To find the Taylor series expansion of $V_{rd,L}$, we need to recover its phase information. However, due to channel fading, this is generally difficult to obtain. Since it is sufficient to exploit the amplitude information of $V_{rd,L}$ for the task of identifying a tag's ID, we temporarily assume that the phase of $V_{rd,L}$ is φ, excluding the phase contained in the complex signal $1/\left(Z_a + R_2 + \frac{Z_1 Z_2}{Z_1 + Z_2 s(t)} \right)$. Therefore, $V_{rd,L}$ can be expressed as follows:

$$V_{rd,L} = \bar{K}_0 \frac{1}{Z_a + R_2 + \frac{Z_1 Z_2}{Z_1 + Z_2 s(t)}} e^{j\varphi}. \tag{2.22}$$

Based on equation (2.22), we have

$$V_{rd,L} = \bar{K}_0 e^{j\varphi} \frac{Z_1 + Z_2 s(t)}{(Z_a + R_2)(Z_1 + Z_2 s(t)) + Z_1 Z_2}$$

$$= \bar{K}_0 e^{j\varphi} \frac{Z_1 + Z_2 s(t)}{(Z_a + R_2)Z_1 + Z_1 Z_2} \cdot \frac{1}{1 + \frac{(Z_a + R_2)Z_2}{(Z_a + R_2)Z_1 + Z_1 Z_2} s(t)}$$

$$\approx \frac{\bar{K}_0 e^{j\varphi}}{(Z_a + R_2 + Z_2)Z_1} [Z_1 + Z_2 s(t)] \left[1 - \frac{(Z_a + R_2)Z_2}{(Z_a + R_2 + Z_2)Z_1} s(t) \right] \tag{2.23}$$

$$\approx \frac{\bar{K}_0 e^{j\varphi}}{Z_a + R_2 + Z_2} + \frac{\bar{K}_0 Z_2^2 e^{j\varphi}}{(Z_a + R_2 + Z_2)^2 Z_1} s(t). \tag{2.24}$$

In deriving equations (2.23) and (2.24), we have used the following conditions

$$\left| \frac{(Z_a + R_2)Z_2}{(Z_a + R_2 + Z_2)Z_1} s(t) \right| \ll 1 \tag{2.25}$$

and

$$\left| \frac{Z_a + R_2}{Z_1} s(t) \right| \ll 1 \tag{2.26}$$

respectively.

When Conditions (2.25) and (2.26) hold true, the input and output of the backscattering modulation RFID system exhibit an affine linear relationship. When the term $\frac{\bar{K}_0 e^{j\varphi}}{Z_a + R_2 + Z_2}$ in equation (2.24) is calibrated, a pure linear relationship for the input and output can be obtained.

2.6 Statistical Channel Models

In the preceding section, we have obtained the input-output relationship from the perspective of radio wave propagation and circuit theory. The obtained relationship can be considered as a deterministic channel model. This kind of channel model is useful for the design of RFID circuits, but in practice, it is not easy to obtain relevant circuit parameters exactly, especially when there are many scatterers in the communication media. In this case, the superposition of the signals coming from different scatterers at the receiver will cause fading and hence, describing the input-output relationship with a statistical model is more suitable. On the other hand, we often need to evaluate the performance of a generic RFID system with a new signal processing algorithm. In this situation, a statistical channel model for relevant RFID systems is also helpful.

In this section, several statistical channel models for RFID systems will be discussed. First the mathematical and physical backgrounds for these statistical channel models will be addressed. Then how to apply these models to RFID systems is elaborated.

2.6.1 Backgrounds of Rayleigh, Ricean and Nakagami Fading

2.6.1.1 Rayleigh Fading

Rayleigh fading is widely used in electrical engineering to characterize the distribution of envelope- or magnitude-related random variables. The probability density function (pdf) of Rayleigh distribution is

$$p_R(x) = \begin{cases} \dfrac{x}{\sigma^2}\, e^{-\frac{x^2}{2\sigma^2}} & x \geq 0, \\[2mm] 0 & x < 0, \end{cases} \tag{2.27}$$

where R denotes Rayleigh random variable and parameter σ is related with the mean value and variance of R in the following way:

$$\mathbb{E}(R) = \sqrt{\frac{\pi}{2}}\sigma, \quad \mathrm{Var}(R) = \frac{4-\pi}{2}\sigma^2, \quad \mathbb{E}(R^2) = 2\sigma^2.$$

Hereafter, we use $R \sim \texttt{Rayleigh}(\sigma^2)$ to denote that R is distributed as in (2.27).

Rayleigh distribution can be derived from Gaussian distribution. Suppose that there are two independent Gaussian-distributed random variables $X \sim \mathcal{N}(0, \sigma^2)$ and $Y \sim \mathcal{N}(0, \sigma^2)$. Let

$$R := |X + {}_J Y| = \sqrt{X^2 + Y^2}.$$

Then we have $R \sim \texttt{Rayleigh}(\sigma^2)$.

This interpretation is widely used in engineering. For example, it gives an explanation why radio channels in wireless communications fit well to Rayleigh distribution.

In practical world, a Gaussian-distributed random variable can be approximately obtained by the superposition of several random variables. This fact is due to the Central Limit Theorem (CLT) in probability theory. Let $\{X_1,...,X_n\}$ be a sequence of independent identically distributed random variables with

$$\mathbb{E}(X_i) = \mu, \quad \mathrm{Var}(X_i) = \sigma^2. \tag{2.28}$$

CLT says that the sum variable

$$X = \sqrt{n}\left(\frac{1}{n}\sum_{i=1}^{n}X_i\right) - \mu = \sum_{i=1}^{n}\xi_i - \mu \quad \text{with} \quad \xi_i = \frac{X_i}{\sqrt{n}} \tag{2.29}$$

approaches Gaussian distribution $\mathcal{N}(0, \sigma^2)$ as n approaches infinity. Condition (2.28) is equivalent to

$$\mathbb{E}(\xi_i) = \frac{\mu}{\sqrt{n}}, \quad \text{Var}(\xi_i) = \frac{\sigma^2}{n}. \tag{2.30}$$

Condition (2.30) requires that the contribution of each individual variable such as ξ_i to the sum variable X in equation (2.29) is very small, i.e., no variable dominates the sum.

In a wireless communication channel, as shown in Figure 2.12, generally there are a large number of scatterers in the communication environments. The overall channel coefficient h is the sum of all the individual path gain $\{h_i\}$ coming from different scatterers that are unresolvable by the receiver in the temporal domain:

$$h = \sum_{i=1}^{n} h_i = \sum_{i=1}^{n} |h_i| e^{j2\pi f_c \tau_i} = \sum_{i=1}^{n} |h_i| e^{j\phi_i}, \tag{2.31}$$

where

$$\phi_i := 2\pi f_c \tau_i \text{ modulo } 2\pi = 2\pi \frac{d_i}{\lambda} \text{ modulo } 2\pi,$$

f_c and λ are the frequency and wavelength of the carrier, respectively, and d_i and τ_i are, respectively, the distance and time that the wave of the ith path travels. When $d_i \gg \lambda$, it is reasonable to assume that ϕ_i is uniformly distributed in the interval $[0, 2\pi]$ and all the phases $\{\phi_i\}$ are independent of each other. When the scatterers are not correlated with each other and no strong scatterers exist in the environments, then it is reasonable to assume that all $\{|h_i|\}$ are independent of each other and follow the same distribution. In this case, both the real part and imaginary part of h can be approximated by a Gaussian distribution and independent of each other. Thus the amplitude fading of the channel coefficient h can be approximated by Rayleigh

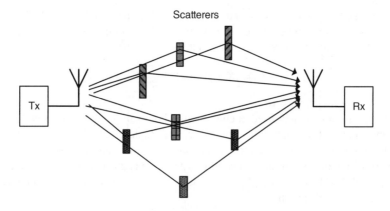

Figure 2.12 An illustration of a Rayleigh channel.

distribution. Note that it is difficult to examine whether or not the condition 'all $\{|h_i|\}$ follow the same distribution' holds true. Therefore, in this argument, this condition is often replaced with that 'all the scatterers are irregularly distributed' in practice.

Based on this argument, it can be concluded that if $\mathbb{E}(|h_i|^2) = \frac{\sigma^2}{n}$, then $|h| \sim$ Rayleigh(σ^2).

In the following chapters, we need the distribution of the function $Z = R^2$, i.e. the distribution of power of the channel fading. By using the theory of functions of random variables, it can be easily found that

$$p_Z(z) = \begin{cases} \dfrac{1}{2\sigma^2} \, e^{-\frac{z}{2\sigma^2}} & z \geq 0, \\ 0 & z < 0. \end{cases} \tag{2.32}$$

2.6.1.2 Ricean Fading

When there is a very strong path (called a specular path), as shown in Figure 2.13, and the path gains from all other scatterers in the propagation environments are relatively small and distributed independently and identically, then the channel coefficient can be modelled by the following equation

$$h = h_0 + \sum_{i=1}^{n} h_i = A e^{j\phi_0} + \sum_{i=1}^{n} |h_i| e^{j\phi_i},$$

where A is a constant, ϕ_0 can be either a constant or a random variable uniformly distributed in the interval $[0, 2\pi]$, and h_i and ϕ_i have the same meaning as in equation (2.31).

If h_i and ϕ_i satisfy the conditions discussed in preceding subsection, then the magnitude of h can be approximated by Ricean distribution, whose pdf is given by

$$p_{|h|}(x) = \begin{cases} \dfrac{x}{\sigma^2} \, e^{-\frac{x^2+A^2}{2\sigma^2}} I_0\left(\dfrac{Ax}{\sigma^2}\right) & x \geq 0, \\ 0 & x < 0, \end{cases} \tag{2.33}$$

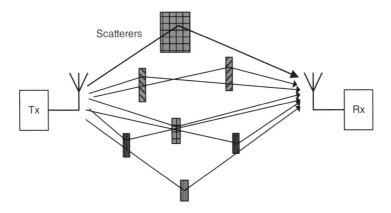

Figure 2.13 An illustration of Ricean channel.

where I_0 denotes the modified Bessel function of the first kind and zeroth order. Some details about function I_0 will be discussed in Appendix 2.A.

In the Ricean fading defined by equations (2.33), the power possessed by the specular path is A^2, and the power possessed by other scatterers are $2\sigma^2$ (both the real part and imaginary part are of $\mathcal{N}(0, \sigma^2)$). The Ricean fading is often characterized by parameter K, which is called Rice factor and defined as the ratio between the power of specular path and the total power of all other paths, i.e.,

$$K = \frac{A^2}{2\sigma^2}.$$

Let the total power of all the paths be Ω. Then we have $A^2 = \frac{K\Omega}{K+1}$ and $2\sigma^2 = \frac{\Omega}{K+1}$. Therefore, the pdf of Ricean fading can be rewritten as

$$p_{|h|}(x) = \begin{cases} \dfrac{2(K+1)x}{\Omega} e^{-K-\frac{(K+1)x^2}{\Omega}} I_0\left(2\sqrt{\dfrac{K(K+1)}{\Omega}} x\right) & x \geq 0, \\ 0 & x < 0. \end{cases}$$

It can be easily seen that Ricean fading reduces to Rayleigh fading when $K = 0$; and it exhibits no fading when $K \to \infty$.

Let $Z = |h|^2$. Then we can find the pdf of Z as follows:

$$p_Z(z) = \begin{cases} \dfrac{1}{2\sigma^2} e^{-\frac{z+A^2}{2\sigma^2}} I_0\left(\dfrac{A\sqrt{z}}{\sigma^2}\right) & z \geq 0, \\ 0 & z < 0. \end{cases}$$

2.6.1.3 Nakagami Fading

The pdf of Nakagami distribution is given by

$$p_X(x) = \begin{cases} \dfrac{2m^m x^{2m-1}}{\Gamma(m)\Omega^m} e^{-\frac{mx^2}{\Omega}} & \text{when} \quad x \geq 0, \\ 0 & \text{when} \quad x < 0, \end{cases} \qquad m \geq \frac{1}{2},$$

where Γ denotes the Gamma function, $\Omega = \mathbb{E}(X^2)$, and $m = [\mathbb{E}(X^2)]^2/\mathrm{Var}(X^2)$. The parameter m is called shape factor. It is apparent that Nakagami fading reduces to Rayleigh fading when $m = 1$, to a one-sided Gaussian distribution when $m = \frac{1}{2}$, and to an impulse (or deterministic channel) when $m \to \infty$.

There is no clear explanation from physics about how the Nakagami distribution is linked with a channel fading. Nakagami first discovered, in the early 1960s, that it can provide a good match to empirical data of rapid fading in long-distance high-frequency channels [17]. A good feature of Nakagami fading is that it can closely approximate Ricean fading by using the following relationship of their parameters [14]:

$$K = \frac{\sqrt{m^2 - m}}{m - \sqrt{m^2 - m}}, \tag{2.34}$$

$$m = \frac{(K + 1)^2}{2K + 1}. \tag{2.35}$$

A good analytical property for Nakagami fading is that it involves with only elementary functions and does *not* contain a Bessel function as Ricean fading does and the squared magnitude of a Nakagami fading channel is of the Gamma distribution, which has some nice analytical properties in deriving closed-form expressions of some system performance. Therefore, Nakagami fading is often used as an alternative channel model for Ricean fading model in system performance analysis.

Let $Z = X^2$. Then we can find the pdf of Z as follows:

$$p_Z(z) = \begin{cases} \frac{m^m z^{m-1}}{\Gamma(m)\Omega^m} e^{-\frac{mz}{\Omega}} & \text{when} \quad z \geq 0, \\ 0 & \text{when} \quad z < 0, \end{cases} \qquad m \geq \frac{1}{2}.$$

A comparison between Rayleigh, Ricean and Nakagami distributions is illustrated in Figure 2.14, where the power of the magnitude of the fading X is chosen to be one, i.e. $\mathbb{E}(X^2) = 1$, and the Ricean factor in Ricean distribution and the shape factor in Nakagami

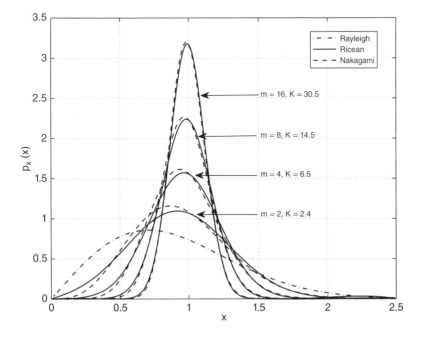

Figure 2.14 A comparison among Rayleigh, Ricean and Nakagami distributions.

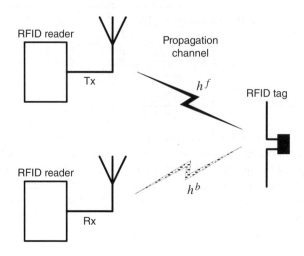

Figure 2.15   An illustration of RFID forward and backward channels.

distribution are chosen according to equations (2.34) and (2.35), respectively. It is seen that Ricean fading and Nakagami fading can be well approximated by each other.

2.6.2   Statistical Channel Models of RFID Systems

The RFID channel differs from a conventional wireless communication channel in that the former is the cascade of two fading channels: the forward and backward links as shown in Figure 2.15. The forward link refers to the propagation channel from the reader's transmitter to the tag, while the backward link refers to the propagation channel from the tag to the reader's receiver. Denote the channel gains of the forward and backward links as h^f and h^b respectively. Then the whole channel gain is

$$h = h^b \gamma h^f, \tag{2.36}$$

where γ denotes the backscattering coefficient of the tag for the case of backscattering modulation or coupling function of the tag for the case of load modulation. In discussing the fading property, γ can be neglected since it is a deterministic function of tag's ID.

From equation (2.36) we can see that, to characterize the fading property of an RFID channel, two issues should be clarified:

- the fading property of h^f and h^b; and
- the relationship between h^f and h^b.

Due to the fact that RFID is typically operated within short distances (within a range of about 10 m) and in the scenario of line of sight (LoS), both forward link and backward link often contain a strong path, in addition to other weak scattering paths. Therefore, Ricean distribution is a proper characterization for the fading property of h^f and h^b. This observation is confirmed by the channel measurement results of [12]. As elaborated on in the preceding subsection, Ricean distribution can be well approximated by Nakagami distribution. Therefore, in theoretical system performance analysis, the Nakagami fading model is also used to characterized both h^f and h^b. For example, report [14] uses Nakagami fading to derive a closed-form expression for the average interrogation range of RFID, and report [15] uses Nakagami fading to characterize the RFID channel and analyse the bit-error rate of a space-time encoded RFID system.

In some cases [10, 11], Rayleigh fading is also used to characterize h^f and h^b. For example, in the testbed developed in [11], non-line-of-sight (NLoS) measurements for RFID channels are obtained at 5.8 GHz. It is found that Rayleigh fading provides a very good fit for the cumulative distribution function of the envelope of the channel.

The relationship between h^f and h^b depends on the relative locations of the transmitter and receiver at the reader. Basically there are two kinds of configuration: a monostatic system and a bistatic system [19]. In a monostatic system, the reader transmit and receive antennas are closely spaced or colocated. The RF power decoupling for this kind of system is realized through an RF isolator [21]. In a bistatic system, the reader transmit and receive antennas are located far apart from each other (with the distance being typically more than half of the wavelength of the carrier) or decoupled via polarization. In the former case, the forward link and the backward link are highly correlated. By applying the reciprocal principle of radio channels, it can be reasonably assumed that

$$h^b = h^f.$$

In the latter case, it is widely assumed that the forward link and the backward link are independent of each other. Let $p_{h^f}(x)$ and $p_{h^b}(y)$ be the pdfs of h^f and h^b, respectively. Let $h = h^f h^b$ (i.e. omitting the backscattering coefficient γ in equation (2.36)). Then the pdf of h is given by [22]

$$p_h(z) = \int_{-\infty}^{\infty} p_{h^f}(x) p_{h^b}\left(\frac{z}{x}\right) \frac{1}{|x|} \, dx. \tag{2.37}$$

2.6.3 Large Scale Path Loss

To fully characterize the statistical model of an RFID channel, we need to have a model for the large scale path loss about the RFID channel. Since only far field is concerned for the large scale path loss, only backscattering-modulated RFID channel should be considered. Let us consider the general case: the reader is bistatic, the distances of transmit and receive antennas to the tag are r_1 and r_2, respectively, and the gain of tag antenna is of different values in the receive mode and transmit mode. Denote the gains to be $G_{tag,r}$ and $G_{tag,t}$, respectively. Then following the same derivation as that of equation (2.12), we can obtain the relationship between transmitted power $P_{rd,0}$ and received power $P_{rd,1}$ in free space as follows:

$$\frac{P_{rd,1}}{P_{rd,0}} = G_{tag,r} G_{tag,t} G_{rd,t} G_{rd,r} K \left(\frac{\lambda}{4\pi r_1}\right)^2 \left(\frac{\lambda}{4\pi r_2}\right)^2,$$

where K is the reflection coefficient of the tag. Therefore, the path loss in decibels can be expressed as

$$L_p := 10 \log \frac{P_{rd,0}}{P_{rd,1}} = 10 \log G_0 + 20 \log r_1 + 20 \log r_2 = L_0 + 20 \log r_1 + 20 \log r_2$$

where

$$G_0 := \frac{1}{G_{\text{tag,r}} G_{\text{tag,t}} G_{\text{rd,t}} G_{\text{rd,r}} K} \left(\frac{4\pi}{\lambda}\right)^4,$$

$$L_0 := 10 \log G_0 = 10 \log \frac{P_{\text{rd,0}}}{P_{\text{rd,1}}|_{r_1=r_2=1 \text{ m}}}. \tag{2.38}$$

In a generalized propagation media, the path loss exponent in one-way propagation, denoted as n, may not equal 2. Then the path loss can be generally expressed as:

$$L_p = L_0 + 10n \log r_1 + 10n \log r_2, \tag{2.39}$$

where parameter L_0 is defined similar to equation (2.38). In equation (2.39), it is assumed that the propagation media is symmetric. Hence the path loss exponents for the forward link and backward link are equal to each other.

The measured results in report [12] show path loss exponents of 2.09 and 2.16 in different room environments.

The results in [12] also confirm that the path loss exponent of the two-way link is approximately twice that of a traditional one-way link in the same environment. It is also pointed out in [19] that in any real propagation environment, no matter how complex it is, the forward and backward channels are always symmetrical: the path losses of the forward and backward channels between the reader and tag antennas are equal to each other. Any measured asymmetry is usually due to RF hardware calibration. This observation suggests that we can calculate link budgets for the backscattering-modulated RFID channels by using many existing propagation models for traditional one-way links.

Some further discussions on propagation model of RFID systems can be found in reports [19] for a general overview, [20] for near-field channel modelling, [2] for wideband RFID channels, [16] for RFID channel measurement and [11, 13, 14] for MIMO RFID channel modelling.

2.7   A Review of RFID Protocol

In recent decades, there have been a few organizations developing their own standards for RFID. Now two leading figures in this competition are the ISO (International Organization for Standardization) and EPCglobal©. EPCglobal is a joint venture between GS1 (formerly known as EAN International) and GS1 US (the former Uniform Code Council (UCC), Inc.). It is an organization set up to achieve worldwide adoption and standardization of Electronic Product Code (EPC) technology. The ISO has developed a series of RFID standards for different purposes, including those for animal identification, Proximity cards, Vicinity cards, item management and so on [1]. EPCglobal stemmed from the Auto-ID Center [2], which was set up in 1999 to develop the EPC and related technologies that could be used to identify products and track them through the global supply chain. Since EAN and UCC have been involved in EPCglobal from the beginning, the EPCglobal standards are down-compatible with the widely used barcode standard.

---

[1] See the link http://rfid.net/basics/186-iso-rfid-standards-a-complete-list
[2] See the link www.autoidlabs.org/

Table 2.2 Five classes of EPCglobal RFID standards. Here WORM stands for 'write once and read many', and WMRM 'write many and read many'

EPC class	passive/active	R/W ability	bits of tag ID	others
Class 0	passive	read only	64 bits	
Class 1	passive	WORM	$\geq$ 96 bits	
Class 1, Gen 2	passive	WMRM	$\geq$ 96 bits	
Class 2	semi-active	WMRM	$\geq$ 96 bits	enhanced Gen 2 not yet completed
Class 3	semi-active	WMRM	$\geq$ 96 bits	data logging capability not fully defined
Class 4	active	WMRM	extended tag ID	in early definition stage

EPCglobal has proposed five classes of RFID standards, some of which have been finalized and some of which are still under developing or in its early definition stage. The main features of these five classes of standards are illustrated in Table 2.2.

Note that Class 0 tags and Class 1 tags use different air interfaces to communicate with the reader, and hence are not inter-operable, while Class 1, Gen 2 (or C1G2 in short) merges Class 0 and Class 1 into an inter-operable EPC standard. In the long run, C1G2 will replace Class 0 and Class 1 tags.

C1G2 operates in the 860 MHz $\sim$ 960 MHz frequency range. However, in no countries can the readers and tags legally operate over the entire bandwidth due to local radio power regulations.

The EPCglobal network now forms a network that allows authorized trading partners of a supply chain to exchange previously registered data. The Discovery Services facilitate to find EPC-related data in the EPCglobal Network. Access to EPC-related data is controlled by EPC Information Services. The companies themselves determine which trading partners are given access to registered data. Through these functionalities, an information network is developed, which can be used for worldwide and real-time tracking of goods through the supply chains [9].

EPCglobal C1G2 represents a major step in standardization, performance and quality. Now it is accepted by ISO and known as ISO/IEC 18000 Type C standard.

In the following, some basic information for the C1G2 protocol will be outlined.

2.7.1 Physical Layer

In RFID applications, there are often many tags and might be several readers in a specific environment, which form a communication network. In this network, only the parameters and protocols for physical (PHY) layer and media access control (MAC) need to be specified.

2.7.1.1 Reader-to-Tag Communications

In the PHY layer, readers communicate with tags using modulations of double sideband (DSB) amplitude-shift keying (ASK), single sideband (SSB) ASK, or phase-reversal (PR)

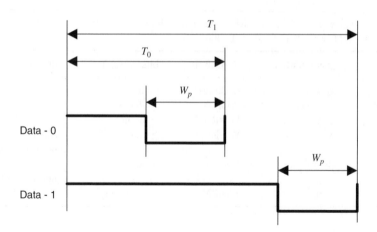

Figure 2.16 An illustration of PIE symbols.

ASK. Pulse-interval encoding (PIE) is used to encode the basic data symbols '0' and '1', which is illustrated in Figure 2.16. The basic timing unit in the PIE is the duration of data symbol '0'. It is denoted by T_0. Other timing durations such as the pulse width W_p and duration of data symbol '1', denoted as T_1, depend on T_0. The ranges for T_0, T_1 and W_p are as follows [8]:

$$6.25 \ \mu s \leq T_0 \leq 25 \ \mu s,$$

$$\max\{0.265T_0, 2 \ \mu s\} \leq W_p \leq 0.525T_0,$$

$$1.5T_0 \leq T_1 \leq 2T_0.$$

The reason for using PIE encoding at a reader is that the RF power density contained in PIE waveform in the corresponding RFID frequency band is ample.

A reader starts the signalling for the reader-to-tag link with either a preamble or a frame-sync, which are illustrated in Figure 2.17. The preamble consists of four parts: a delimiter with duration T_d, data '0', reader-to-tag (R→T) calibration with duration T_{RTcal}, and tag-to-reader (T→R) calibration with duration T_{TRcal}. These parameters are specified as follows [8]:

$$T_d = 12.5 \ \mu s \pm 5\%,$$

$$2.5T_0 \leq T_{\text{RTcal}} = T_0 + T_1 \leq 3T_0,$$

$$1.1T_{\text{RTcal}} \leq T_{\text{TRcal}} \leq 3T_{\text{RTcal}}.$$

The R→T calibration provides a basis for a tag to decode the reader's symbols. A tag measures the length of R→T calibration T_{RTcal} and computes $p = T_{\text{RTcal}}/2$. The tag interprets subsequent reader symbols shorter than p as symbol '0', longer than p as symbol '1', and further longer than $4T_{\text{RTcal}}$ to be an invalid symbol.

The T→R calibration, together with the command *divide ratio* (DR), specifies a tag's backscatter link frequency (BLF) or backscatter-link pulse-repetition frequency. DR can have

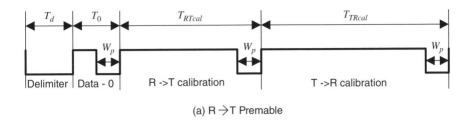

(a) R →T Premable

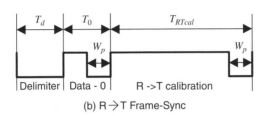

(b) R →T Frame-Sync

Figure 2.17   An illustration of reader-to-tag signalling.

two possible values: DR = 8 or DR = 64/3. The tag measures T_{TRcal} and then computes BLF as [8]

$$\text{BLF} = \frac{\text{DR}}{T_{\text{TRcal}}}.$$

In addition to operating in a fixed band in the 860 MHz ~ 960 MHz frequency range, the Gen 2 standard also uses frequency-hopping spread-spectrum (FHSS) modulation to transmit. In FHSS, the reader rapidly switches its carrier frequency among a number of channels during the radio signal transmission, and reads tags at slightly different frequencies to get the best possible reading result from the tag. Finally, the reader compares the results to determine if the reading is successful or not.

2.7.1.2   Tag-to-Reader Communications

A tag communicates with a reader using backscattering modulation. The backscatter uses ASK or phase-shift keying (PSK) modulation scheme.

A tag encodes the backscattered data using either FM0 or Miller baseband encoding scheme. In the FM0 baseband encoding scheme, data '1' changes its pulse phase at the beginning of every symbol; while data '0' changes its pulse phase not only at the beginning of every symbol but also in the middle of a bit period. In the Miller encoding scheme, data '0' causes no change in signal level unless it is followed by another '0' in which case a transition to the other level takes place at the end of the first bit period; while data '1' causes a transition from one level to the other in the middle of the bit period. An illustration for Miller and FM0 baseband encoding schemes is shown in Figure 2.18, where NRZ code denotes the typical non-return-to-zero line code. It can be seen from Figure 2.18 that both FM0 or Miller schemes, not like NRZ code, guarantee that the signal level will change at least for a period of two-symbol time.

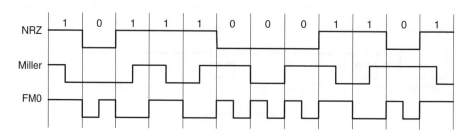

Figure 2.18   An illustration of Miller and FM0 encoding schemes.

In the communications from the tag to reader, the FM0 or Miller signalling begins with one or two preambles with a specific pulse pattern. For the details, readers are referred to Figure 6.11 and Figure 6.15 of [8].

2.7.2   MAC Layer

2.7.2.1   Tag Memory Structure

The information of a tag is stored in tag memory, some of which is writable by a reader. In this way, the tag and reader can communicate mutually. The tag memory consists of four banks: reserved memory bank, EPC memory bank, TID memory bank and user memory bank, as shown in Figure 2.19 [8].

The reserved memory contains the 'Kill Password' and 'Access Password', if the password functionality is implemented on the tag. Through the 'Kill Password', a reader can recommission a tag or kill a tag by rendering it nonresponsive to readers permanently. A tag with a nonzero 'Access Password' requires a reader to issue this password before transitioning to the secured state, in which the tag can execute all access commands, for exemple *Read, Write, Lock* etc.

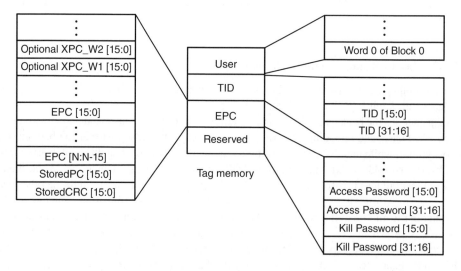

Figure 2.19   The structure of tag memory.

The EPC memory contains a StoredCRC, StoredPC, EPC words and optional XPC (extended protocol control). The StoredCRC denotes cyclic redundancy check (CRC) code. The tag's backscattered CRC code is used by the reader to verify the integrity of a received protocol control (PC) word, XPC word or words and EPC words, thus identifying possible data collision event.

The StoredPC comprises EPC-word length, a UMI (user-memory indicator), an XI (XPC_W1 indicator), and an NSI (numbering system identifier). If a tag supports XPC functionality, it will also implement a PacketPC, which differs from StoredPC in its EPC-word length field. The UMI consists of one-bit information, which is fixed by the tag chip manufacturer. If the tag does not have user memory, then this bit is set to binary 0; if the tag does support user memory, then this bit is set to binary 1. The XI bit identifies whether an extended protocol control word 1 (XPC_W1) exists in the EPC Memory. The NSI bits identify whether the data in the EPC data area conforms to an EPCglobal tag data standard (with the first bit being set to binary 0) or other non-EPCglobal applications such as ISO/IEC 15961 coding scheme (with the first bit being set to binary 1).

The EPC words are the code that identifies the object to which a tag is affixed. After several rounds of 'hand-shaking' communications between a tag and reader starting from the power up of the tag, the reader may issue a command that will cause the tag to backscatter the code contained in the EPC words.

The XPC word or words are used for a tag to support recommissioning. A reader may issue a *Select* command with a mask that covers all or part of XPC_W1 (the first word of XPC) and/or XPC_W2 (the second word of XPC) and may read a tag's XPC_W1 and XPC_W2 by using a *Read* command.

The TID (tag's ID, different from the ID of tag-affixed product or object) memory is used to identify the tag's model and manufacturer. It contains one of two ISO/IEC 15963 class-identifier values and vendor-specific data (e.g. a tag serial number). It includes sufficient identifying information for a reader to uniquely identify the custom commands and optional features that the tag supports. The codes in the TID memory are written on the chips during fabrication and are protected against rewriting.

The user memory is optional. It is used to store user-specific data. It is rewritable. During recommissioning a reader may render a tag's user memory inaccessible, as if it no longer exists.

2.7.2.2 Sessions and Inventoried Flags

Readers and tags in C1G2 are designed to support four independent sessions, denoted as S0, S1, S2 and S3, respectively. Each session is of two states called 'A' and 'B', as illustrated in Figure 2.20. The states 'A' and 'B' can be used as a flag to label whether a tag is inventoried by a reader. For example, a reader can mark by state 'B' as inventoried tags. When a tag is not inventoried, it is put in state 'A'; when inventoried, it is put in state 'B'. In the case of multiple readers, different readers can inventory the same group of tags by using different sessions independently.

A parameter *persistence* is used to control the transition from state 'B' to state 'A'. Different sessions have different values for the parameter *persistence*. However, these values cannot be set by users. Instead, they are set by manufacturers of tags. The range of *persistence* for different sessions is listed in Table 6.16 of [8].

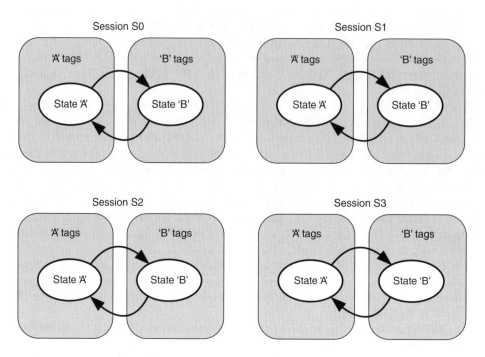

Figure 2.20   An illustration of sessions and states.

During an inventory round, a tag participates in one and only one session corresponding to one reader. A reader can use the command *Select* to set up the state of a session.

2.7.2.3   Tag States and Slot Counter

A tag is of seven states named 'Ready', 'Arbitrate', 'Reply', 'Acknowledged', 'Open', 'Secured' and 'Killed'. These states are used to control the whole process of inventory, and the transition among these states depends on the condition of the reading process. The 'Ready' state can be viewed as a 'holding' state for energized tags that are neither killed nor participating in an inventory round. The 'Arbitrate' state represents the status that a tag is participating in an inventory round but whose slot counter holds a nonzero value. The 'Reply' state is an intermediate state to represent the status that the tag is backscattering its ID to the reader. The state 'Acknowledged' is also an intermediate state meaning that the tag is backscattering a reply signal containing the tag's protocol-control words. The states 'Open' and 'Secured' can be viewed as two 'holding' states for further processing depending on the reader's new command. The difference between 'Open' and 'Secured' lies in that the former needs no password, i.e., when the parameter 'access password' $\neq 0$, while the latter needs a password, i.e., when the parameter 'access password' $= 0$. The state 'Killed' refers to the state of a tag that it is permanently disabled.

A sketch of the transition of these states is illustrated in Figure 2.21. For the detailed state transition diagram and the relevant transition conditions, readers are referred to Figure 6.19 of [8].

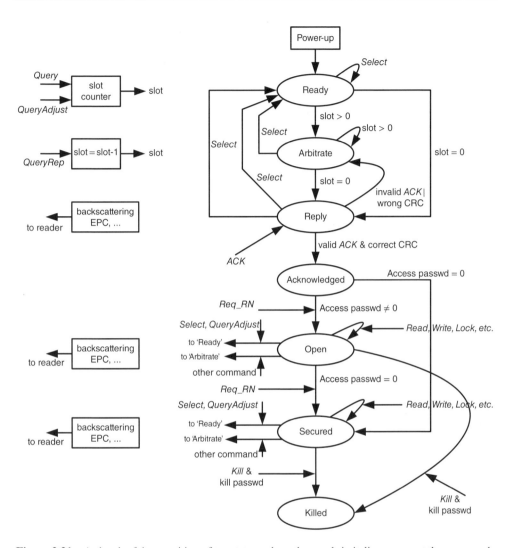

Figure 2.21   A sketch of the transition of tag states, where the words in italics represent the commands from the reader.

When a tag is in the 'Ready' state, it will receive a *Query* command. The *Query* command sets up a slot-count parameter Q. In terms of this value, the tag's chip will generate a random integer uniformly distributed over $[0, 2^Q - 1]$ and load this integer into the tag's slot counter. When a tag is in the 'Arbitrate' state, its slot counter will decrease its value by '1' every time when the tag receives a *QueryRep* command. When the slot counter reaches zero, the tag will transition to the state 'Reply' and backscatter its ID to the reader. When finished, the tag will transition to the state 'Acknowledged' and at the same time the reader will send back an acknowledgement signal *ACK* to the tag containing the tag's ID. Depending on the received *ACK* information, the tag will transition to different states. For example, the tag will transition to the state 'Arbitrate' if it receives no *ACK* within a specified duration, an invalid *ACK* or an

ACK with an erroneous tag ID, and to the state 'Open' or 'Secured' if it receives a valid *ACK* with a correct tag ID. When a tag is in the 'Open' or 'Secured' state, it can communicate with the reader indefinitely and transition to the state 'Ready' if a new command '*Select*' is received or to the state 'Killed' if a new command '*Kill*' is received.

The mechanism of random 'slot counter' is designed for a single reader to access multiple tags in an inventory round. This mechanism is borrowed from contention-based CSMA (carrier sense multiple accesses) media-access mechanism in WLAN.

In summary, it is seen that according to EPCglobal C1G2 protocol, one can use multiple (up to 4) readers to inventory multiple tags (the upper limit depends on the parameter Q) simultaneously.

2.8   Challenges in RFID

The main challenges of RFID come from two aspects corresponding to its applications. From the perspective of low-end applications, the long-term goal of RFID is to replace the barcode technology in the mass market. However, a barcode label is very cheap. There is still a long way to go to manufacture RFID tags with the price sufficiently close to that of barcode labels.

From the perspective of high-end applications, it is highly desirable for a reader to iden-tify the identities of multiple tags almost simultaneously without human's interference. For example, one often sees shopping trolleys carrying several or many products passing through casher gates in almost all super-markets. If the reader can reliably identify all the products in a shopping trolley simultaneously without bringing the products one-by-one to the scanner, a great benefit in saving the labor force for the super market and in reducing the time for the cus-tomer will be gained. Another high-end application scenario is that several readers inventory thousands of products in the same site such as in a store or a manufacturer. Each reader will experience strong interfering signals coming from other readers and the tags that have been already inventoried by the reader itself but not by other readers. In this case, to solve the prob-lem within a reasonable speed is challenging. The foreseeable techniques to solve this kind of problem are to employ multiple antennas in the PHY layer and advanced collision-avoidance techniques in the MAC layer. Another challenge in this perspective is, how a reader can reli-ably identify tag ID under NLoS scenario. For barcode technology, this problem is neglected since, simply, it cannot be solved. RFID has the potential to solve this problem with some advanced signal processing techniques. For example, using multiple antennas can partly solve this problem.

2.9   Summary

In this chapter, the following issues have been discussed:

- the basic operating principle of RFID systems – Inductive coupling, backscattering coupling and capacitive coupling are addressed.
- classification of RFID systems – Passive, semi-passive and active RFID are discussed.
- an example circuit of RFID tag – An equivalent circuit for this example circuit is presented, based on which many new modulation schemes can be proposed.

- circuit analysis for signal transfer in an RFID system – Here, how the tag ID signal is trans-ferred to the reader has been *quantitatively* analysed.
- signal analysis for RFID systems – Here, the relationship between the tag's transmitted ID signal (as input signal) and the received signal at the reader's receiver (as output signal) has been analysed both *qualitatively* and *quantitatively*, based on which a design guideline for RFID tags are drawn and further signal processing models for more complicated RFID systems can be developed.
- statistical channel models for RFID – Three statistical channel models, namely, Rayleigh, Ricean and Nakagami fading, for RFID systems have been presented. The mathematical and physical backgrounds for these statistical channel models are addressed and how to apply these models to RFID systems is elaborated.
- RFID protocols – A review of RFID protocols, especially on EPCglobal Class 1 Gen 2 standard from both PHY and MAC layers, has been presented.

The aforementioned discussions provide a basis for the following chapters.

Appendix 2.A Modified Bessel Function of the First Kind

The modified Bessel function of the first kind can be defined by either a series or an integral as follows:

$$I_0(x) = \sum_{k=0}^{\infty} \frac{\left(\frac{x}{2}\right)^{2k}}{(k!)^2},$$

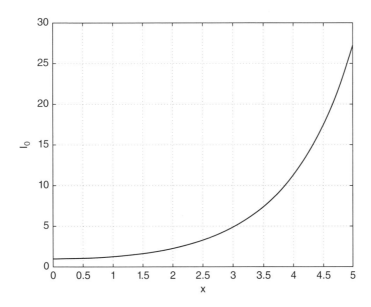

Figure 2.22 An illustration of the modified Bessel function of the first kind $I_0(x)$.

$$I_0(x) = \frac{1}{\pi} \int_{-1}^{1} \frac{e^{-xt}}{\sqrt{1 - t^2}} dt.$$

The function is illustrated in Figure 2.22.

References

[1] C. Angerer, R. Langwieser, G. Maier, and M. Rupp. Maximal ratio combining receivers for dual antenna RFID readers. In *IEEE 2009 Int. Microwave Workshop on Wireless Sensing, LocalPositioning, and RFID*, Cavtat, Croatia, 24–25 Sept. 2009.

[2] D. Arnitz, U. Muehlmann, and K. Witrisal. Characterization and modeling of UHF RFID channels for rangingand localization. *IEEE Trans. Antennas Propag.*, 60:2491–2501, 2012.

[3] C. A. Balanis. *Antenna Theory: Analysis and Design,* 3rd ed. John Wiley & Sons, Inc., Hoboken, NJ, 2005.

[4] J. F. Dickson. On-chip high-voltage generation in MNOS integrated circuits usingan improved voltage multiplier technique. *IEEE J. Solid-State Circuits*, 11:374–378, 1976.

[5] D. Dobkin and T. Wandinger. A radio-oriented introduction to radio frequency identification. *High Frequency Electronics*, 4(6):46–54, 2005.

[6] D. M. Dobkin and S. M. Weigand. UHF RFID and tag antenna scattering, part I: Experimentalresults. *Microwave Journal, Euro-Global Edition*, 49:170–190, 2007.

[7] EPCglobal. EPC radio-frequency identity protocols – Class-1 Generation-2UHF RFID protocol for communications at 860 MHz - 960 MHz, version1.0.9. 2005. Available:www.gs1.org/gsmp/kc/epcglobal/uhfc1g2/uhfc1g2_1_0_9-standard-20050126.pdf.

[8] EPCglobal. EPC radio-frequency identity protocols – Class-1 Generation-2UHF RFID protocol for communications at 860 MHz - 960 MHz, version1.2.0. 2008. Available:www.gs1.org/gsmp/kc/epcglobal/uhfc1g2/uhfc1g2_1_2_0-standard-20080511.pdf.

[9] K. Finkenzeller. *RFID Handbook,* 3rd ed. John Wiley & Sons, Ltd, Chichester, 2010.

[10] J. D. Griffin and G. D. Durgin. Gains for RF tags using multiple antennas. *IEEE Trans. Antennas Propag.*, 56:563–570, 2008.

[11] J. D. Griffin and G. D. Durgin. *Multipath fading measurements for multi-antenna backscatter RFID at5.8 GHz.* In *Proc. 2009 IEEE Int. Conf. RFID*, p. 322–329, Orlando,USA, 27–28 Apr. 2009.

[12] D. Kim M. A. Ingram and W. W. Smith. Measurements of small-scale fading and path loss for long range RFtags. *IEEE Trans. Antennas Propag.*, 51:1740–1749, 2003.

[13] M. A. Ingram, M. F. Demirkol, and D. Kim. Transmit diversity and spatial multiplexing for RF links usingmodulated backscatter. In *Int. Symp. Signals, Systems, and Electronics*, Tokyo, Japan, 24–27 July 2001.

[14] D.-Y. Kim, H.-S. Jo, H. Yoon, C. Mun, B.-J. Jang, and J.-G. Yook. Reverse-link interrogation range of a UHF MIMO-RFID system inNakagami-m fading channels. *IEEE Trans. Industrial Electronics*, 57:1468–1477, 2010.

[15] K. Maichalernnukul, S. Buaroong, F. Zheng, and T. Kaiser. BER analysis of a space-time coded RFID system in Nakagami-mfading channels. In. *Proc. Asia-Pacific Radio Science Conf.*, Taipei, Taiwan, 3–7 Sept. 2013.

[16] L. W. Mayer, M. Wrulich, and S. Caban. Measurements and channel modeling for short range indoor UHFapplications. In *1st European Conf. Antennas and Propagation*, Nice, France, 6–10 Nov. 2006.

[17] M. Nakagami. The *m*-distribution: A general formula of intensity distributionof rapid fading. In *Statistical Methods in Radio Wave Propagation*, W. C. Hoffman, Ed., p. 3–36, Oxford, UK: Pergamon, 1960.

[18] P. V. Nikitin and K. V. S. Rao. Theory and measurement of backscattering from RFID tags. *IEEE Antennas and Propagation Mag.*, 48(6):212–218, 2006.

[19] P. V. Nikitin and K. V. S. Rao. Antennas and propagation in UHF RFID systems. In *2008 IEEE Int. Conf. on RFID*, p. 277–288, Las Vegas,USA, 16–17 Apr. 2008.

[20] P. V. Nikitin, K. V. S. Rao, and S. Lazar. An overview of near field UHF RFID. In *2007 IEEE Int. Conf. on RFID*, p. 167–174, Grapevine,Texas, USA, 26–28 Mar. 2007.

[21] J. Polivka. Wideband UHF/microwave active isolators. *High Frequency Electronics*, pages 70–72, Dec. 2006.

[22] K. F. Riley, M. P. Hobson, and S. J. Bence. *Mathematical Methods for Physics and Engineering,* 2nd ed. Cambridge University Press, Cambridge, UK, 2002.

[23] A. Sample, D. Yeager, P. Powledge, A. Mamishev, and J. Smith. Design of an RFID-based battery-free programmable sensing platform. *IEEE Trans. Instrumentation and Measurement*, 57:26082615,2008.

[24] S. Thomas and M. S. Reynolds. QAM backscatter for passive UHF RFID tags. In *2010 IEEE Int. Conf. on RFID*, pages 210–214, Orlando,Florida, USA, 14–16 Apr. 2010.

[25] M. S. Trotter and G. D. Durgin. Survey of range improvement of commercial RFID tags with poweroptimized waveforms. In *2010 IEEE Int. Conf. on RFID*, p. 195–202, Orlando,Florida, USA, 14–16 Apr. 2010.

[26] R. Weinstein. RFID: A technical overview and its application to the enterprise. *IT Professional*, 7(3):27–33, 2005.

[27] B. York. Fundamental properties of antennas, *Lecture notes on* 'AntennaTheory', University of California in Santa Barbara, 2003. Available:`http://my.ece.ucsb.edu/York/Bobsclass/201C/Handouts/Chap3.pdf`.

3

Basic Signal Processing for RFID

For a passive RFID system, the signal received by the reader's receiver is a superposition of the backscattered signal from the tag, the leaked signal from the transmitter of the reader, and environmental or thermal noise. Since the tag has no active transmitter, the magnitude of the leaked signal might be several orders of magnitude stronger than that of the backscattered signal, and the ratio between the power of the backscattered signal and that of the noise might be also very low. The backscattered signal can be separated from the leaked signal by using a subharmonic modulation technique. However, the separation might not be clean enough. Therefore, further bandpass filtering in frequency domain for the separated signal is necessary. The low signal-to-noise power ratio (SNR) problem can be solved with a matching filter in time domain. In this chapter, we will discuss the design of bandpass filters and matching filters.

Besides, in RFID applications such as localization, several kinds of optimal estimators are often used to process obtained data in order to give the best estimate for the locations of tags or objects. In this chapter, the basic principles and algorithms for optimal estimators will be presented.

3.1 Bandpass Filters and Their Applications to RFID

3.1.1 Lowpass Filter Performance Specification

The design of a bandpass filter is mainly derived from the design of a corresponding lowpass filter. An ideal lowpass filter is of the following frequency response:

$$|H(j\omega)| = \begin{cases} H_0 & \text{for } |\omega| \le \omega_c, \\ 0 & \text{for } |\omega| > \omega_c, \end{cases} \tag{3.1}$$

$$\angle H(j\omega) = -\omega t_0, \tag{3.2}$$

where $H(j\omega)$ denotes the frequency response of the lowpass filter, ω_c is called the cut-off frequency, and H_0 and t_0 are constants. The frequency interval $[0, \omega_c]$ is called the pass-band. Equation (3.1) says that only the signal whose frequency is lower than ω_c can pass through the filter, while equation (3.2), i.e. the linear-phase property in the frequency response,

---

Digital Signal Processing for RFID, First Edition. Feng Zheng and Thomas Kaiser.
© 2016 John Wiley & Sons, Ltd. Published 2016 by John Wiley & Sons, Ltd.

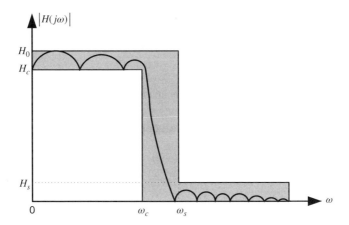

Figure 3.1   An illustration for the specification of lowpass filters.

requires that no distortion to the input signal is caused by the filter. Actually, equations (3.1) and (3.2) imply that the output signal is a pure time-delayed version of the input signal with an amplification if all its frequencies are lower than ω_c.

However, signal analysis theory shows that a filter will be noncausal if the magnitude of its frequency response is discontinuous. Therefore, the filter (3.1)–(3.2) cannot be realized in the real world. Hence, we need to use a smooth function to approximate the discontinuous function in the right-hand side of equation (3.1). The approximation[1] should satisfy some requirements. These requirements are mainly applied to the magnitude of the frequency response of the filter, while no restriction is applied on the phase of the frequency response. This greatly simplifies the design of the filter. Typically, the requirements are expressed as follows[2]:

$$\begin{cases} H_c \leq |H(j\omega)| \leq H_0 & \text{when } 0 \leq \omega \leq \omega_c, \\ |H(j\omega)| \leq H_s & \text{when } \omega \geq \omega_s, \end{cases} \tag{3.3}$$

where ω_s is called the stopband edge frequency, and H_c and H_s are constants. The frequency interval $[\omega_s, +\infty)$ is called the stopband. Generally, H_c should approach H_0 as much as possible, H_s should be much smaller than H_0 and ω_s $(> \omega_c)$ should approach ω_c as much as possible. No restriction is pre-specified on the magnitude of the frequency response for the frequency interval $\omega \in (\omega_c, \omega_s)$, but it is generally taken for granted that $|H(j\omega)|$ should be smaller than H_c when $\omega \in (\omega_c, \omega_s)$. This relaxation should facilitate the filter design.

The above requirements are illustrated in Figure 3.1.

To reduce unnecessary design parameters, two procedures are popularly adopted: magnitude normalization and frequency scaling. In magnitude normalization, the highest gain of the filter (normally it is the gain at the direct-current frequency) is set to be unity, that is, $H_0 = 1$. In frequency scaling, the frequency of the filter is scaled as

$$\Omega = \frac{\omega}{\omega_c}.$$

---

[1] Approximation is the heart of a realizable filter design [4].
[2] The requirements are only shown for the positive frequency part. The counterpart for the negative frequencies can be inferred from the symmetric property of the frequency response of the filter.

With these two procedures, requirements (3.3) can be rewritten as

$$
\begin{cases}
H_c \le |\bar{H}(j\Omega)| \le 1 & \text{when } 0 \le \Omega \le 1, \\
|\bar{H}(j\Omega)| \le H_s & \text{when } \Omega \ge \Omega_s,
\end{cases}
\tag{3.4}
$$

where $\Omega_s = \frac{\omega_s}{\omega_c}$, and $\bar{H}$ denotes the new filter designed in the new frequency domain according to the new requirements in (3.4).

Once the filter $\bar{H}$ is worked out, the original filter can be obtained via

$$
H(j\omega) = \bar{H}(j\Omega)\big|_{\Omega=\frac{\omega}{\omega_c}}.
$$

3.1.2   Lowpass Filter Design

Generally, we choose $|H(j\Omega)|^2$ to be a rational function of Ω and let $|H(j\Omega)|$ satisfy requirements in (3.4). By such a choice, the transfer function of the filter is also a rational function. Hence the filter can be realized by using either a passive RLC circuit or an active RC network (with the help of operational amplifiers). Theoretically, there are infinite approximations for the rational functions that satisfy requirements in (3.4), but four kinds of approximation methods are commonly used. These four methods are: Butterworth, Chebyshev, Pascal and elliptic approximations [4, 5]. In the following, the Butterworth and Chebyshev approximation design methods will be briefly introduced.

For simplicity, let us consider the case that the filter is of an all-pole transfer function, i.e., the transfer function of the filter is of the form

$$
H(s) = \frac{b}{A(s)} = \frac{b}{s^N + a_{N-1}s^{N-1} + \cdots + a_1 s + a_0},
$$

where the filter order N and coefficients $b, a_{N-1}, \ldots, a_0$ are to be decided in terms of the filter specifications.

The squared amplitude of the frequency response function is

$$
|H(j\Omega)|^2 = \frac{b^2}{|A(j\Omega)|^2} := \frac{b^2}{1 + \beta^2 K^2(\Omega)},
\tag{3.5}
$$

where parameter β and polynomial $K(\Omega)$ (real) are chosen to satisfy

$$
\beta^2 |K(\Omega)|^2 = |A(j\Omega)|^2 - 1.
$$

It is conventional in filter design to consider polynomial $K(\Omega)$ instead of $A(j\Omega)$ directly. Note that $K(\Omega)$ is the function of Ω instead of $j\Omega$. Furthermore, it is clear that $K(\Omega)$ is an even function of Ω if the filter parameters are restricted to be real.

3.1.2.1   Butterworth Filter

For the Butterworth filter, the polynomial $K(\Omega)$ is simply chosen as

$$
K(\Omega) = \Omega^N.
\tag{3.6}
$$

With this choice, the amplitude of the frequency response $|H(j\Omega)|$ is a monotonously decreasing function of Ω. Therefore, specification (3.4) can be translated to

$$
\begin{cases}
|\bar{H}(j0)| \leq 1 & \Leftarrow b = 1, \\
|\bar{H}(j1)| \geq H_c & \Leftrightarrow \dfrac{b}{\sqrt{1+\beta^2}} \geq H_c \quad \Leftrightarrow \beta \leq \sqrt{\dfrac{1}{H_c^2} - 1} := \beta_0, \\
|\bar{H}(j\Omega_s)| \leq H_s & \Leftrightarrow \dfrac{b}{\sqrt{1+\beta^2 \Omega_s^{2N}}} \leq H_s \Leftrightarrow N \geq \dfrac{\log\left(\dfrac{1}{H_s^2}-1\right)-2\log\beta}{2\log\Omega_s} := f(\beta).
\end{cases}
$$

The choice of N depends on the choice of β. Since the complexity of a filter increases with N, N should be as small as possible. Therefore, β and N can be designed as follows:

$$
\beta = \beta_0,
$$
$$
N = \lceil f(\beta_0) \rceil := N_0, \tag{3.7}
$$

where the notation $\lceil x \rceil$ stands for the minimum integer that is larger than or equal to x.

Once the filter order N is solved in terms of equation (3.7), there is some additional design freedom to choose parameter β. Let us define

$$
\beta_1 := f^{-1}(N)\big|_{N=N_0}.
$$

It is clear that $\beta_1 \leq \beta_0$. Then β can be chosen to be any number in the interval $[\beta_1, \beta_0]$. For example, if we choose $\beta = \beta_1$ with $N = N_0$, the performance of the filter in the passband will be improved, that is, the amplitude of its frequency response will be higher in the passband than the case when $\beta = \beta_0$ is chosen.

To realize the filter as designed here, we need to solve its transfer function from its amplitude frequency response. This can be generally carried out by using the following relationship

$$
|H(j\Omega)|^2 = H(s)H(-s)\big|_{s=j\Omega} \tag{3.8}
$$

and the fact that $H(s)$ is a stable transfer function. From equation (3.8) we can choose $H(s)$ such that

$$
H(s)H(-s) = |H(j\Omega)|^2\big|_{\Omega=-js}. \tag{3.9}
$$

The right-hand side of equation (3.9) can be obtained from equation (3.5). Based on the obtained rational function $H(s)H(-s)$, factorizing it and assigning the poles on the left-hand half plane to $H(s)$, one obtains a stable transfer function for the filter.

From equations (3.5), (3.6) and (3.9) it can be seen that $H(s)H(-s)$ is of the following form

$$
H(s)H(-s) = \frac{1}{1 + \beta^2(-s^2)^N} = \frac{1}{(-1)^N\beta^2} \times \frac{1}{s^{2N} + \dfrac{1}{(-1)^N\beta^2}},
$$

from which we can immediately solve the poles of $H(s)H(-s)$ as follows

$$
p_k = \left(\frac{1}{\beta^2}\right)^{\frac{1}{2N}} e^{j\left(\frac{2k+1}{2N}\pi + \frac{\pi}{2}\right)}, \quad k = 0, 1, \ldots, 2N-1.
$$

The poles on the left-hand half plane are

$$p_k^- = \frac{1}{\beta^{\frac{1}{N}}} e^{j\left(\frac{2k+1}{2N}\pi + \frac{\pi}{2}\right)}, \quad k = 0, 1, \dots, N-1. \tag{3.10}$$

Therefore, the transfer function of the Butterworth filter is

$$H(s) = \frac{1}{\beta} \times \frac{1}{\prod_{k=0}^{N-1}(s - p_k^-)}.$$

Based on this transfer function, one can use either passive RLC circuits or active RC networks to implement the filter.

Equation (3.10) shows that the poles of the Butterworth filter are located on a circle.

3.1.2.2 Chebyshev Filter

For the Chebyshev filter, the function $K(\Omega)$ is chosen as

$$K(\Omega) = T_N(\Omega), \tag{3.11}$$

where $T_N(\Omega)$ denotes the Chebyshev polynomial of order N. The Chebyshev polynomial can be defined in different ways. Two popular ways are [10]:

$$T_N(\Omega) = \begin{cases} \cos(N\cos^{-1}(\Omega)) & \text{when } |\Omega| \le 1, \\ \cosh(N\cosh^{-1}(\Omega)) & \text{when } |\Omega| > 1, \end{cases} \tag{3.12}$$

or

$$\begin{cases} T_N(\Omega) = 2\Omega T_{N-1}(\Omega) - T_{N-2}(\Omega), \\ T_0(\Omega) = 1 \quad \text{and} \quad T_1(\Omega) = \Omega. \end{cases} \tag{3.13}$$

From the recursive definition (3.13), one can easily find the polynomial expression for a specific order N. Some examples can be found in [1] for N up to 12.

A nice property of the Chebyshev polynomial is that it oscillates in the range $[-1, 1]$ when $|\Omega| \le 1$ and approaches to infinity monotonically and rapidly when $|\Omega| > 1$ for a large N, as shown in Figure 3.2. This property is very useful for filter approximation.

Taking the aforementioned property of the Chebyshev polynomial into consideration, we can translate specification (3.4) to

$$\begin{cases} |\bar{H}(j\Omega)|\big|_{0\le\Omega\le1} \le 1 & \Leftarrow b = 1, \\ |\bar{H}(j\Omega)|\big|_{0\le\Omega\le1} \ge H_c & \Leftrightarrow \frac{1}{\sqrt{1+\beta^2}} \ge H_c \quad\quad \Leftrightarrow \beta \le \sqrt{\frac{1}{H_c^2} - 1} := \beta_0, \\ \bar{H}(j\Omega)|\big|_{\Omega\ge\Omega_s} \le H_s & \Leftrightarrow \frac{1}{\sqrt{1+\beta^2 T_N^2(\Omega_s)}} \le H_s \Leftrightarrow T_N(\Omega_s) \ge \frac{1}{\beta}\sqrt{\frac{1}{H_s^2} - 1}. \end{cases} \tag{3.14}$$

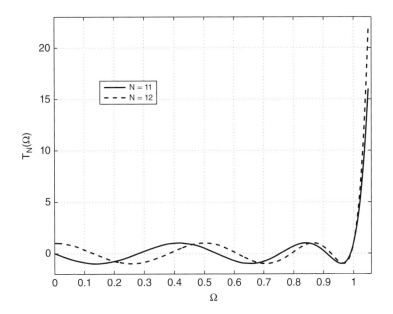

Figure 3.2 An illustration of Chebyshev polynomial, where the curves for $T_{11}(\Omega) = 1024\,\Omega^{11} - 2816\,\Omega^9 + 2816\,\Omega^7 - 1232\,\Omega^5 + 220\,\Omega^3 - 11\Omega$ and $T_{12}(\Omega) = 2048\,\Omega^{12} - 6144\,\Omega^{10} + 6912\,\Omega^8 - 3584\,\Omega^6 + 840\,\Omega^4 - 72\,\Omega^2 + 1$, as two examples, are plotted.

In terms of definition (3.12), the last inequality in the third line of equation (3.14) is equivalent to

$$N\cosh^{-1}(\Omega_s) \geq \cosh^{-1}\left(\frac{1}{\beta}\sqrt{\frac{1}{H_s^2} - 1}\right)$$

$$\Updownarrow$$

$$N \geq \frac{\cosh^{-1}\left(\frac{1}{\beta}\sqrt{\frac{1}{H_s^2} - 1}\right)}{\cosh^{-1}(\Omega_s)} := g(\beta).$$

Based on the same argument as in the Butterworth filter, we should make N as small as possible. Since the function $\cosh^{-1}(x)$ monotonically increases with x when $x > 0$, one should make β as large as possible in determining the filter order N. Therefore, β and N can be designed as follows:

$$\beta = \beta_0,$$
$$N = \lceil g(\beta_0) \rceil := N_0. \tag{3.15}$$

Once the filter order N is solved in terms of equation (3.15), there is some additional design freedom for parameter β. Let us define

$$\beta_2 := g^{-1}(N)\big|_{N=N_0}.$$

It is clear that $\beta_2 \leq \beta_0$. Then β can be chosen to be any number in the interval $[\beta_2, \beta_0]$. For example, if we choose $\beta = \beta_2$ with $N = N_0$, the gain ripple in the passband will be minimized.

For the realization of the filter, we need to find the transfer function $H(s)$. Substituting the definition of Chebyshev polynomial (3.12) into equation (3.9), we can solve the poles of $H(s)H(-s)$ as follows[3]:

$$p_k = \pm \sin\left[(2k+1)\frac{\pi}{2N}\right] \sinh\left[\frac{1}{N}\sinh^{-1}\left(\frac{1}{\beta}\right)\right]$$

$$+ j\cos\left[(2k+1)\frac{\pi}{2N}\right] \cosh\left[\frac{1}{N}\sinh^{-1}\left(\frac{1}{\beta}\right)\right], \quad k = 0, 1, \ldots, N-1. \quad (3.16)$$

The poles on the left-hand half plane are assigned to $H(s)$, that is, $H(s)$ has the following poles:

$$p_k^- = -\sin\left[(2k+1)\frac{\pi}{2N}\right] \sinh\left[\frac{1}{N}\sinh^{-1}\left(\frac{1}{\beta}\right)\right]$$

$$+ j\cos\left[(2k+1)\frac{\pi}{2N}\right] \cosh\left[\frac{1}{N}\sinh^{-1}\left(\frac{1}{\beta}\right)\right], \quad k = 0, 1, \ldots, N-1.$$

Summarizing the results obtained so far, we get $H(s)$ as follows:

$$H(s) = \frac{1}{\beta c_N} \times \frac{1}{\prod_{k=0}^{N-1}(s - p_k^-)}, \quad (3.17)$$

where $c_N = 2^{N-1}$ is the coefficient of the highest-order term Ω^N of $T_N(\Omega)$. Based on equation (3.17), the filter can be implemented by either a passive RLC circuit or an active RC network.

From equation (3.16) it is easily seen that

$$\left[\frac{\mathrm{Re}(p_k)}{\sinh\left[\frac{1}{N}\sinh^{-1}\left(\frac{1}{\beta}\right)\right]}\right]^2 + \left[\frac{\mathrm{Im}(p_k)}{\cosh\left[\frac{1}{N}\sinh^{-1}\left(\frac{1}{\beta}\right)\right]}\right]^2 = 1. \quad (3.18)$$

Equation (3.18) says that the poles of the Chebyshev filter are located on an ellipse.

Another approximation method is to use an elliptic rational function, which leads to an elliptic filter. Since its design is involved with the elliptic integral and Jocobi elliptic sine function [1, 4], which is beyond the scope of this introductory chapter, we omit this design method.

Some brief comparison about these filters is in order at this point. The Butterworth filter has poor gain performance, but shows good phase performance. The Chebyshev filter has better gain performance than the Butterworth filter, but is of poorer phase performance than the Butterworth filter. Of all the filters for the same gain performance specification, the elliptic filter can use a transfer function of the lowest order to realize the specification, i.e., the elliptic filter is of the best gain performance. However, its phase performance is the worst: the nonlinearity of the phase of the frequency response is highest among all the four kinds of filters. For more discussions, readers are referred to [4, 5].

---

[3] For details about the solution, readers are referred to Appendix 3.A

3.1.3 Bandpass Filter Design

An ideal bandpass filter is of the following amplitude frequency response:

$$|H_B(j\omega)| = \begin{cases} H_0 & \text{for } \omega_{c1} \le \omega \le \omega_{c2}, \\ 0 & \text{otherwise,} \end{cases} \tag{3.19}$$

where $H_B(j\omega)$ denotes the frequency response of the bandpass filter, and ω_{c1} and ω_{c2} are called lower cut-off frequency and higher cut-off frequency, respectively. In terms of signal analysis theory, filter (3.19) cannot be realized in the real world. Therefore, the amplitude frequency response $H_B(j\omega)$ as shown in equation (3.19) should be approximated by some smooth functions. The approximation should satisfy some requirements, namely

$$\begin{cases} H_c \le |H_B(j\omega)| \le H_0 & \text{when } \omega_{c1} \le \omega \le \omega_{c2}, \\ |H(j\omega)| \le H_s & \text{when } \omega \le \omega_{s1} \text{ or } \omega \ge \omega_{s2}, \end{cases}$$

where ω_{s1} and ω_{s2} are called the lower stopband edge frequency and higher stopband edge frequency, respectively. These requirements are illustrated in Figure 3.3, where ω_0 is the 'centre' frequency of the bandpass filter.

The basic way of designing a bandpass filter is to move the frequency response of a lowpass filter to a targeted frequency range by using a frequency transformation. A prerequisite for this frequency transformation is that the transformed transfer function can be implemented by using RLC circuits or active RC networks. Therefore, rational functions are typically used for this frequency transformation, since the new transfer function by applying this transformation is also a rational function.

Now let $H_L(s)$ denote the transfer function of a lowpass filter designed by any methods presented in the preceding subsection, and let $H_B(\bar{s})$ be the transfer function of a bandpass filter. The frequency transformation is carried out through the following substitution

$$s = \frac{1}{B} \frac{\bar{s}^2 + \omega_0^2}{\bar{s}}. \tag{3.20}$$

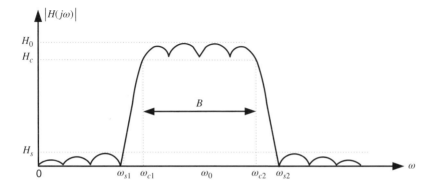

Figure 3.3 An illustration of the specification of bandpass filters.

That is to say, the transfer function of the bandpass filter is given by

$$H_{\mathrm{B}}(\bar{s}) = H_{\mathrm{L}}(s)\Big|_{s=\frac{1}{B}\frac{\bar{s}^2+\omega_0^2}{\bar{s}}}. \tag{3.21}$$

Through equation (3.21) we can see that the form of the amplitude of the frequency response is not changed, while the frequency of the original lowpass filter is shifted via a nonlinear mapping. The frequency mapping function can be found through the relationship (3.21). Let $s = j\Omega$ and $\bar{s} = j\omega$. Then substituting s and $\bar{s}$ into both sides of equation (3.20) yields

$$\omega^2 - B\Omega\omega - \omega_0^2 = 0. \tag{3.22}$$

For any a given frequency $\Omega = \Omega_0$ in the s-plane, there are two frequencies $\omega = \{\omega_1, \omega_2\}$, which solves equation (3.22):

$$\begin{cases} \omega_1 = \dfrac{B}{2}\Omega_0 - \sqrt{\dfrac{B^2}{4}\Omega_0^2 + \omega_0^2} < 0, \\[4mm] \omega_2 = \dfrac{B}{2}\Omega_0 + \sqrt{\dfrac{B^2}{4}\Omega_0^2 + \omega_0^2} > 0. \end{cases} \tag{3.23}$$

From equation (3.23) we can see that for the frequency $\Omega = -\Omega_0$, the two frequencies $\omega = \{-\omega_1, -\omega_2\}$ solve equation (3.22). Let us define

$$\begin{cases} \omega_{c1} := -\dfrac{B}{2}\Omega_c + \sqrt{\dfrac{B^2}{4}\Omega_c^2 + \omega_0^2}, \\[4mm] \omega_{c2} := \dfrac{B}{2}\Omega_c + \sqrt{\dfrac{B^2}{4}\Omega_c^2 + \omega_0^2}, \end{cases} \tag{3.24}$$

where Ω_c is the cut-off frequency of the lowpass filter. Then the frequency mapping relationship between the lowpass filter and bandpass filter can be illustrated in Figure 3.4 [4].

From equation (3.24) it is found that

$$\omega_0 = \sqrt{\omega_{c1}\omega_{c2}}. \tag{3.25}$$

Equation (3.25) says that the centre frequency of the bandpass filter is the geometric mean value of ω_{c1} and ω_{c2} instead of their arithmetic mean value. Notice that once ω_0 is calculated in terms of equation (3.25), Ω_c will be normalized to the unity, i.e.,

$$\Omega_c = \frac{\omega_0^2 - \omega_{c1}^2}{B\omega_{c1}} = \frac{\omega_{c2}^2 - \omega_0^2}{B\omega_{c2}} = 1. $$

Similar to equation (3.24), let us define the mapping regarding the stopband edge frequencies between the lowpass filter and bandpass filter as follows:

$$\omega_{s1} := \sqrt{\frac{B^2}{4}\Omega_s^2 + \omega_0^2} - \frac{B}{2}\Omega_s \quad\Longleftrightarrow\quad \Omega_s = \frac{\omega_0^2 - \omega_{s1}^2}{B\omega_{s1}}, \tag{3.26}$$

$$\omega_{s2} := \sqrt{\frac{B^2}{4}\Omega_s^2 + \omega_0^2} + \frac{B}{2}\Omega_s \quad\Longleftrightarrow\quad \Omega_s = \frac{\omega_{s2}^2 - \omega_0^2}{B\omega_{s2}}. \tag{3.27}$$

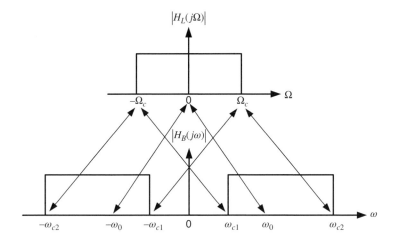

Figure 3.4 An illustration of the frequency mapping between a lowpass filter and a bandpass filter.

Since the relationship $\sqrt{\omega_{s1}\omega_{s2}} = \omega_0$ might *not* hold true, the values for the parameter Ω_s calculated in terms of equations (3.26) and (3.27) might *not* be equal to each other. In this case, we choose the tighter design specification for the lowpass filter, i.e.,

$$\Omega_s = \min \left\{ \frac{\omega_0^2 - \omega_{s1}^2}{B\omega_{s1}}, \frac{\omega_{s2}^2 - \omega_0^2}{B\omega_{s2}} \right\}. \tag{3.28}$$

From these discussions, we can summarize the design procedure of a bandpass filter as follows:

1. Given the specification of a bandpass filter as shown in Figure 3.3, namely $\{\omega_{c1}, \omega_{c2}, \omega_{s1}, \omega_{s2}, H_0, H_c, H_s\}$.
2. Recalculate the centre frequency ω_0 according to equation (3.25).
3. Set $\Omega_c = 1$ and calculate Ω_s according to equation (3.28).
4. Calculate the lowpass filter $H_L(s)$ in terms of specification $\{\Omega_c, \Omega_s, H_0, H_c, H_s\}$.
5. Calculate the bandpass filter $H_B(\bar{s})$ in terms of transformation (3.21).

3.1.4 Bandpass Filters for RFID Systems

As mentioned in Chapter 2, a reader often uses subcarrier modulations to avoid the strong leakage interference signal from reader's transmitter to reader's receiver. Therefore, a well-designed bandpass filter at the reader is very helpful to reduce the interference caused by the leakage signal.

In this subsection, the filter design methods presented in subsections 3.1.2 and 3.1.3 will be applied to RFID systems. We will show how the filter complexity is affected by the filter specification via two examples: one is for a high-frequency RFID system, and another is for a microwave-frequency RFID system.

3.1.4.1 Bandpass Filters for High-Frequency RFID Systems

Let us consider a reader with a carrier frequency of 13.56 MHz. The tag uses a subcarrier of 212 kHz to modulate the tag-ID signal. Therefore, the centre carrier frequency of the tag signal is $f_0' = 13.56 \pm 0.212$ MHz. We consider the case of $f_0' = 13.56 + 0.212 = 13.712$ MHz. The bandwidth of received tag-backscattered signal is coordinated by the reader's R→T preamble. Assume that the parameter divide ratio is setting to be 8. The signal bandwidth can be calculated as $\text{BLF} = \dfrac{8}{T_{\text{TRcal}}}$, where the parameter T_{TRcal} is specified in Figure 2.17. According to Section 2.7.1, we can choose the following parameters

$$T_0 = 25 \ \mu s, \ T_{\text{RTcal}} = 3T_0 = 75 \ \mu s, \ T_{\text{TRcal}} = 3T_{\text{RTcal}} = 225 \ \mu s \qquad (3.29)$$

as a design example. This setting gives

$$B = 2\pi\text{BLF} = 2\pi \frac{8}{T_{\text{TRcal}}} = 2.2340 \times 10^5 \ 1/s,$$

which corresponds to a bandwidth of 35.6 kHz. The lower and higher cut-off frequencies are given by

$$\omega_{c1} = 2\pi f_0' - \frac{B}{2} = 8.6044 \times 10^7 \ 1/s, \qquad (3.30)$$

$$\omega_{c2} = 2\pi f_0' + \frac{B}{2} = 8.6267 \times 10^7 \ 1/s, \qquad (3.31)$$

respectively.

In the passband, it is hoped that the gain ripple is less than 1 dB. In the stopband, it is hoped that the filter gain falls down to at least 20 dB lower than the passband nominal gain outside $\pm 5\%B$ of the passband. These requirements give

$$H_c = 0.7943 \quad \text{with} \quad H_0 = 1,$$

$$H_s = 0.01,$$

$$\omega_{s1} = \omega_{c1} - 0.05B = 8.6032 \times 10^7 \ 1/s,$$

$$\omega_{s2} = \omega_{c2} + 0.05B = 8.6278 \times 10^7 \ 1/s.$$

Since $2\pi f_0'$ is the arithmetic mean value of ω_{c1} and ω_{c2}, we need to calculate the geometric mean value of ω_{c1} and ω_{c2}, denoted as ω_0, to compute other parameters of the filter. From equations (3.30) and (3.31) we get

$$\omega_0 = \sqrt{\omega_{c1}\omega_{c2}} = 8.6155 \times 10^7 \ 1/s.$$

Then from equation (3.28) we obtain

$$\Omega_s = 1.0999.$$

Having obtained the values of parameters $\{\Omega_c, \Omega_s, H_0, H_c, H_s\}$, we can calculate the lowpass Butterworth filter and Chebyshev filter as follows:

Butterworth filter:

$$H_{L,\,\text{Butt}}(s) = \frac{1}{\beta} \times \frac{1}{\prod\limits_{k=0}^{N-1}(s - p_k^-)}, \tag{3.32}$$

$$N = 52, \tag{3.33}$$

$$\beta \in [\beta_1, \beta_0], \quad \text{with} \quad \beta_0 = 0.7648, \ \beta_1 = 0.7085, \tag{3.34}$$

$$p_k^- = \frac{1}{\beta^{\frac{1}{52}}} e^{J\left(\frac{2k+1}{104}\pi + \frac{\pi}{2}\right)}, \quad k = 0, 1, \ldots, 51. \tag{3.35}$$

Chebyshev filter:

$$H_{L,\,\text{Cheb}}(s) = \frac{1}{2^{N-1}\beta} \times \frac{1}{\prod\limits_{k=0}^{N-1}(s - p_k^-)}, \tag{3.36}$$

$$N = 13, \tag{3.37}$$

$$\beta \in [\beta_2, \beta_0], \quad \text{with} \quad \beta_0 = 0.7648, \ \beta_2 = 0.6286, \tag{3.38}$$

$$p_k^- = -\sin\left[(2k+1)\frac{\pi}{26}\right]\sinh\left[\frac{1}{13}\sinh^{-1}\left(\frac{1}{\beta}\right)\right]$$

$$+J\cos\left[(2k+1)\frac{\pi}{26}\right]\cosh\left[\frac{1}{13}\sinh^{-1}\left(\frac{1}{\beta}\right)\right], \quad k = 0, 1, \ldots, 12. \tag{3.39}$$

Substituting equation (3.20) into (3.32) and (3.36) gives bandpass Butterworth filter and Chebyshev filter as follows:

$$H_{B,\,\text{Butt}}(\bar{s}) = \frac{1}{\beta} \times \frac{(B\bar{s})^{52}}{\prod\limits_{k=0}^{51}(\bar{s}^2 - Bp_k^-\bar{s} + \omega_0^2)}, \tag{3.40}$$

$$H_{B,\,\text{Cheb}}(\bar{s}) = \frac{1}{2^{12}\beta} \times \frac{(B\bar{s})^{13}}{\prod\limits_{k=0}^{12}(\bar{s}^2 - Bp_k^-\bar{s} + \omega_0^2)}, \tag{3.41}$$

with the corresponding parameters being given by equations (3.33)–(3.35) and (3.37)–(3.39), respectively.

The amplitudes of the frequency response of these two bandpass filters are shown in Figures 3.5 and 3.6, respectively. It can be seen that, to achieve the same performance within the passband and at the stopband edge frequency, the Butterworth filter (with the filter order being 52) is much more complicated than the Chebyshev filter (with the filter order being 13). As a compensation for this drawback, the Butterworth filter offers a much deeper fall within the stopband than the Chebyshev filter does.

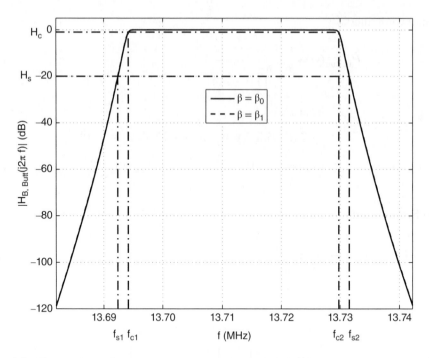

Figure 3.5 The amplitude of the frequency response of the bandpass filter (3.40), where $f_{c1} = \omega_{c1}/(2\pi)$, $f_{c2} = \omega_{c2}/(2\pi), f_{s1} = \omega_{s1}/(2\pi)$ and $f_{s2} = \omega_{s2}/(2\pi)$.

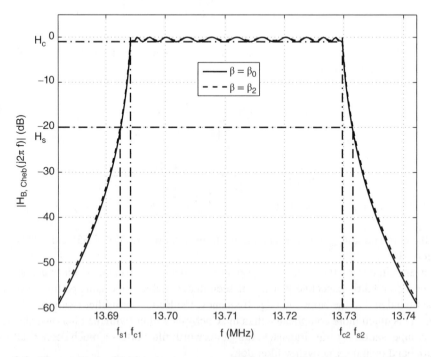

Figure 3.6 The amplitude of the frequency response of the bandpass filter (3.41), where $f_{c1} = \omega_{c1}/(2\pi)$, $f_{c2} = \omega_{c2}/(2\pi), f_{s1} = \omega_{s1}/(2\pi)$ and $f_{s2} = \omega_{s2}/(2\pi)$.

3.1.4.2 Bandpass Filters for Microwave-Frequency RFID Systems

Now let us consider a reader with a carrier frequency of 5.8 GHz. The tag also uses a subcarrier of 212 kHz to modulate the tag-ID signal. Therefore, the centre carrier frequency of the tag signal is $f_0' = 5800 \pm 0.212$ MHz. We consider the case of $f_0' = 5800 + 0.212 = 5800.212$ MHz. Choosing the same parameter for the reader's R→T preamble as in equation (3.29) gives the bandwidth of $B = 2.2340 \times 10^5$ 1/s. Then the lower and higher cut-off frequencies are given by

$$\omega_{c1} = 2\pi f_0' - \frac{B}{2} = 3.64437803 \times 10^{10} \text{ 1/s,}$$

$$\omega_{c2} = 2\pi f_0' + \frac{B}{2} = 3.64440037 \times 10^{10} \text{ 1/s,}$$

respectively.

Similar to the preceding subsection, it is required that the gain ripple in the passband is less than 1 dB and the filter gain falls down in the stopband to at least 20 dB lower than the passband nominal gain outside $\pm 5\%B$ of the passband. These requirements give

$$H_c = 0.7943 \quad \text{with} \quad H_0 = 1,$$

$$H_s = 0.01,$$

$$\omega_{s1} = 3.64437692 \times 10^{10} \text{ 1/s,}$$

$$\omega_{s2} = 3.64440149 \times 10^{10} \text{ 1/s,}$$

$$\omega_0 = 3.64438920 \times 10^{10} \text{ 1/s,}$$

$$\Omega_s = 1.1000.$$

Given these values of parameters $\{\Omega_c, \Omega_s, H_0, H_c, H_s\}$, we can calculate the lowpass Butterworth filter and Chebyshev filter as follows:

Butterworth filter:

$$H_{\text{L, Butt}}(s) = \frac{1}{\beta} \times \frac{1}{\prod_{k=0}^{N-1} (s - p_k^-)}, \tag{3.42}$$

$$N = 52, \tag{3.43}$$

$$\beta \in [\beta_1, \beta_0], \text{ with } \beta_0 = 0.7648, \ \beta_1 = 0.7040, \tag{3.44}$$

$$p_k^- = \frac{1}{\beta^{\frac{1}{52}}} e^{j\left(\frac{2k+1}{104}\pi + \frac{\pi}{2}\right)}, k = 0, 1, \ldots, 51. \tag{3.45}$$

Chebyshev filter:

$$H_{L,\,\text{Cheb}}(s) = \frac{1}{2^{N-1}\beta} \times \frac{1}{\displaystyle\prod_{k=0}^{N-1}(s - p_k^-)}, \tag{3.46}$$

$$N = 13, \tag{3.47}$$

$$\beta \in [\beta_2, \beta_0], \text{ with } \beta_0 = 0.7648, \ \beta_2 = 0.6262, \tag{3.48}$$

$$p_k^- = -\sin\left[(2k+1)\frac{\pi}{26}\right]\sinh\left[\frac{1}{13}\sinh^{-1}\left(\frac{1}{\beta}\right)\right]$$

$$+ j\cos\left[(2k+1)\frac{\pi}{26}\right]\cosh\left[\frac{1}{13}\sinh^{-1}\left(\frac{1}{\beta}\right)\right], k = 0, 1, \ldots, 12. \tag{3.49}$$

Substituting equation (3.20) into (3.42) and (3.46) gives the bandpass a Butterworth filter and Chebyshev filter whose expressions are the same as equations (3.40) and (3.41), respectively, with the corresponding parameters being given by equations (3.43)–(3.45) and (3.47)–(3.49), respectively.

The amplitude of the frequency response of these two bandpass filters are shown in Figures 3.7 and 3.8, respectively.

Comparing equations (3.32)–(3.35) and (3.42)–(3.45), equations (3.36)–(3.39) and (3.46)–(3.49) and Figures 3.5–3.6 and Figures 3.7–3.8, we can see that shifting carrier frequency does not affect considerably the performance and complexity of a bandpath filter. Notice that the performance of the designed bandpass filter varies only marginally caused by the change of the parameter β in the pre-specified range. This is due to the fact that the bandwidth of the filter is very narrow compared to the carrier frequencies.

3.2 Matching Filters and their Applications to RFID

Matching filters play a crucial role in optimal receivers of radio systems and other kinds of systems such as acoustic systems. To show the basic principle of a matching filter, let us consider a linear system with impulse response $h(t)$ and frequency response $H(j\omega)$, subject to the constraint

$$\int_{-\infty}^{\infty} h^2(t)dt = 1, \tag{3.50}$$

i.e., no power amplification will be provided by the system. Assume that a finite-support and finite-energy signal $x(t)$ is applied to the system:

$$x(t) = 0 \quad \text{when } t < 0 \text{ or } t > T, \text{ and } \int_0^T x^2(t)dt < \infty.$$

Our question is: how to design $h(t)$ such that the system will produce a maximal output at some given time t_0? The question can be easily answered by using the Cauchy–Schwarz inequality, while the result has changed some design ideas of many systems and affected the course of some events dramatically when its principle was first applied to practical radio systems in World War II.

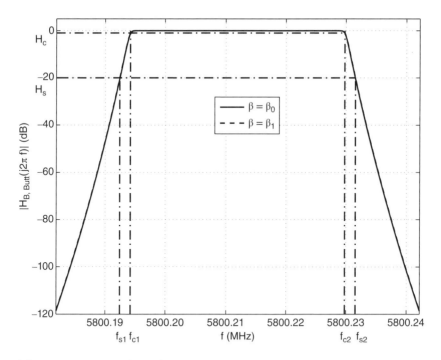

Figure 3.7 The amplitude of the frequency response of the Butterworth bandpass filter for a microwave-frequency RFID system.

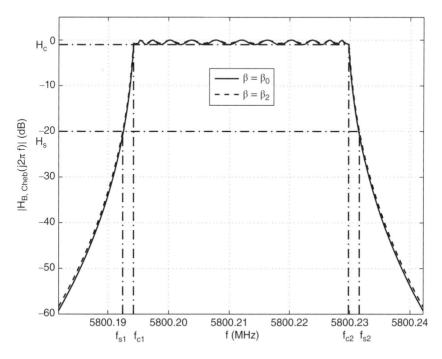

Figure 3.8 The amplitude of the frequency response of the Chebyshev bandpass filter for a microwave-frequency RFID system.

Denote the system output with $y(t)$. Then we have

$$y(t_0) = \int_{-\infty}^{\infty} h(t)x(t_0 - t)dt \leq \sqrt{\int_{-\infty}^{\infty} h^2(t)dt} \sqrt{\int_{-\infty}^{\infty} x^2(t_0 - t)dt}$$

with the equality occurring if and only if

$$h(t) = cx(t_0 - t) = cx(-(t - t_0)), \tag{3.51}$$

where c is a constant. The system characterized by the impulse response (3.51) is the optimal filter that maximizes the system output at time t_0. It is seen that the optimal filter consists of two procedures: time shift and time reversal of the input signal. It is due to the time reversal property that the obtained optimal filter is called matching filter. An illustration of the matching filter is shown in Figure 3.9. One can see from Figure 3.9 that time t_0 must be chosen to satisfy $t_0 \geq T$ to guarantee the causality of the filter.

Substituting equation (3.51) into (3.50) yields

$$c = \frac{1}{\sqrt{\int_{-\infty}^{\infty} x^2(t)dt}}.$$

Let $X(j\omega)$ and $Y(j\omega)$, respectively, denote the Fourier transforms of input signal $x(t)$ and output signal $y(t)$. Then from equation (3.51) we get

$$H(j\omega) = cX^*(j\omega)e^{j\omega t_0}.$$

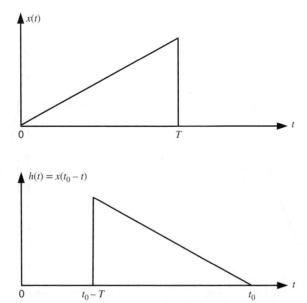

Figure 3.9   An illustration of a matching filter.

Since $Y(j\omega) = H(j\omega)X(j\omega)$, one can see that the matching filter weights the input signal in frequency domain in such a way: it reduces the weaker spectral components of $x(t)$ by assigning a proportionally lower gain and enhances the stronger spectral components of $x(t)$ by assigning a proportionally higher gain in frequency domain. This gives another explanation of the word 'matching' for the filter: which is actually the principle of frequency-domain maximal ratio combining.

Another derivation of matching filter comes from the maximization of the output SNR of a linear filter. Consider the system as shown in Figure 3.10, where the input signal $x(t)$ is corrupted by an additive white stationary noise $n(t)$. Suppose that the waveform of $x(t)$ is known and has finite support, as in the cases of the signalling signals in RFID systems and baseband waveforms of the transmitted symbols in communication systems. Now the question is: how to design a linear processor (filter) $h(t)$ so that the output SNR at some time t_0 is maximized? By using the Cauchy–Schwarz inequality it can be easily seen that the optimal filter is the one as shown in equation (3.51), i.e. the matching filter.

From these discussions, we can easily obtain the matching filter at the tag for reader-to-tag communications. Figure 3.11 shows the matching filters for the PIE symbol detection, where $h_0(t)$ is the impulse response of the matching filter for detecting data '0', and $h_1(t)$ for detecting data '1'.

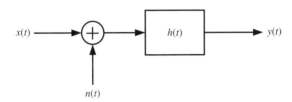

Figure 3.10   An illustration of a linear system to maximize the output SNR.

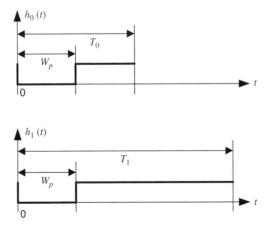

Figure 3.11   Matching filters for PIE symbol detection at RFID tags.

Similarly, we can design matching filters at the reader for detecting data symbols coming form the tag. The challenge lies in that the waveforms of the tag data symbols '0' or '1' might vary considerably due to the simplicity of modulation circuits in some RFID tags.

3.3   A Review of Optimal Estimation

In Chapter 8, we will discuss the localization problem with the help of RFID systems. The localization problem can be basically formulated as a parameter estimation problem, where the position of an object, e.g., two or three-dimensional coordinates of the object (even the speed of an object in some applications), should be estimated based on measured or observed information. Therefore, we will discuss some basic estimation algorithms in this section, to provide a preliminary for the sequel chapters.

Suppose that there is a set of parameters that are associated to an entity and yet independent of each other. For example, the coordinates of an object, the tap coefficients of a finite impulse response filter, and the mean value and variance of a random variable are such sets of parameters, while the current I passing through a resistor R and the voltage V across R do not belong to this class of parameters since I and V are simply constrained by the relationship $V = IR$. Let θ be the set of the concerned parameters. We have also a group of measurements that are associated with θ. Denote the measured information with $\mathbf{y}$. Suppose that $\mathbf{y}$ and θ are associated with each other by the following equation

$$\mathbf{y} = \mathbf{f}(\theta) + \mathbf{w} \tag{3.52}$$

or

$$\mathbf{y} = \mathbf{H}\theta + \mathbf{w}, \tag{3.53}$$

where $\mathbf{f}$ is a nonlinear vector-valued function, $\mathbf{H}$ is a matrix with an appropriate dimension, and $\mathbf{w}$ is unknown and called measurement noise, which corresponds to measurement errors and makes equation (3.52) or (3.53) hold true for every measurement. Our objective is to get a proper estimate for the true value of θ from $\mathbf{y}$ and $\mathbf{f}$ or $\mathbf{H}$ (if known). Equation (3.52) is called nonlinear measurement model, while equation (3.53) is a linear measurement model.

Clearly, a proper estimator depends on the knowledge about the measurement noise. If one has more detailed knowledge about the noise, one can expect an estimator of better performance. For example, if one knows nothing about the noise, then the least square (LS) estimation is a proper estimator; if one knows the probability distribution of the noise, then the maximum likelihood (ML) estimator is one of the best choices for the estimator; if one knows the statistical properties of the noise and parameter[4] up to the second order, then minimum mean square error (MMSE) estimation is a good choice for the estimator.

3.3.1   Linear Least Square Estimation

In LS estimation, we do not know anything about measurement noise. Therefore, the LS estimation is formulated as a deterministic optimization problem. Let $\hat{\theta}_{LS}$ denote the LS estimate

---

[4] In this case, the parameter as a whole is considered as a random variable and the parameter embedded in the obtained measurements is considered as a realization of this random variable.

of θ. The estimation criterion is to make $\mathbf{f}(\hat{\theta}_{LS})$ or $\mathbf{H}\hat{\theta}_{LS}$ approach to $\mathbf{y}$ as much as possible, i.e.,

$$\hat{\theta}_{LS} = \arg\min_{\hat{\theta}} \|\mathbf{y} - \mathbf{f}(\hat{\theta})\|$$

or

$$\hat{\theta}_{LS} = \arg\min_{\hat{\theta}} \|\mathbf{y} - \mathbf{H}\hat{\theta}\|, \tag{3.54}$$

where $\| \cdot \|$ denotes the l_2 (Euclidean) norm [6] in the corresponding vector space. The solution to problem (3.54) is given by [7]:

$$\hat{\theta}_{LS} = (\mathbf{H}^T\mathbf{H})^{-1}\mathbf{H}^T\mathbf{y}. \tag{3.55}$$

From equation (3.55) we can see that the matrix $\mathbf{H}^T\mathbf{H}$ should be invertible to guarantee that the LS estimation is well-posed. When the dimension of $\mathbf{y}$ is higher than the dimension of θ and the entries of $\mathbf{y}$ come from different sources, the matrix $\mathbf{H}^T\mathbf{H}$ is generally invertible.

The measurements contained in vector $\mathbf{y}$ can come from different kinds of sensors, and each sensor might be of different accuracy. To take this possibility into account, the LS estimation criterion is often changed to

$$\hat{\theta}_{WLS} = \arg\min_{\hat{\theta}} (\mathbf{y} - \mathbf{f}(\hat{\theta}))^T\mathbf{W}(\mathbf{y} - \mathbf{f}(\hat{\theta})),$$

or

$$\hat{\theta}_{WLS} = \arg\min_{\hat{\theta}} (\mathbf{y} - \mathbf{H}\hat{\theta})^T\mathbf{W}(\mathbf{y} - \mathbf{H}\hat{\theta}), \tag{3.56}$$

where $\mathbf{W}$ is a positive definite matrix, and $\hat{\theta}_{WLS}$ denotes the weighted least square (WLS) estimate of θ. For example, when $\mathbf{W}$ is a diagonal matrix, we can choose $\mathbf{W}_{ii}$ to be inversely proportional to the variance of $\mathbf{y}_i$, so that the measurements of higher accuracy are evaluated more importantly in the performance constraint. The solution to problem (3.56) is given by [7]:

$$\hat{\theta}_{WLS} = (\mathbf{H}^T\mathbf{W}\mathbf{H})^{-1}\mathbf{H}^T\mathbf{W}\mathbf{y}. \tag{3.57}$$

The LS algorithm (3.57) is called WLS estimation, which is widely used in practice.

Since measurement model (3.53) is linear in θ, the LS estimator (3.55) and WLS estimator (3.57) are called linear least square (LLS) estimator and linear weighted least square (LWLS) estimator.

For the nonlinear measurement model (3.52), there are no closed-form solutions for the LS or WLS estimation. Various kinds of approximation methods can be used to obtain approximated LS or WLS solution.

3.3.2   Linear Minimum Mean Square Error Estimation

In MMSE estimation, we know the statistical properties of the measurement noise up to the second order, and hence the problem is formulated as a stochastic optimization problem. In this regard, the true parameter θ is also considered as a random variable (vector) and its statistical properties up to the second order are also assumed to be known. The general MMSE estimation can be conducted based on a general correlation model (in the sense of probability) between measurement $\mathbf{y}$ and parameter θ. Let $\mathbf{C}_{\mathbf{xz}}$ denote the covariance matrix between random vectors $\mathbf{x}$ and $\mathbf{z}$: $\mathbf{C}_{\mathbf{xz}} = \mathbb{E}\{[\mathbf{x} - \mathbb{E}(\mathbf{x})][\mathbf{z} - \mathbb{E}(\mathbf{z})]^T\}$.

The MMSE estimation criterion is

$$\hat{\theta}_{\mathrm{MMSE}} = \arg \min_{\hat{\theta}} \mathbb{E}\{(\hat{\theta} - \theta)^T (\hat{\theta} - \theta)\}, \qquad (3.58)$$

where $\hat{\theta}_{\mathrm{MMSE}}$ stands for the MMSE estimate of θ. The linear MMSE (LMMSE) estimation further restricts the estimate of θ, denoted as $\hat{\theta}_{\mathrm{LMMSE}}$, to be a linear function of $\mathbf{y}$. Suppose that the correlation of $\mathbf{y}$ and θ is characterized by the covariance matrix $\mathbf{C}_{\theta\mathbf{y}}$. Then the LMMSE estimate of θ is given by [7]

$$\hat{\theta}_{\mathrm{LMMSE}} = \mathbf{C}_{\theta\mathbf{y}} \mathbf{C}_{\mathbf{yy}}^{-1} [\mathbf{y} - \mathbb{E}(\mathbf{y})] + \bar{\theta}, \qquad (3.59)$$

where $\bar{\theta} = \mathbb{E}(\theta)$.

Specializing equation (3.59) to the linear measurement model (3.53) and supposing that $\mathbf{C}_{\theta\mathbf{w}} = 0$ and $\mathbb{E}(\mathbf{w}) = 0$, we have [7]:

$$\hat{\theta}_{\mathrm{LMMSE}} = \mathbf{\Psi}(\mathbf{y} - \mathbf{H}\bar{\theta}) + \bar{\theta}, \qquad (3.60)$$

$$\mathbf{\Psi} = \mathbf{C}_{\theta\theta} \mathbf{H}^T (\mathbf{H}\mathbf{C}_{\theta\theta}\mathbf{H}^T + \mathbf{C}_{\mathbf{ww}})^{-1} \qquad (3.61)$$

$$= (\mathbf{H}^T \mathbf{C}_{\mathbf{ww}}^{-1} \mathbf{H} + \mathbf{C}_{\theta\theta}^{-1})^{-1} \mathbf{H}^T \mathbf{C}_{\mathbf{ww}}^{-1}.$$

From equation (3.61) we can see that the matrix $\mathbf{H}\mathbf{C}_{\theta\theta}\mathbf{H}^T + \mathbf{C}_{\mathbf{ww}}$ should be invertible to guarantee that the MMSE estimation is well-posed. When the measurements contained in $\mathbf{y}$ are independent of each other, matrix $\mathbf{C}_{\mathbf{ww}}$, and hence $\mathbf{H}\mathbf{C}_{\theta\theta}\mathbf{H}^T + \mathbf{C}_{\mathbf{ww}}$, are invertible. Therefore, it is not required in MMSE estimation that the dimension of $\mathbf{y}$ is higher than the dimension of θ. This is because some prior information, such as $\bar{\theta}$, about θ is available. Then even without any measurements, some 'proper' estimate about θ can be made to make (3.58) satisfied. For example, we can simply let $\hat{\theta}_{\mathrm{MMSE}} = \bar{\theta}$ when no measurements are available.

Generally, the statistical information about the measurement noise is easy to get, but in many applications such as the case of identification of a black-box system, the (a prior) statistical information about the parameter is difficult to obtain. In this case, one often assumes that $\mathbf{C}_{\theta\theta} = \sigma_\theta^2 \mathbf{I}$ with $\sigma_\theta^2 \to \infty$. Thus we have $\mathbf{C}_{\theta\theta}^{-1} \to 0$ and

$$-\mathbf{\Psi}\mathbf{H}\bar{\theta} + \bar{\theta} = (\mathbf{H}^T \mathbf{C}_{\mathbf{ww}}^{-1} \mathbf{H} + \mathbf{C}_{\theta\theta}^{-1})^{-1} \mathbf{C}_{\theta\theta}^{-1} \bar{\theta} \to 0.$$

Substituting the above result into equation (3.60) yields

$$\hat{\theta}_{\mathrm{LMMSE}} \to (\mathbf{H}^T \mathbf{C}_{\mathbf{ww}}^{-1} \mathbf{H})^{-1} \mathbf{H}^T \mathbf{C}_{\mathbf{ww}}^{-1} \mathbf{y},$$

which is the WLS estimate with $\mathbf{W} = \mathbf{C}_{\mathbf{ww}}^{-1}$.

Similar to the case of LS estimation, there are no closed-form solutions to the MMSE estimation of θ for nonlinear measurement model (3.52).

3.3.3   Maximum Likelihood Estimation

In ML estimation, the parameter vector θ is considered as a deterministic vector. Since the probability distribution of the measurement noise is available, the probability distribution of the measurement vector $\mathbf{y}$ for a given θ can be computed based on equation (3.52) or (3.53). Let $p_{\mathbf{w}}(\mathbf{w})$ be the probability density function (pdf) of the measurement noise $\mathbf{w}$, and let $p_{\mathbf{y}}(\mathbf{y}; \theta)^5$ be the pdf of the measurement vector $\mathbf{y}$ for a given θ. If the pdf of $p_{\mathbf{w}}(\mathbf{w})$ is known, then the pdf $p_{\mathbf{y}}(\mathbf{y}; \theta)$ can be readily obtained:

$$p_{\mathbf{y}}(\mathbf{y}; \theta) = p_{\mathbf{w}}(\mathbf{w})\big|_{\mathbf{w}=\mathbf{y}-\mathbf{f}(\theta)} \tag{3.62}$$

or

$$p_{\mathbf{y}}(\mathbf{y}; \theta) = p_{\mathbf{w}}(\mathbf{w})\big|_{\mathbf{w}=\mathbf{y}-\mathbf{H}\theta} \tag{3.63}$$

by considering the measurement equation (3.52) or (3.53), respectively. The pdf $p_{\mathbf{y}}(\mathbf{y}; \theta)$ can be investigated from two perspectives. The first perspective is to change $\mathbf{y}$ for a fixed θ — this is a natural way. In this perspective, $p_{\mathbf{y}}(\mathbf{y}; \theta)d\mathbf{y}$ represents the probability that $\mathbf{y}$ falls in the region $[\mathbf{y}, \mathbf{y} + d\mathbf{y}]$. The second perspective is to change θ for a fixed $\mathbf{y} = \mathbf{y}_0$. Let us consider two possible values of $p_{\mathbf{y}}(\mathbf{y}_0; \theta)$ in this perspective, say $p_{\mathbf{y}}(\mathbf{y}_0; \theta_1)$ and $p_{\mathbf{y}}(\mathbf{y}_0; \theta_2)$. Suppose that

$$p_{\mathbf{y}}(\mathbf{y}_0; \theta_1) > p_{\mathbf{y}}(\mathbf{y}_0; \theta_2). \tag{3.64}$$

Equation (3.64) means that when parameter θ takes value θ_1, the event $\mathbf{y} \in [\mathbf{y}_0, \mathbf{y}_0 + d\mathbf{y}]$ would be more likely to happen than when parameter θ takes value θ_2. Therefore, a natural estimation for parameter θ is to choose those θ that maximize $p_{\mathbf{y}}(\mathbf{y}_0; \theta)$ for an obtained measurement $\mathbf{y} = \mathbf{y}_0$. This is the basic idea of ML estimation, which is formally defined as

$$\hat{\theta}_{\text{ML}} = \arg \max_{\theta} p_{\mathbf{y}}(\mathbf{y}; \theta), \tag{3.65}$$

where $\hat{\theta}_{\text{ML}}$ denotes the ML estimate of θ. The pdf $p_{\mathbf{y}}(\mathbf{y}; \theta)$, when viewed as a function of θ, is called maximum likelihood function of θ.

In practice, the ML function is often defined as $\ln p_{\mathbf{y}}(\mathbf{y}; \theta)$ and the constant contained in $\ln p_{\mathbf{y}}(\mathbf{y}; \theta)$ is often dropped out. This is because the pdf of $\mathbf{y}$ is often related with some expressions of an exponential function. Therefore, the ML estimation is often recast as

$$\hat{\theta}_{\text{ML}} = \arg \max_{\theta} \ln p_{\mathbf{y}}(\mathbf{y}; \theta). \tag{3.66}$$

The ML estimate of θ can be directly obtained by combining equation (3.66) (or (3.65)) with equation (3.62) or (3.63) for nonlinear measurement model and linear measurement model, respectively.

---

[5] Note the subtle difference between the notations $p_{\mathbf{y}}(\mathbf{y}; \theta)$ and $p_{\mathbf{y}}(\mathbf{y} \mid \theta)$: $p_{\mathbf{y}}(\mathbf{y}; \theta)$ represents the pdf of $\mathbf{y}$ for a fixed deterministic parameter θ, while $p_{\mathbf{y}}(\mathbf{y} \mid \theta)$ represents the conditional pdf of $\mathbf{y}$ for a given realization of random θ.

3.3.4 Comparison of the Three Estimation Algorithms

The performance of an estimation algorithm is often characterized by the covariance matrix of the estimate. For an unbiased estimate of the parameters, the smaller[6] the covariance matrix of the estimate, the better the estimator. However, the covariance matrix of any unbiased estimate cannot be smaller than a matrix characterized by the distribution of the measurements when a very loose condition is satisfied. More accurately, suppose that the pdf $p(\mathbf{y}; \theta)$ satisfies the condition

$$\mathbb{E}\left[\frac{\partial \ln \, p(\mathbf{y}; \theta)}{\partial \theta}\right] = 0, \text{for all } \theta \quad \Leftrightarrow \quad \int \frac{\partial \ln \, p(\mathbf{y}; \theta)}{\partial \theta} d\mathbf{y} = 0, \text{for all } \theta, \tag{3.67}$$

and $\hat{\theta}$ is an unbiased estimate of θ. Then the covariance matrix of $\hat{\theta}$ is lower bounded by

$$C_{\hat{\theta}\hat{\theta}} \geq \mathbf{F}^{-1}(\theta), \tag{3.68}$$

where $\mathbf{F}$ denotes the Fisher information matrix defined by

$$[\mathbf{F}(\theta)]_{ij} = -\mathbb{E}\left[\frac{\partial^2 \ln \, p(\mathbf{y}; \theta)}{\partial \theta_i \partial \theta_j}\right].$$

The lower bound specified by inequality (3.68) is called Cramer–Rao bound (CRB), while condition (3.67) is called regularity condition. It is seen from the second equality of (3.67) that the regularity condition will hold true if the order of the partial derivative with respect to θ and the expectation with respect to $\mathbf{y}$ in (3.67) is interchangeable.

It is proved [7, 9] that the covariance matrix of the ML estimate of θ converges to the CRB asymptotically when the sample size of measurements is sufficiently large. In general, among the estimators as discussed in this section, the MMSE estimator performs best, followed by the ML estimator, LMMSE estimator, and WLS estimator, respectively, and the LS estimator performs worst. However, the MMSE estimator is difficult to realize for the case of non-Gaussian measurement noise. When the measurement noise is Gaussian-distributed and the measurement equation is linear in θ, ML and WLS (with an appropriate weighting matrix) estimators produce the same estimate for θ.

In the following, we will use a simple example to illustrate the comparison among these estimators.

Example Consider a black-box system, whose input x and output y are related by

$$y = \theta_1 + \theta_2 x + \theta_3 x^2 + \theta_4 x^3, \tag{3.69}$$

where $\theta_1, \theta_2, \theta_3$ and θ_4 are constants and independent parameters, which are unknown. Define $\theta = [\theta_1 \ \theta_2 \ \theta_3 \ \theta_4]^T$. Our purpose is to estimate θ based on a group of measurements to system (3.69).

Since the measurement to the output y cannot be exact, the practical measurement model to system (3.69) can be revised to

$$y = \theta_1 + \theta_2 x + \theta_3 x^2 + \theta_4 x^3 + w, \tag{3.70}$$

---

[6] The words 'smaller' and 'greater' are understood in the sense of positive definiteness of matrices.

where w stands for measurement noise. We use a subscript to index the input, measurement, and measurement noise at a specific time instant. Let

$$
\mathbf{H} = \begin{bmatrix} 1 & x_0 & x_0^2 & x_0^3 \\ 1 & x_1 & x_1^2 & x_1^3 \\ \vdots & \vdots & \vdots & \vdots \\ 1 & x_{N-1} & x_{N-1}^2 & x_{N-1}^3 \end{bmatrix},
$$

$$
\mathbf{y} = [y_0 \ y_1 \ \cdots \ y_{N-1}]^T,
$$

$$
\mathbf{w} = [w_0 \ w_1 \ \cdots \ w_{N-1}]^T,
$$

where N is the number of measurement samples. With these notations, equation (3.70) can be rewritten as

$$
\mathbf{y} = \mathbf{H}\theta + \mathbf{w}.
$$

It is assumed that all the measurement noises are independent of each other and have zero mean. Therefore, we have $\mathbf{C_{ww}} = \sigma_w^2 \mathbf{I}$, where σ_w^2 denotes the variance of each measurement noise.

In the WLS estimation, we choose the weighting matrix $W = \mathbf{C_{ww}^{-1}} = \frac{1}{\sigma_w^2}\mathbf{I}$. For this choice, the WLS estimate of θ reduces to the LS estimate of θ.

From the very definition of MMSE estimation, parameter θ is considered to be random. For a deterministic parameter θ, which implies that $\mathbf{C_{\theta\theta}} = 0$, this leads to a trivial estimate of the true value of θ, i.e., $\hat{\theta}_{\text{MMSE}} = \bar{\theta} = \theta$. Since now the system to be identified is treated as a black box, *no a prior* knowledge about θ is available, except that the four parameters θ_1, θ_2, θ_3 and θ_4 are mutually independent. Therefore, it is reasonable to assume that $\mathbf{C_{\theta\theta}} = \sigma_\theta^2 \mathbf{I}$ with $\sigma_\theta^2 \to \infty$. Under this assumption the MMSE estimate of θ coincides with its LS estimate.

We will examine how the aforementioned estimators perform when the measurement noise undergoes two different distributions: Gaussian distribution and Laplace distribution.

For the case of Gaussian noise, from equation (3.63) one can find the pdf $p_\mathbf{y}(\mathbf{y}; \theta)$ as follows

$$
p_\mathbf{y}(\mathbf{y}; \theta) = \frac{1}{\sqrt{(2\pi)^N \sigma_w^{2N}}} \exp\left[-\frac{1}{2\sigma_w^2}(\mathbf{y} - \mathbf{H}\theta)^T(\mathbf{y} - \mathbf{H}\theta)\right]. \tag{3.71}
$$

From equation (3.71) it is seen that the ML estimate of θ is the θ, which minimizes $(\mathbf{y} - \mathbf{H}\theta)^T(\mathbf{y} - \mathbf{H}\theta)$. This coincides with the LS estimate of θ.

For the case of Laplace noise, the pdf of the measurement noise w_k at a specific time instant k reads

$$
p_{w_k}(w_k) = \frac{1}{2\breve{\sigma}} \exp\left(-\frac{|w_k|}{\breve{\sigma}}\right),
$$

where $\breve{\sigma} = \frac{\sqrt{2}}{2}\sigma_w$. Considering the independence among $w_k, k = 0, 1, \ldots, N$, we obtain the pdf of $\mathbf{w}$ as follows:

$$
p_\mathbf{w}(\mathbf{w}) = \frac{1}{(2\breve{\sigma})^N} \exp\left(-\frac{\sum_{k=0}^{N-1} |w_k|}{\breve{\sigma}}\right). \tag{3.72}
$$

From equation (3.72) we get the pdf of $\mathbf{y}$ as follows:

$$p_{\mathbf{y}}(\mathbf{y};\theta) = \frac{1}{(2\breve{\sigma})^N} \exp\left[-\frac{\sum_{k=0}^{N-1} |(\mathbf{y} - \mathbf{H}\theta)_k|}{\breve{\sigma}}\right], \quad (3.73)$$

where $(\cdot)_k$ denotes the kth entry of the concerned vector. From equation (3.73) it is seen that the ML estimate of θ is

$$\hat{\theta}_{\mathrm{ML}} = \arg\min_{\hat{\theta}} \sum_{k=0}^{N-1} |(\mathbf{y} - \mathbf{H}\theta)_k| = \arg\min_{\hat{\theta}} \|\mathbf{y} - \mathbf{H}\theta\|_1, \quad (3.74)$$

where $\|\cdot\|_1$ denotes the l_1 norm [6] in the corresponding vector space. Problem (3.74) is an l_1-minimization problem, which can be solved by the Matlab routine `l1decode_pd` in ℓ_1-MAGIC [3] developed by Candes and Romberg.

In the numerical results, the true value of θ is set to be $\theta = [1 \ \ 2 \ \ 3 \ \ 4]^T$, and $x_0, x_1, \ldots, x_{N-1}$ are chosen randomly (uniformly distributed) and independently from the interval $[-1, 1]$. In total, 300 runs of simulations are executed for each case to get relatively smooth results. We use the average estimation error (AEE)

$$\epsilon = \frac{1}{M} \sum_{m=1}^{M} \sqrt{\frac{\|\hat{\theta} - \theta\|_2}{4}}$$

to evaluate the estimation performance, where m stands for the index of Monte Carlo run (the error $\hat{\theta} - \theta$ depends on m implicitly), and M is the total runs (here $M = 300$).

The simulation results are depicted in Figures 3.12–3.13. From the two figures it is seen that the AEE ϵ decreases with N, the number of measurements in a single Monte Carlo run, but it is not a strictly monotonously decreasing function of N. In contrast, ϵ increases with σ_w^2 monotonously. When N is large enough, say $N \geq 400$, the decreasing rate of ϵ with respect to N is very small, which means that it does not make too much sense to further increase the number of measurements when N is greater than 400.

From Figure 3.13 it is seen that the ML estimation algorithm yields better performance than the LS and MMSE estimation algorithms. Note that the MMSE estimation algorithm is not a true *MMSE* estimation method since the parameter vector to be identified is considered deterministic here and no *a prior* information about θ is exploited in the MMSE estimate.

3.4 Summary

In this chapter, we have reviewed some basic signal processing techniques and their applications to RFID systems. Three main techniques are addressed. They are

- analogue bandpass filters,
- matching filters, and
- optimal estimation.

The power of leakage interference signal from reader's transmitter to reader's receiver is very strong. To reduce the effect of this interference signal to backscattered tag-ID signal is a critical issue for RFID systems. Using the RF isolator [8] and polarization technique in the RF frontend

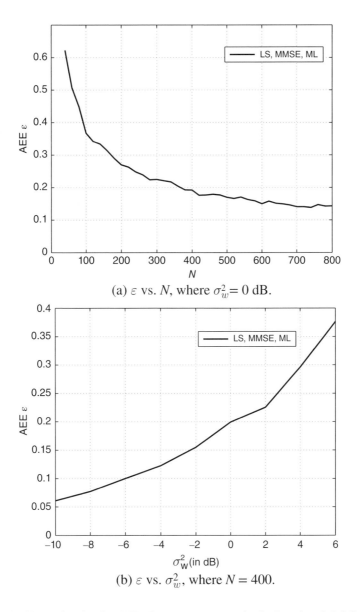

(a) ε vs. N, where $\sigma_w^2 = 0$ dB.

(b) ε vs. σ_w^2, where $N = 400$.

Figure 3.12 An illustration for the AEE: the measurement noise is Gaussian. LS, MMSE and ML estimates of θ coincide with each other.

is the first step to combat against the leakage signal. Using an analogue bandpass filtering technique in the subharmonic modulation system is the second step to reduce the effect of the leakage signal. Both RF isolator and bandpass filtering techniques are used to improve the quality of the received raw signal from the tag. On the other hand, matching filters should be used to improve the SNR of the received signal from the perspective of symbol level. This means that, to decode a symbol, a filter matched to the waveform of that symbol should be adopted.

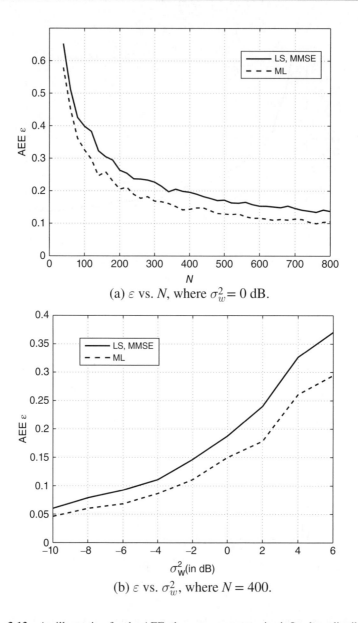

(a) ε vs. N, where $\sigma_w^2 = 0$ dB.

(b) ε vs. σ_w^2, where $N = 400$.

Figure 3.13 An illustration for the AEE: the measurement noise is Laplace distributed.

Three optimal estimation algorithms, namely, LS, MMSE and ML estimators, are discussed in this chapter. They can be used to estimate locations of objects attached with RFID tags and to decode symbols in space-time coding RFID systems. When the parameters to be estimated are considered deterministic variables, both LS and ML estimators can be applied; when the parameters to be estimated are random variables and the statistical properties up to the second order are available, the MMSE estimator can be applied. For both cases, the LS estimator can be applied with some sacrifice of estimation performance. If the speed of the object is also needed to estimate, Kalman filters [2] can be used. Kalman filtering is a kind of MMSE optimal estimation algorithm with the concerned system being modeled by a dynamic equation instead of a static equation such as in model (3.52) or (3.53).

Appendix 3.A Derivation of Poles of the Chebyshev Filter

This derivation follows basically the method presented in [5, Chapter 2]. From equations (3.5), (3.9) and (3.11) we can see that the poles of $H(s)H(-s)$ are those s that satisfies

$$T_N^2(\Omega)\big|_{\Omega=s/J} = -\frac{1}{\beta^2}.$$

Using the first equation of definition (3.12)[7], the above equation is equivalent to

$$\cos\left[N\cos^{-1}\left(\frac{s}{J}\right)\right] = \pm J\frac{1}{\beta}. \tag{3.75}$$

Let

$$\cos^{-1}\left(\frac{s}{J}\right) = \zeta + J\eta, \tag{3.76}$$

where ζ and η are real numbers to be determined. Substituting equation (3.76) into (3.75) yields

$$
\begin{aligned}
&\cos[N(\zeta + J\eta)] \\
&= \frac{e^{JN(\zeta+J\eta)} + e^{-JN(\zeta+J\eta)}}{2} \\
&= \frac{e^{JN\zeta-N\eta} + e^{-JN\zeta+N\eta}}{2} \\
&= \frac{1}{2}[\cos(N\zeta) + J\sin(N\zeta)]e^{-N\eta} + \frac{1}{2}[\cos(N\zeta) - J\sin(N\zeta)]e^{N\eta} = \pm J\frac{1}{\beta}. \tag{3.77}
\end{aligned}
$$

Equating the real and imaginary parts of both sides of equation (3.77) gives

$$\cos(N\zeta)(e^{-N\eta} + e^{N\eta}) = 0, \tag{3.78}$$

$$\sin(N\zeta)\frac{e^{-N\eta} - e^{N\eta}}{2} = \pm\frac{1}{\beta}. \tag{3.79}$$

From equation (3.78) we have

$$\cos(N\zeta) = 0 \quad \Longleftrightarrow \quad \zeta = (2k+1)\frac{\pi}{2N}, \; k = 0, 1, \dots, N-1. \tag{3.80}$$

Substituting equation (3.80) into (3.79) yields

$$(-1)^k \sinh(N\eta) = \pm\frac{1}{\beta} \quad \Longleftrightarrow \quad \eta = \pm\frac{1}{N}\sinh^{-1}\left(\frac{1}{\beta}\right). \tag{3.81}$$

---

[7] If the second equation of definition (3.12) is used, the same expression for the poles can be obtained.

Substituting equations (3.80) and (3.81) back into (3.76) yields:

$$s = j \cos(\zeta + j\eta)$$

$$= j \frac{e^{j(\zeta + j\eta)} + e^{-j(\zeta + j\eta)}}{2}$$

$$= \frac{j}{2}[(\cos \zeta + j \sin \zeta)e^{-\eta} + (\cos \zeta - j \sin \zeta)e^{\eta}]$$

$$= \sin \zeta \sinh \eta + j \cos \zeta \cosh \eta$$

$$= \pm \sin \left[(2k + 1)\frac{\pi}{2N} \right] \sinh \left[\frac{1}{N}\sinh^{-1} \left(\frac{1}{\beta} \right) \right]$$

$$+ j \cos \left[(2k + 1)\frac{\pi}{2N} \right] \cosh \left[\frac{1}{N}\sinh^{-1} \left(\frac{1}{\beta} \right) \right].$$

This is equation (3.16).

References

[1] M. Abramowitz and I. A. Stegun (eds.). *Handbook of Mathematical Functions with Formulas, Graphs, and Mathematical Tables*. Dover Publications, 1965.

[2] B. D. O. Anderson and J. B. Moore. *Optimal Filtering*. Prentice-Hall, Englewood Cliffs, NJ, 1979.

[3] E. Candes and J. Romberg. ℓ_1-MAGIC: Recovery of sparse signals via convex programming, October, 2005. Available: http://users.ece.gatech.edu/ justin/l1magic/.

[4] H. G. Dimopoulos. *Analog Electronic Filters: Theory, Design and Synthesis*. Springer, 2012.

[5] W. K. Chen (ed.). *Passive, Active, and Digital Filters*. CRC Press, Boca Raton, 2009.

[6] R. A. Horn and C. R. Johnson. *Matrix Analysis*. Cambridge University Press, Cambridge, UK, 1986.

[7] S. M. Kay. *Fundamentals of Statistical Signal Processing: Vol. 1:Estimation Theory*. Prentice-Hall, Upper Saddle River, New Jersey, 1993.

[8] J. Polivka. Wideband UHF/microwave active isolators. *High Frequency Electronics*, pages 70–72, Dec. 2006.

[9] C. R. Rao. *Linear Statistical Inference and Its Applications*. John Wiley & Son, Inc., New York, 1973.

[10] T. J. Rivlin. *The Chebyshev Polynomials*. John Wiley & Sons, Inc., New York, 1974.

4

RFID-Oriented Modulation Schemes

4.1 A Brief Review of Analogue Modulation

In general, the frequency range of information-bearing signals or message signals is in the baseband, which is not suitable for transmission in a communication channel. To make the message signals propagate satisfactorily from the transmitter to the receiver, a possible way is to use the message signals to modulate a high-frequency sinusoidal wave, which is called the carrier, so that the frequency range of the transmitted signal is shifted to a proper band. This process is called modulation. Considering the scarcity of radio spectrum resources nowadays, the ultimate goal of a modulation technique is to transport the message signal through a radio channel at the best possible quality while occupying the least amount of radio spectrum.

Since a sinusoidal wave $A\cos(2\pi ft + \varphi)$ has three variables: amplitude A, frequency f and phase φ, a message-bearing signal $m(t)$ can be embedded into these three variables, yielding three basic modulations:

- amplitude modulation (AM) – the modulated signal can be generally represented as $s_{AM}(t) = A(m(t))\cos(2\pi f_c t + \varphi_0)$, where f_c and φ_0 are constants.
- frequency modulation (FM) – the modulated signal can be generally represented as $s_{FM}(t) = A_0 \cos(2\pi f(f_c, m(t))t + \varphi_0)$, where A_0 and φ_0 are constants, and
- phase modulation (PM) – the modulated signal can be generally represented as $s_{PM}(t) = A_0 \cos(2\pi f_c t + \varphi(m(t)))$, where f_c and A_0 are constants.

For the modulation of analogue signals, frequency modulation and phase modulation are mathematically equivalent to each other under the mild condition that the function $\varphi(m(t))$ is differentiable with respect to time t. Therefore, both frequency and phase modulations are also called angle modulation. However, for the modulation of digital signals, frequency modulation and phase modulation are different due to the fact that $\varphi(m(t))$ is discontinuous with respect to time t.

Frequency modulation has better noise immunity than amplitude modulation. Since the information message is embedded in the phase of the modulated signal, the amplitude fading

Digital Signal Processing for RFID, First Edition. Feng Zheng and Thomas Kaiser.
© 2016 John Wiley & Sons, Ltd. Published 2016 by John Wiley & Sons, Ltd.

of the wireless channel has little effect on the performance of FM systems. Due to this advantage, FM is most popularly used in wireless radio systems [7]. However, FM signals occupy a wider radio spectrum than amplitude modulation signals do.

For a more extensive comparison between FM and AM, readers are referred to [7].

Traditionally, the amplitude signal in AM is expressed as $A(m(t)) = A_c(1 + m(t))$, where A_c is a constant. Let $M(f)$ be the spectrum of $m(t)$. The spectrum of AM signal $A(m(t))$ $\cos(2\pi f_c t + \varphi_0)$ can be expressed as

$$S_{AM}(f) = \frac{A_c}{2}[\delta(f - f_c) + \delta(f + f_c) + M(f - f_c) + M(f + f_c)],$$

where $\delta(\cdot)$ denotes Dirac δ function. An illustration of the spectra $M(t)$ and $S_{AM}(f)$ is illustrated in Figure 4.1.

Generally, $m(t)$ is a real signal. Due to this fact, $M(f)$ is symmetric with respect to the axis $f = 0$. Hence the upper sideband spectrum and lower sideband spectrum of $S_{AM}(f)$ are symmetric with respect to the vertical line $f = f_c$, as shown in Figure 4.1. Therefore, it is wasteful in both power and bandwidth if AM signal $s_{AM}(t)$ is directly transmitted. In practical systems, either the upper sideband signal or lower sideband signal is transmitted. This results in single sideband (SSB) modulation. One of its realization schemes is illustrated in Figure 4.2, where $s_{SSB}(t)$ stands for the SSB-modulated signal and the passband of the bandpass filter is either $[f_c, f_c + f_m]$ or $[f_c - f_m, f_c]$.

SSB modulation has been used in the reader-to-tag communications of RFID systems.

The AM signals can be demodulated in two ways. The first one is the product detector, which is a coherent demodulator as illustrated in Figure 4.3, where φ_r stands for the phase of

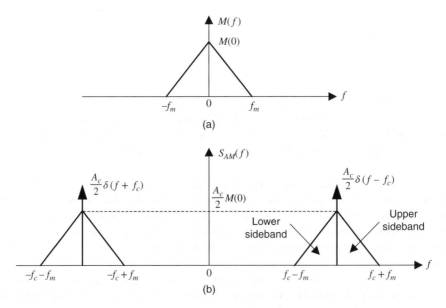

Figure 4.1 An illustration of the spectra of message signal and AM modulated signal. (a) The spectrum of message signal; (b) the spectrum of AM signal, where f_m stands for the maximal frequency of the message signal.

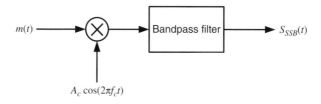

Figure 4.2   A realization scheme of SSB modulation.

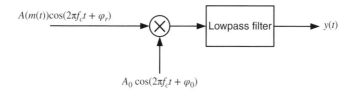

Figure 4.3   Block diagram of a product demodulator.

the received signal at the RF front end, φ_0 is the phase of the local oscillator, and the amplitude fading of the channel is not taken into account.

Assuming an ideal lowpass filter, it can be easily found that the output of the product detector reads

$$y(t) = \frac{1}{2}A(m(t))A_0 \cos(\varphi_r - \varphi_0),$$

which is proportional to the modulated message signal. It is seen that the output of the detector achieves maximum when the phase of the local oscillator is synchronized to the phase of the received RF signal. However, to realize this synchronization, a voltage-controlled oscillator (VCO) is needed to allow a phase-locked loop (PLL) system to track the phase of the input signal. This is not easy for passive tags.

Another method of demodulation of AM signals is the envelope detector, which is a noncoherent demodulator, as illustrated in Figure 4.4.

In Figure 4.4, the diode rectifies the input RF signal and the capacitor, if properly designed, will only allow the baseband message signal to pass through. Actually, the capacitor and resistor constitute a lowpass filter.

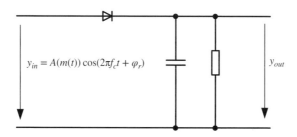

Figure 4.4   An envelope demodulator.

The advantage of envelope demodulators is that they can be easily implemented. Due to this fact, they are widely used in passive tags. The envelope detector works well when the input signal power is at least 10 dB greater than the noise power, whereas the product detector is able to process AM signals even when the power ratio between the input signal and noise is well below 0 dB [7].

For the generation and detection of frequency-modulated and phase-modulated signals, readers are referred to the textbooks [1, 3, 6, 9]. Since the detection of FM or PM signals typically needs a PLL system, which is not easy to implement in passive tags, amplitude modulation is generally used at the reader for the reader-to-tag communication link.

4.2 Amplitude- and Phase-Shift Keying and Performance Analysis

RFID tags use digital modulation techniques. Digital modulation offers many advantages over analogue modulation, for example, greater noise immunity and robustness to channel impairments and more flexibility in supporting complex signal conditioning and processing techniques such as source coding and encryption. While many new digital modulation techniques, e.g., variable-rate variable-power modulation and adaptive coded modulation [2], have been recently developed to improve the performance of overall communication links, RFID tags can only use very basic digital modulation schemes due to simple chips resident in RFID tags. Therefore, these simple digital modulation schemes will be discussed and their performance will be analysed in this section.

Because the change in amplitude, phase or frequency with the message is discontinuous in digital modulations, the corresponding digital amplitude, phase or frequency modulations are called amplitude-, phase- or frequency-shift keying, respectively.

4.2.1 M-ary Quadrature Amplitude Modulation

For M-ary quadrature amplitude modulation (M-QAM), the information bits are encoded in both amplitude and phase of the transmitted signal. The message signal is represented by a constellation in the complex plane, as shown in Figure 4.5. In Figure 4.5, a message indexed by m_i is represented by a point $A_i e^{j\theta_i}$ in the constellation, where A_i and θ_i are the amplitude and phase, respectively, corresponding to message m_i. If message m_i is transmitted, the transmitted signal is:

$$s_i(t) = \text{Re}[A_i e^{j\theta_i} g(t) e^{j2\pi f_c t}], \quad i = 1, 2, \dots .M, \tag{4.1}$$

where f_c is the carrier frequency, and $g(t)$ is a pulse shaping waveform, which is designed to make the spectrum of the transmitted signal satisfy specific requirements.

In practical systems, the modulator (4.1) is generally implemented by using in-phase and quadrature structure. To show this implementation, we write

$$\alpha_{Ii} = A_i \cos(\theta_i),$$

$$\alpha_{Qi} = A_i \sin(\theta_i).$$

It is seen that message m_i can be equivalently represented by the point $(\alpha_{Ii}, \alpha_{Qi})$ in the complex plane. Based on these notations, equation (4.1) can be rewritten as

$$s_i(t) = \alpha_{Ii} g(t) \cos(2\pi f_c t) - \alpha_{Qi} g(t) \sin(2\pi f_c t), \quad i = 1, 2, \dots .M. \tag{4.2}$$

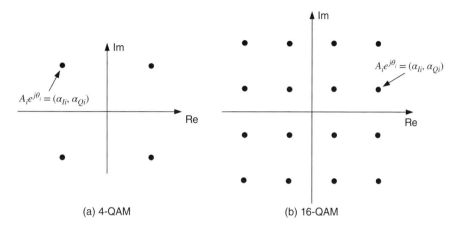

Figure 4.5 Constellations of 4-QAM and 16-QAM.

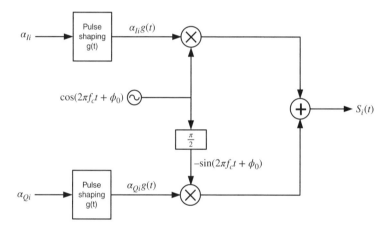

Figure 4.6 QAM modulator (coherent).

Based on equation (4.2), the QAM modulator (4.1) can be implemented with in-phase and quadrature structure shown in Figure 4.6. The constant phase ϕ_0 in Figure 4.6 is the initial phase of the oscillator. In the demodulator, this phase must be compensated to correctly demodulate the transmitted message.

The demodulator of QAM is shown in Figure 4.7. The purpose of the carrier phase recovery is to recover the initial phase ϕ_0 of the oscillator at the transmitter, which is one of the most challenging task for the demodulator. Since the information about the initial phase ϕ_0 is exploited in Figure 4.7, it is called a coherent demodulator. The variable h contained in the input of the demodulator $y(t)$ denotes the RFID channel fading, which has been described in Chapter 2.

In digital modulations, another issue is how to map information bits to message points in the constellation. For an M-QAM modulation, where $M = 2^k$ with k being a positive integer, a message m_i can carry k information bits. These k information bits can be designed in such a

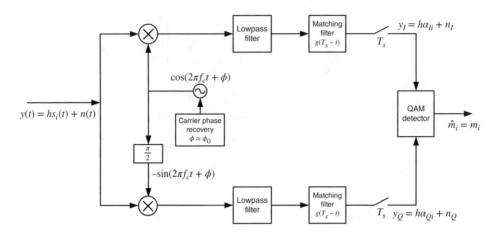

Figure 4.7   QAM demodulator.

way that the adjacent message points differ by only one binary digit. This mapping is called Gray encoding [10]. The detection errors that most likely happen due to noises in the detection are those that are caused by erroneous selection of an adjacent message point. In such a case, only a single bit error occurs in one decoded k-bit sequence.

4.2.2   Symbol Error Rate Analysis of M-QAM

Now let us analyse the symbol error rate (SER) of the communication link with M-QAM. Suppose that $M = L^2$ and the constellation points are in a square with both horizontal and vertical distances between any two neighbouring points being $2d$, as shown in Figure 4.8. The decision maker at the demodulator makes the decision based on the principle of shortest distance selection in the in-phase and quadrature axes, respectively, i.e.,

$$\hat{\alpha}_I(y_I) = \arg\min_{\alpha_{Ii}}\{|\alpha_{Ii} - y_I|, \quad i = 1, 2, \dots, M\},$$

$$\hat{\alpha}_Q(y_Q) = \arg\min_{\alpha_{Qi}}\{|\alpha_{Qi} - y_Q|, \quad i = 1, 2, \dots, M\},$$

where $\hat{\alpha}_I$ and $\hat{\alpha}_Q$ denote, respectively, the in-phase and quadrature coordinates of the estimated message.

Suppose that the lowpass filter has the ideal frequency response as shown in equations (3.1)–(3.2) with the gain being unity and the pulse shaping waveform or the matching filter has an energy of 2. Then the sampler outputs in the in-phase and quadrature branches of the QAM demodulator read, respectively:

$$y_I = \frac{1}{2}h\alpha_I \int_0^{T_s} g^2(t)dt + n_I = h\alpha_I + n_I, \tag{4.3}$$

$$y_Q = \frac{1}{2}h\alpha_Q \int_0^{T_s} g^2(t)dt + n_Q = h\alpha_Q + n_Q, \tag{4.4}$$

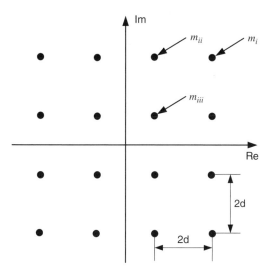

Figure 4.8   An illustration of SER calculation for 16-QAM system.

where n_I and n_Q stand for the noises of in-phase and quadrature branches, respectively, and the coefficient $\frac{1}{2}$ is due to the mixer. Suppose that n_I and n_Q are mutually independent and that $n_I \sim \mathcal{N}(0, \sigma_0^2)$, $n_Q \sim \mathcal{N}(0, \sigma_0^2)$.

First let us consider the case of no fading, that is, $h = 1$. Therefore, the communication link becomes an AWGN (additive white Gaussian noise) channel. Then erroneous decisions can be categorized into the following three cases:

1. (Ci) When the sending message corresponds to the corner points in the constellation such as m_i, an error will occur when either $n_I < -d$ or $n_Q < -d$. This results in an error probability

$$P_{Ci} = P\{\{n_I < -d\} \cup \{n_Q < -d\}\}$$
$$= P\{n_I < -d\} + P\{n_Q < -d\} - P\{\{n_I < -d\} \cap \{n_Q < -d\}\}$$
$$= P\{n_I < -d\} + P\{n_Q < -d\} - P\{n_I < -d\}P\{n_Q < -d\}$$
$$= 2Q\left(\frac{d}{\sigma_0}\right) - \left[Q\left(\frac{d}{\sigma_0}\right)\right]^2,$$

where $Q(\cdot)$ denotes the Gaussian Q-function defined by [5, p. 40]:

$$Q(x) = \frac{1}{\sqrt{2\pi}} \int_x^\infty e^{-\frac{z^2}{2}} \, dz. \tag{4.5}$$

2. (Cii) When the sending message corresponds to the edge points in the constellation such as m_{ii}, an error will occur when either $|n_I| > d$ or $n_Q < -d$. This results in an error probability

$$P_{Cii} = P\{\{|n_I| > d\} \cup \{n_Q < -d\}\}$$
$$= P\{|n_I| > d\} + P\{n_Q < -d\} - P\{\{|n_I| > d\} \cap \{n_Q < -d\}\}$$

$$= P\{|n_I| > d\} + P\{n_Q < -d\} - P\{|n_I| > d\}P\{n_Q < -d\}$$

$$= 3Q\left(\frac{d}{\sigma_0}\right) - 2\left[Q\left(\frac{d}{\sigma_0}\right)\right]^2.$$

3. (Ciii) When the sending message corresponds to the inner points in the constellation such as m_{iii}, an error will occur when either $|n_I| > d$ or $|n_Q| > d$. This results in an error probability

$$P_{Ciii} = P\{\{|n_I| > d\} \cup \{|n_Q| > d\}\}$$

$$= P\{|n_I| > d\} + P\{|n_Q| > d\} - P\{\{|n_I| > d\} \cap \{|n_Q| > d\}\}$$

$$= P\{|n_I| > d\} + P\{|n_Q| > d\} - P\{|n_I| > d\}P\{|n_Q| > d\}$$

$$= 4Q\left(\frac{d}{\sigma_0}\right) - 4\left[Q\left(\frac{d}{\sigma_0}\right)\right]^2.$$

Supposing that all the messages are sent out with equal probability, the average symbol-error probability is given by

$$P_S = \frac{1}{M}[4P_{Ci} + 4(L - 2)P_{Cii} + (M - 4L + 4)P_{Ciii}]$$

$$= \frac{1}{M}\left\{4(M - L)Q\left(\frac{d}{\sigma_0}\right) - 4(M - 2L + 1)\left[Q\left(\frac{d}{\sigma_0}\right)\right]^2\right\}$$

$$= 4\frac{\sqrt{M} - 1}{\sqrt{M}}Q\left(\frac{d}{\sigma_0}\right) - 4\frac{(\sqrt{M} - 1)^2}{M}\left[Q\left(\frac{d}{\sigma_0}\right)\right]^2. \qquad (4.6)$$

Equation (4.6) holds true for $M = L^2$ with $L \geq 2$. For the case of $L = 2$, the QAM modulation reduces to quadrature phase-shift keying (QPSK) and the average symbol-error probability reads

$$P_{S,QPSK} = 2Q\left(\frac{d}{\sigma_0}\right) - \left[Q\left(\frac{d}{\sigma_0}\right)\right]^2. \qquad (4.7)$$

Now let us consider the situation of fading channels. Without sacrificing generality we can assume that $h > 0$, that is, only amplitude fading is considered. Two cases should be distinguished: (1) the channel fading h, i.e. channel state information (CSI), is known at the receiver and (2) the channel fading h, i.e. CSI, is unknown at the receiver.

For the first case, the signal constellation at the receiver is also scaled by h at the receiver. The receiver can make optimal demodulation based on this new signal constellation. For a given h, following the same procedure as in the AWGN case, we can obtain the SER for cases (Ci), (Cii), (Ciii) and hence the total conditional SER:

$$P_{Ci|h} = P\{\{n_I < -hd\} \cup \{n_Q < -hd\}\}$$

$$= 2Q\left(\frac{hd}{\sigma_0}\right) - \left[Q\left(\frac{hd}{\sigma_0}\right)\right]^2,$$

$$P_{\text{Cii}|h} = P\{\{|n_I| > hd\} \cup \{n_Q < -hd\}\}$$

$$= 3Q\left(\frac{hd}{\sigma_0}\right) - 2\left[Q\left(\frac{hd}{\sigma_0}\right)\right]^2,$$

$$P_{\text{Ciii}|h} = P\{\{|n_I| > hd\} \cup \{|n_Q| > hd\}\}$$

$$= 4Q\left(\frac{hd}{\sigma_0}\right) - 4\left[Q\left(\frac{hd}{\sigma_0}\right)\right]^2,$$

$$P_{S|h} = \frac{1}{M}[4P_{\text{Ci}|h} + 4(L-2)P_{\text{Cii}|h} + (M - 4L + 4)P_{\text{Ciii}|h}]$$

$$= 4\frac{\sqrt{M}-1}{\sqrt{M}}Q\left(\frac{hd}{\sigma_0}\right) - 4\frac{(\sqrt{M}-1)^2}{M}\left[Q\left(\frac{hd}{\sigma_0}\right)\right]^2.$$

Based on these equations, the average SER can be obtained as

$$P_{S,\text{CSI}} = \int_0^\infty P_{S|h} P_h(h)\mathrm{d}h$$

$$= \int_0^\infty \left\{4\frac{\sqrt{M}-1}{\sqrt{M}}Q\left(\frac{hd}{\sigma_0}\right) - 4\frac{\left(\sqrt{M}-1\right)^2}{M}\left[Q\left(\frac{hd}{\sigma_0}\right)\right]^2\right\}P_h(h)\mathrm{d}h. \quad (4.8)$$

Here CSI is attached in the subscript to indicate that equation (4.8) holds true for the case that the CSI is available at the receiver.

Substituting the RFID channel fading $h = h^b h^f$ into equation (4.8), we can get the average SER. For example, let h^f and h^b be distributed identically and independently in Rayleigh:

$$p_{h^f}(x) = p_{h^b}(x) = \begin{cases} \frac{x}{\sigma^2}e^{-\frac{x^2}{2\sigma^2}} & \text{when } x \geq 0, \\ 0 & \text{when } x < 0. \end{cases} \quad (4.9)$$

Then the pdf of h is given by

$$p_h(x) = \int_0^\infty \left(2\lambda \upsilon e^{-\lambda \upsilon^2}\right)\left(2\lambda \frac{x}{\upsilon}e^{-\lambda \frac{x^2}{\upsilon^2}}\right)\frac{1}{\upsilon}\mathrm{d}\upsilon \, \texttt{sign}(x)$$

$$= 4\lambda^2 x \int_0^\infty \frac{1}{\upsilon}e^{-\lambda\left(\upsilon^2 + \frac{x^2}{\upsilon^2}\right)}\mathrm{d}\upsilon \, \texttt{sign}(x), \quad (4.10)$$

where parameter λ and function $\texttt{sign}(\cdot)$ are defined by

$$\lambda := \frac{1}{2\sigma^2}, \quad (4.11)$$

$$\texttt{sign}(x) := \begin{cases} 1 & \text{when } x \geq 0, \\ 0 & \text{when } x < 0. \end{cases}$$

For the second case, the receiver cannot make use of the scaled signal constellation. It makes a decision based on the original signal constellation. For a sent message corresponding to an inner point in the signal constellation, say $\alpha_I = k_I d$, and $\alpha_Q = k_Q d$, a decision error will occur when

$$|h\alpha_I + n_I - k_I d| > d \quad \Leftrightarrow \quad \{n_I < -(h-1)k_I d - d\} \cup \{n_I > -(h-1)k_I d + d\} \tag{4.12}$$

or

$$|h\alpha_Q + n_Q - k_Q d| > d \quad \Leftrightarrow \quad \{n_Q < -(h-1)k_Q d - d\} \cup \{n_Q > -(h-1)k_Q d + d\},$$

where k_I and k_Q are corresponding integers: $\pm 1, \pm 3, \ldots, \pm(L-3)$.

Similarly, for a sent message corresponding to a corner point in the signal constellation, say $\alpha_I = (L-1)d$, and $\alpha_Q = (L-1)d$, a decision error will occur when

$$h\alpha_I + n_I - (L-1)d < -d \quad \Leftrightarrow \quad n_I < -(h-1)(L-1)d - d$$

or

$$h\alpha_Q + n_Q - (L-1)d < -d \quad \Leftrightarrow \quad n_Q < -(h-1)(L-1)d - d. \tag{4.13}$$

For a sent message corresponding to an edge point in the signal constellation, say $\alpha_Q = (L-1)d$ and $\alpha_I = k_I d$ with $k_I = \pm 1, \pm 3, \ldots, \pm(L-3)$, a decision error will occur when either event (4.12) or event (4.13) happens.

Based on these discussions, we can obtain the SER for cases (Ci), (Cii) and (Ciii) as follows:

$$\bar{P}_{\text{Ci}|h} = P\{\{n_I < -(h-1)(L-1)d - d\} \cup \{n_Q < -(h-1)(L-1)d - d\}\}$$

$$= 2Q\left(\frac{(h-1)(L-1)d + d}{\sigma_0}\right) - \left[Q\left(\frac{(h-1)(L-1)d + d}{\sigma_0}\right)\right]^2, \tag{4.14}$$

$$\bar{P}_{\text{Cii}|h} = P\left\{\{\{n_I < -(h-1)k_I d - d\} \cup \{n_I > -(h-1)k_I d + d\}\}\right.$$

$$\left. \cup \{n_Q < -(h-1)(L-1)d - d\}\right\}$$

$$= Q\left(\frac{(h-1)k_I d + d}{\sigma_0}\right) + Q\left(\frac{-(h-1)k_I d + d}{\sigma_0}\right) + Q\left(\frac{(h-1)(L-1)d + d}{\sigma_0}\right)$$

$$- \left[Q\left(\frac{(h-1)k_I d + d}{\sigma_0}\right) + Q\left(\frac{-(h-1)k_I d + d}{\sigma_0}\right)\right] Q\left(\frac{(h-1)(L-1)d + d}{\sigma_0}\right), \tag{4.15}$$

$$\bar{P}_{\text{Ciii}|h} = P\left\{\{\{n_I < -(h-1)k_I d - d\} \cup \{n_I > -(h-1)k_I d + d\}\}\right.$$

$$\left. \cup \{\{n_Q < -(h-1)k_Q d - d\} \cup \{n_Q > -(h-1)k_Q d + d\}\}\right\}$$

$$= Q\left(\frac{(h-1)k_I d + d}{\sigma_0}\right) + Q\left(\frac{-(h-1)k_I d + d}{\sigma_0}\right)$$

$$+ Q\left(\frac{(h-1)k_Q d + d}{\sigma_0}\right) + Q\left(\frac{-(h-1)k_Q d + d}{\sigma_0}\right)$$

$$- \left[Q\left(\frac{(h-1)k_I d + d}{\sigma_0}\right) + Q\left(\frac{-(h-1)k_I d + d}{\sigma_0}\right)\right]$$

$$\times \left[Q\left(\frac{(h-1)k_Q d + d}{\sigma_0}\right) + Q\left(\frac{-(h-1)k_Q d + d}{\sigma_0}\right)\right]. \tag{4.16}$$

The total conditional SER is:

$$
\bar{P}_{S|h} = \frac{1}{M}\left[4\bar{P}_{Ci|h} + 4 \sum_{k_I = \pm 1, \pm 3, \dots, \pm(L-3)} \bar{P}_{Cii|h} + \sum_{k_I, k_Q = \pm 1, \pm 3, \dots, \pm(L-3)} \bar{P}_{Ciii|h} \right]
$$

$$
= \frac{1}{M}\left[4\bar{P}_{Ci|h} + 8 \sum_{k_I = 1, 3, \dots, L-3} \bar{P}_{Cii|h} + 4 \sum_{k_I, k_Q = 1, 3, \dots, L-3} \bar{P}_{Ciii|h} \right]. \tag{4.17}
$$

In this derivation, we have used the symmetric property of the signal constellation and communication channel. The average SER can be obtained as

$$
P_{S,\overline{CSI}} = \int_0^\infty \bar{P}_{S|h} \, p_h(h) \mathrm{d}h. \tag{4.18}
$$

Here $\overline{CSI}$ is attached in the subscript to indicate that equation (4.18) holds true for the case that the CSI is not available at the receiver.

Specializing this result to the case of QPSK, we have $L = 2$, which means that the second and third entries in equation (4.17) become void and thus we have

$$
P_{S,\overline{CSI},QPSK} = \int_0^\infty \left\{ 2Q\left(\frac{hd}{\sigma_0}\right) - \left[Q\left(\frac{hd}{\sigma_0}\right)\right]^2 \right\} p_h(h)\mathrm{d}h. \tag{4.19}
$$

Comparing equations (4.19) and (4.8), we can see that the SER performance of the system with QPSK modulation is not affected by the knowledge of the amplitude fading of the channel. This observation is helpful for RFID system design.

For the case of 16-QAM, we have $L = 4$, and k_I and k_Q in equation (4.17) can only take the value of one. Substituting L, k_I and k_Q into equations (4.14), (4.15) and (4.16) gives

$$
\bar{P}_{Ci|h,16-QAM} = 2Q\left(\frac{3hd - 2d}{\sigma_0}\right) - \left[Q\left(\frac{3hd - 2d}{\sigma_0}\right)\right]^2, \tag{4.20}
$$

$$
\bar{P}_{Cii|h,16-QAM} = Q\left(\frac{hd}{\sigma_0}\right) + Q\left(\frac{-hd + 2d}{\sigma_0}\right) + Q\left(\frac{3hd - 2d}{\sigma_0}\right)
$$
$$
- \left[Q\left(\frac{hd}{\sigma_0}\right) + Q\left(\frac{-hd + 2d}{\sigma_0}\right)\right] Q\left(\frac{3hd - 2d}{\sigma_0}\right), \tag{4.21}
$$

$$
\bar{P}_{Ciii|h,16-QAM} = 2Q\left(\frac{hd}{\sigma_0}\right) + 2Q\left(\frac{-hd + 2d}{\sigma_0}\right) - \left[Q\left(\frac{hd}{\sigma_0}\right) + Q\left(\frac{-hd + 2d}{\sigma_0}\right)\right]^2. \tag{4.22}
$$

Substituting equations (4.20), (4.21) and (4.22) into (4.17) gives

$$
P_{S,\overline{CSI},16-QAM|h} = Q\left(\frac{3hd - 2d}{\sigma_0}\right) + Q\left(\frac{hd}{\sigma_0}\right) + Q\left(\frac{-hd + 2d}{\sigma_0}\right)
$$
$$
- \frac{1}{4}\left[Q\left(\frac{3hd - 2d}{\sigma_0}\right)\right]^2 - \frac{1}{4}\left[Q\left(\frac{hd}{\sigma_0}\right) + Q\left(\frac{-hd + 2d}{\sigma_0}\right)\right]^2
$$
$$
- \frac{1}{2}\left[Q\left(\frac{hd}{\sigma_0}\right) + Q\left(\frac{-hd + 2d}{\sigma_0}\right)\right] Q\left(\frac{3hd - 2d}{\sigma_0}\right). \tag{4.23}
$$

Combining equation (4.23) and (4.18), we can get the average SER for 16-QAM system under the case of no CSI at the receiver.

In the above, the average SER is derived mainly for QAM of square constellations. For other kind of constellations, e.g., a rectangular constellation, a similar approach can be used to derive the corresponding average SER.

4.2.3 Numerical Results for M-QAM

The average SER can be generally calculated based on equations (4.8) and (4.18), for the cases of known CSI and unknown CSI, respectively, by combining the corresponding pdf of RFID channels (e.g. equation (4.10)). For some scenarios, nearly closed-form expressions of the average SER can be obtained. This is the case of known CSI at the receiver and double-Rayleigh channel fading. For other scenarios, we can only resort to equations (4.8) and (4.18) to get the average SER.

4.2.3.1 CSI Known at the Receiver and Double-Rayleigh Fading

In this scenario, channel fading is assumed to be distributed in double Rayleigh, i.e., the fading of both forward and backward channels is Rayleigh distributed. Note that $h = h^f h^b$. It is assumed that h^f and h^b are independent of each other and have the same distribution characterized by (4.9).

Substituting equation (4.10) into (4.8), we can get the average SER as follows:

$$P_{S,CSI} = 4\frac{\sqrt{M} - 1}{\sqrt{M}}I_1 - 4\frac{(\sqrt{M} - 1)^2}{M}I_2, \tag{4.24}$$

where

$$I_1 := \int_0^\infty Q\left(\frac{hd}{\sigma_0}\right) p_h(h) dh$$

$$= -\frac{1}{\pi}\int_0^{\pi/2} \eta\sin^2\theta e^{\eta\sin^2\theta} \texttt{Ei}(-\eta\sin^2\theta) d\theta, \tag{4.25}$$

$$I_2 := \int_0^\infty \left[Q\left(\frac{hd}{\sigma_0}\right)\right]^2 p_h(h) dh$$

$$= -\frac{1}{\pi}\int_0^{\pi/4} \eta\sin^2\theta e^{\eta\sin^2\theta} \texttt{Ei}(-\eta\sin^2\theta) d\theta, \tag{4.26}$$

$$\eta := \frac{2\lambda^2\sigma_0^2}{d^2},$$

λ is defined by (4.11), and $\texttt{Ei}$ denotes the exponential integral given by

$$\texttt{Ei}(x) = \int_{-\infty}^x \frac{e^z}{z} dz, \quad x < 0. \tag{4.27}$$

The derivation of equation (4.24) is presented in Appendix 4.A.

In the numerical results, the average SNR is used to characterize the quality of the received signal. It is defined as the ratio between the average power of the transmitted signal and the power of the noise at the receiver. Consider the square constellation specified in Subsection 4.2.2. From the signal constellation we can find the average power of the transmitted signal as follows:

$$
\begin{aligned}
P_{\text{av}} &= \frac{4}{M} \sum_{k_{\text{I}}=1,3,\ldots,L-1} \sum_{k_{\text{Q}}=1,3,\ldots,L-1} \left[(k_{\text{I}}d)^2 + (k_{\text{Q}}d)^2 \right] \\
&= \frac{4}{M} \sum_{k_{\text{I}}=1,3,\ldots,L-1} \left[\frac{L}{2}(k_{\text{I}}d)^2 + \sum_{k_{\text{Q}}=1,3,\ldots,L-1} (k_{\text{Q}}d)^2 \right] \\
&= \frac{4}{M} \left[\frac{L}{2} \sum_{k_{\text{I}}=1,3,\ldots,L-1} (k_{\text{I}}d)^2 + \frac{L}{2} \sum_{k_{\text{Q}}=1,3,\ldots,L-1} (k_{\text{Q}}d)^2 \right] \\
&= \frac{4}{M} \cdot \left\{ \frac{L}{2} \cdot \frac{1}{3} \cdot \frac{L}{2} \left[4\left(\frac{L}{2}\right)^2 - 1 \right] d^2 + \frac{L}{2} \cdot \frac{1}{3} \cdot \frac{L}{2} \left[4\left(\frac{L}{2}\right)^2 - 1 \right] d^2 \right\} \\
&= \frac{2}{3}(M-1)d^2.
\end{aligned}
$$

Therefore, the average SNR, denoted by $\overline{\text{SNR}}$, is given by

$$
\overline{\text{SNR}} = \frac{P_{\text{av}}}{\sigma_{n_{\text{I}}}^2 + \sigma_{n_{\text{Q}}}^2} = \frac{\frac{2}{3}(M-1)d^2}{2\sigma_0^2} = \frac{1}{3}(M-1)\frac{d^2}{\sigma_0^2}.
$$

Figure 4.9 and Figure 4.10 show the average SER for a double-Rayleigh RFID channel with QPSK and 16-QAM, respectively. To compare, the average SER of the same system with Rayleigh fading channel is also included. The RFID double-Rayleigh fading channel and the one-way Rayleigh fading channel are of the same fading power, i.e., $\mathbb{E}(h_{\text{RFID}}^2) = \mathbb{E}(h_{\text{Rayleigh}}^2) = 1$. It is seen from Figure 4.9 and Figure 4.10 that the average SER of the RFID system with QPSK is acceptable in the high SNR range. However, for 16-QAM, the performance of the system depends on whether or not the CSI is known at the receiver. If the CSI is known, the average SER of the system is reasonably low when the SNR is high; if the CSI is unknown, the average SER of the system is very high even in the high SNR range. This result suggests that it is better to use the modulation scheme with a lower cardinality of constellation, such as QPSK, at RFID tags when the CSI is not available at the receiver of the reader.

4.3   Phase-Shift Keying and Performance Analysis

In phase-shift keying (PSK), the information bits are encoded only in the phase of the transmitted signal. The message signal can be represented by a constellation in the complex plane as shown in Figure 4.11. In Figure 4.11, message indexed by m_i is represented by a point $Ae^{j\theta_i}$ in the constellation, where A is a constant and θ_i depends on message m_i. If message m_i is transmitted, the transmitted signal is:

$$
s_i(t) = \text{Re}[Ae^{j\theta_i}g(t)e^{j2\pi f_c t}], \quad i = 1, 2, \ldots M, \tag{4.28}
$$

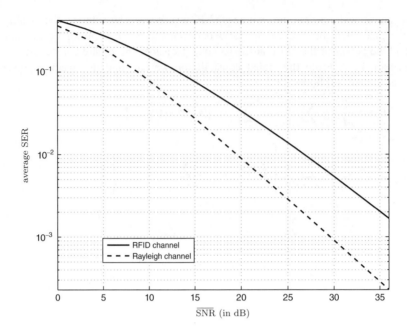

Figure 4.9 The average SER P_S of QPSK modulation for RFID and Rayleigh channels with $\sigma^2 = 1/2$ and $d = 1$.

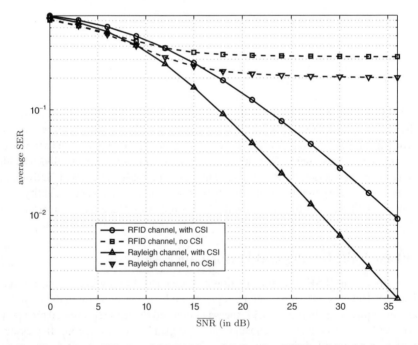

Figure 4.10 The average SER P_S of 16-QAM modulation for RFID and Rayleigh channels with $\sigma^2 = 1/2$ and $d = 1$.

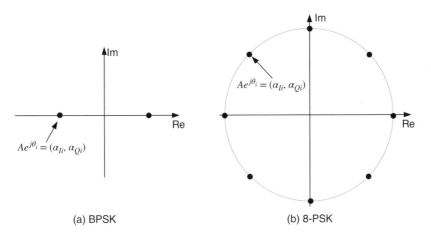

(a) BPSK   (b) 8-PSK

Figure 4.11   Constellations of BPSK and 8-PSK.

where f_c and $g(t)$ have the same meaning as those in M-QAM (4.1). Generally, θ_i, $i = 1, 2, \ldots M$, are distributed on a circle with equal arc distance between any two neighboring points. An M-ary PSK is denoted by M-PSK. In the case of 2-PSK, this is often called BPSK (binary PSK).

In practical systems, the modulator (4.28) is also implemented by using in-phase and quadrature structure as shown in Figure 4.6. The only difference is that both α_{Ii} and α_{Qi} for M-PSK are changed to

$$\alpha_{Ii} = A \cos(\theta_i),$$

$$\alpha_{Qi} = A \sin(\theta_i), i = 1, 2, \ldots, M.$$

Therefore, the demodulator of PSK is of the same structure as that of QAM demodulator, as shown in Figure 4.12. The only difference lies in the decision maker.

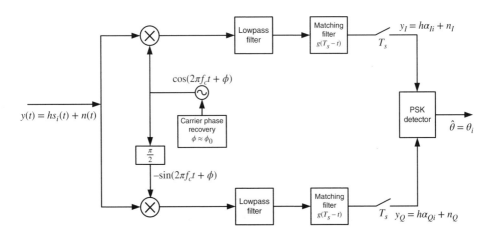

Figure 4.12   PSK demodulator.

The received signals (y_I and y_Q) can be expressed as (4.3) and (4.4), respectively.

Clearly, the optimal decision maker will decode the symbol based on the argument of the received signal (y_I, y_Q) on the complex plane [5, Section 5.2.7]. Specifically, the optimal estimate of θ, denoted by $\hat{\theta}$, is

$$\hat{\theta} = \arg\min_{\theta_i} \left\{ \left| \theta_i - \arctan\left(\frac{y_Q}{y_I}\right) \right|, \quad i = 1, 2, \ldots, M \right\}. \tag{4.29}$$

Now let us analyse the performance of M-PSK systems with RFID channel fading. First consider the AWGN channel. Then $h = 1$ should be substituted into equations (4.3) and (4.4). For this case, the SER of M-PSK with the optimal decoder (4.29) is given by [5, Section 5.2.7]

$$P_{S,\text{PSK,AWGN}} = 1 - \frac{1}{2\pi} \int_{-\pi/M}^{\pi/M} e^{-\rho \sin^2 \vartheta} \int_0^\infty z \exp\left[-\frac{(z - \sqrt{2\rho}\cos\vartheta)^2}{2} \right] dz d\vartheta,$$

$$= 1 - \frac{1}{2\pi} e^{-\rho} \int_0^\infty z \exp\left(-\frac{z^2}{2} \right) \int_{-\pi/M}^{\pi/M} \exp\left(\sqrt{2\rho} z \cos\vartheta \right) d\vartheta dz, \tag{4.30}$$

where parameter ρ is the SNR per symbol given by

$$\rho = \frac{A^2}{2\sigma_0^2}. \tag{4.31}$$

When $M > 4$, equation (4.30) does not admit a simple form. When $M = 2$, equation (4.30) reduces to [5, Section 5.2.7]

$$P_{S,\text{BPSK,AWGN}} = Q(\sqrt{2\rho}). \tag{4.32}$$

When $M = 4$, equation (4.30) reduces to (4.7) [5, Section 5.2.7].

For an RFID fading channel, the conditioned SER for a given realization of the channel h is given by

$$P_{S,\text{PSK}|h} = 1 - \frac{1}{2\pi} \exp\left(-\frac{A^2 h^2}{2\sigma_0^2} \right) \int_0^\infty z \exp\left(-\frac{z^2}{2} \right) \int_{-\pi/M}^{\pi/M} \exp\left(\sqrt{\frac{A^2 h^2}{\sigma_0^2}} z \cos\vartheta \right) d\vartheta dz.$$

The above equation is obtained by simply substituting ρ with ρh^2 into equation (4.30).

Therefore, the average SER of PSK with an RFID channel is given by

$$\bar{P}_{S,\text{PSK}} = \int_0^\infty P_{S,\text{PSK}|h} p_h(h) dh$$

$$= 1 - \frac{1}{2\pi} \int_0^\infty \exp\left(-\frac{A^2 h^2}{2\sigma_0^2} \right) \int_0^\infty z \exp\left(-\frac{z^2}{2} \right)$$

$$\times \int_{-\pi/M}^{\pi/M} \exp\left(\sqrt{\frac{A^2 h^2}{\sigma_0^2}} z \cos\vartheta \right) d\vartheta dz \, p_h(h) dh. \tag{4.33}$$

A closed-form expression for $\bar{P}_{S,\text{PSK}}$ does not exist. Substituting the pdf of a specific distributed RFID channel (e.g. (4.10)) into equation (4.33), we can get numerical results for the average SER.

For the case of BPSK and double-Rayleigh RFID channel, the average SER (reducing to bit-error rate (BER)) reads

$$\bar{P}_{S,BPSK} = -\frac{1}{\pi} \int_0^{\pi/2} \bar{\eta} \sin^2\theta e^{\bar{\eta} \sin^2\theta} \mathrm{Ei}(-\bar{\eta}\sin^2\theta)\mathrm{d}\theta, \qquad (4.34)$$

where $\bar{\eta} = \frac{2\lambda^2\sigma_0^2}{A^2}$. The derivation of equation (4.34) follows directly that of equation (4.25).

Numerical results are obtained via simulations. The results are shown in Figure 4.13, where (a) is for RFID channel and (b) for Rayleigh channel as a comparison. It is seen from Figure 4.13 that the average SERs of the RFID system for BPSK, QPSK and 8-PSK are acceptable in the high SNR range. However, to achieve the same SER, the BPSK RFID system requires an SNR of about 15 dB less than that which the 16-PSK RFID system needs. Since the received SNR at the reader is generally low and data rates in RFID systems are typically not high, BPSK or QPSK are favoured modulation schemes for RFID systems.

Note that the CSI knowledge is not required at the receiver of the reader for the PSK system. This is an advantage of PSK system. Note also that the results shown in Figures 4.9, 4.10 and 4.13 are obtained under the implicit condition that the phase rotation caused by the multipath fading of the channel has been compensated for.

4.4 Frequency-Shift Keying and Performance Analysis

In frequency-shift keying (FSK), the information symbols are embedded in the frequency of the transmitted signal. In a general form, corresponding to message m_i, the transmitted signal can be expressed as

$$s_i(t) = A\cos(2\pi f_i t + \varphi_i), \quad 0 \le t < T_s,$$

where f_i is the frequency corresponding to message m_i, φ_i[1] is the phase offset of the carrier corresponding to message m_i, and T_s is the period of a symbol.

Typically, f_i is set to be $f_i = f_c + i\Delta f$, $i = 1, 2, \ldots, M$ for an M-ary FSK system, where f_c is the fundamental carrier frequency (which should be the integer multiple of $\frac{1}{T_s}$) and Δf the minimal frequency separation among different messages. To ease the design of demodulators, it is expected that the signals corresponding to different messages are orthogonal. To maintain the orthogonality, it is required that

$$\Delta f = \begin{cases} \dfrac{1}{2T_s} & \text{if } \varphi_i = \varphi_j, \forall i,j \in \{1,2,\ldots,M\}, \\ \dfrac{1}{T_s} & \text{if } \varphi_i \ne \varphi_j, \exists i,j \in \{1,2,\ldots,M\}. \end{cases} \qquad (4.35)$$

The FSK modulator can be simply implemented by using a multiple oscillator succeeded by a multiplexor. A coherent demodulator can be implemented by using the scheme shown in Figure 4.14.

From Figure 4.14 it is clear that when condition (4.35) holds true, all the output branches other than the one corresponding to the transmitted message contain only noise.

---

[1] The phase offset φ_i is not a message carrier, even though it is connected with message m_i

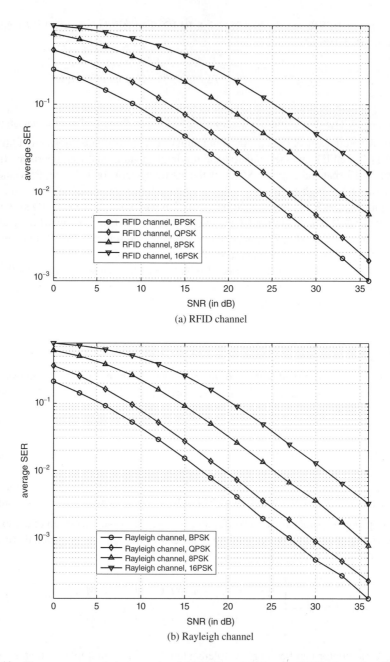

(a) RFID channel

(b) Rayleigh channel

Figure 4.13   The average symbol error rate of M-PSK modulation for RFID channels with $\sigma^2 = \frac{1}{2}$: (a) RFID channel, (b) Rayleigh channel. The results for BPSK can be obtained by either (4.34) or via simulation, which are coincidental.

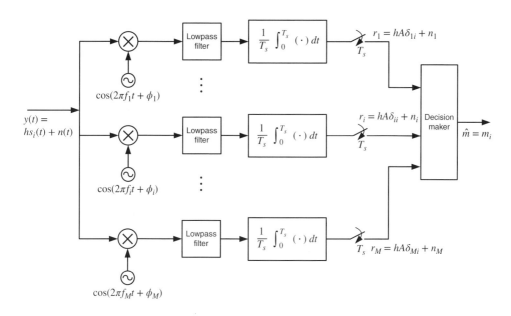

Figure 4.14   FSK demodulator (coherent), where h stands for RFID channel fading.

Assume that the lowpass filter has the ideal frequency response as shown in equations (3.1)–(3.2) with the gain being 2 (to compensate for the gain loss from the mixer) and the bandwidth of the passband being less than, but approaching, Δf. Suppose that the transmitted message is m_i. Then the sampler output for the kth branch reads:

$$r_k = hA\delta_{ki} + n_k, \tag{4.36}$$

where n_k stands for the receiver noise of the kth branch, and δ_{ki} denotes the Kronecker delta function. Note that noise n_k is contributed to by two components: the noise from the RF front end (this component is common for all branches) and the noise from the mixer and low-pass filter at the kth branch. Despite this fact, it is commonly assumed that all the noises n_k, $k = 1, 2, \ldots, M$, are independent of each other in performance analysis of the system. In the following, it is assumed that $n_k \sim \mathcal{N}(0, \sigma_0^2)$.

From equation (4.36) it is seen that only one sampler output, which corresponds to the transmitted message, produces an output whose mean is of a positive value, while all other outputs give noises of zero mean. Therefore, the decision maker at the demodulator selects the branch index whose output is maximal among all the sampler outputs as its optimal estimate of the index of the transmitted message, i.e.,

$$\begin{cases} \hat{m} = m_{i_0}, \\ i_0 = \arg\max_i \{r_i, \ i = 1, 2, \ldots, M\}. \end{cases} \tag{4.37}$$

The SER of FSK system can be easily found from equation (4.37) for the case of $M = 2$, that is, binary FSK (BFSK). Supposing that when message m_1 is transmitted, a decision error will appear if $hA + n_1 < n_2$. A similar event happens when message m_2 is transmitted. By assuming an equally likely message transmission, we obtain the probability of erroneous detection (conditioned on a given channel h) as follows:

$$P_{S,BFSK|h} = \frac{1}{2}P\{hA + n_1 < n_2\} + \frac{1}{2}P\{hA + n_2 < n_1\}$$

$$= P\{n_1 - n_2 > hA\}$$

$$= Q\left(\frac{hA}{\sqrt{2}\sigma_0}\right),$$

where we have used the fact that $\sigma_{n_1-n_2}^2 = \sigma_{n_1}^2 + \sigma_{n_2}^2 = 2\sigma_0^2$.

The average SER of BFSK is given by

$$\bar{P}_{S,BFSK} = \int_0^\infty P_{S,BFSK|h}\, p_h(h)dh = \int_0^\infty Q\left(\frac{hA}{\sqrt{2}\sigma_0}\right) p_h(h)dh.$$

For the case of a double-Rayleigh RFID channel (4.10), the average SER reads

$$\bar{P}_{S,BFSK} = -\frac{1}{\pi}\int_0^{\pi/2} \tilde{\eta}\sin^2\theta\, e^{\tilde{\eta}\sin^2\theta}\, \mathrm{Ei}(-\tilde{\eta}\sin^2\theta)d\theta, \tag{4.38}$$

where $\tilde{\eta} = \frac{4\lambda^2\sigma_0^2}{A^2}$. The derivation of equation (4.38) follows directly that of equation (4.25).

When $M > 2$, the average SER can be derived as follows. From the assumption of an equally likely message transmission and the property of channel symmetry, it is sufficient to consider just the case that message m_1 is transmitted. Then the probability of erroneous detection (conditioned on a given channel h) is given by

$$P_{S,MFSK|h}$$

$$= 1 - P\{(r_1 > r_2)\cap(r_1 > r_3)\cap\cdots\cap(r_1 > r_M)\} \tag{4.39}$$

$$= \frac{1}{\sqrt{(2\pi)^{M-1}\sigma_0^{2(M-1)}\det(\mathbf{C})}}$$

$$\times \int_{-hA}^\infty \int_{-hA}^\infty \cdots \int_{-hA}^\infty \exp\left\{-\frac{1}{2\sigma_0^2}\left[\mathbf{z}^T\mathbf{z} - \frac{1}{M}\left(\sum_{i=1}^{M-1}z_i\right)^2\right]\right\} dz_1 dz_2 \cdots dz_{M-1}, \tag{4.40}$$

where $\mathbf{z}$ is an $(M-1)$-dimensional vector defined in Appendix 4.B and $\mathbf{C}$ is defined by

$$\mathbf{C} = \begin{bmatrix} 2 & 1 & \cdots & 1 \\ 1 & 2 & \cdots & 1 \\ \vdots & \vdots & \ddots & \vdots \\ 1 & 1 & \cdots & 2 \end{bmatrix} = \mathbf{I}_{M-1} + \mathbf{cc}^T \in \mathbf{R}^{(M-1)\times(M-1)},$$

with $\mathbf{c} := [1\ 1\ \cdots\ 1]^T \in \mathbf{R}^{M-1}$.

The average SER of the system is given by

$$\bar{P}_{S,MFSK} = \int_0^\infty P_{S,MFSK|h}\, p_h(h)dh.$$

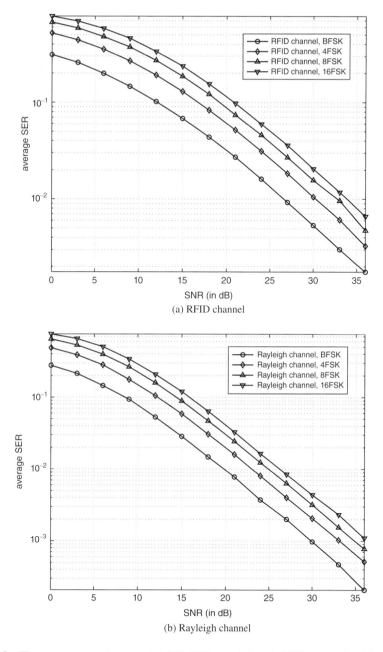

(a) RFID channel

(b) Rayleigh channel

Figure 4.15 The average symbol error rate of M-FSK modulation for RFID channels with $\sigma^2 = 1$ and coherent detection. The results for BFSK can be obtained by either (4.38) or via simulation, which are coincidental.

Numerical results are shown in Figure 4.15, where the SNR is defined as $\frac{A^2}{\sigma_0^2}$. Figure 4.15 (a) is for RFID channel and (b) for Rayleigh channel as a comparison. It is seen from Figure 4.15 that the performance (average SER) gap between BFSK and 16-FSK RFID system is not as great as the gap between BPSK and 16-PSK RFID system: to achieve the same SER, the BFSK RFID system needs an SNR of about 8 dB less than what the 16-FSK RFID system does. Therefore, if high data-rate communications from tags to readers are needed, high-ary FSK, such as 16-FSK, modulations can be adopted at tags. However, the spectral efficiency of high-ary FSK is low. For example, an RFID system using 16-FSK modulation requires a bandwidth of $15\Delta f$, compared to a bandwidth of Δf required by a system using BFSK modulation. Therefore, it is the system designer's task to find compromise between power resource and spectrum resource.

Another important issue of FSK is the phase discontinuity. This problem happens when $\varphi_i \neq \varphi_j$ for $i \neq j$. The phase discontinuity poses several problems, such as spectral spreading and spurious transmissions [7]. To avoid this problem, one can use continuous-phase FSK (CPFSK). A general form of CPFSK is given by

$$s(t) = A \cos \left[2\pi f_c t + 2\pi\beta \int_{-\infty}^{t} u(\tau) \mathrm{d}\tau \right], \qquad (4.41)$$

where

$$u(t) = \sum_k \mathrm{m}_k g(t - kT_s),$$

m_k stands for message sequence, and $g(t)$ the pulse shaping waveform for a single symbol. All the message information is embedded into the waveform $u(t)$. It is clear from (4.41) that the phase of the transmitted signal is a continuous function of the message m_k. Since it is not easy to implement CPFSK in passive tags, we will not discuss further the performance of CPFSK for RFID systems.

The general FSK where the frequency separations between two different neighbouring frequency tones are different, can be easily implemented in RFID tags. For example, in the tag circuit as shown in Figure 2.5, if all the output pins, namely $Q_1, \ldots, Q_7$, of the ripple counter IC 4024 are used and the input of the logic gate IC 7400 is connected to one of them, a system of 7-FSK is realized. The frequency tone for message m_k, being connected to pin Q_k, is $f_0/2^k$, where f_0 is the frequency of the reader's transmitted sinusoidal wave.

4.5   Summary

In this chapter, a brief introduction to analogue modulations and a thorough discussion about digital modulations, tailored to RFID systems, have been presented. Since RFID tags can only implement very simple demodulation circuits, the transmitters of RFID readers generally use analogue AM for communications from readers to tags; while, since the receivers of RFID readers can be equipped with almost all advanced demodulation techniques, typically digital modulations are used in RFID tags.

For all the discussed digital modulations, the average SER performance of the system, based on the double-Rayleigh RFID channel fading model, has been analysed. Numerical results based on either analysis or simulations are provided. Based on these results, it is found that:

- When CSI is known at the reader, high-ary QAM, such as 16-QAM, can be used at the tag with acceptable average SER performance in high SNR range; On the other hand, when CSI is not available at the reader, high-ary QAM will yield very poor SER performance even in high SNR range – in this case, QPSK is recommended since it does not require CSI to decode the symbol.
- The average SER performance of high-ary PSK, such as 16-PSK, is poor. If the available SNR is not high enough, it is recommended to use low-ary PSK, such as BPSK and QPSK.
- The average SER performance of high-ary FSK, such as 8-FSK or 16-FSK, is acceptable in high SNR range. If the available bandwidth is high and the required data rates are also high, one can adopt high-ary FSK.

Although the above conclusions are drawn based on the double-Rayleigh RFID channel, it is believed that other RFID channel models, as discussed in Chapter 2, will bring about the similar results for the corresponding modulation schemes.

Appendix 4.A Derivation of SER Formula (4.24)

Let $V_1 = (h^f)^2$ and $V_2 = (h^b)^2$. Then the pdfs of V_1 and V_2 are given by

$$p_{V_1}(v_1) = \begin{cases} \lambda e^{-\lambda v_1} & v_1 \geq 0, \\ 0 & v_1 < 0, \end{cases} \tag{4.42}$$

$$p_{V_2}(v_2) = \begin{cases} \lambda e^{-\lambda v_2} & v_2 \geq 0, \\ 0 & v_2 < 0, \end{cases} \tag{4.43}$$

respectively.

Let $U = h^2$. From equations (4.42) and (4.43) we can find the pdf of U as follows

$$p_U(u) = \int_0^\infty (\lambda e^{-\lambda v}) \left(\lambda e^{-\lambda \frac{u}{v}} \right) \frac{1}{v} dv \, \texttt{sign}(u)$$

$$= \int_0^\infty \frac{\lambda^2}{v} e^{-\lambda \left(v + \frac{u}{v} \right)} dv \, \texttt{sign}(u).$$

To proceed, we need the following expressions for the Gaussian Q-function and its square [8, p. 85, p. 88]:

$$Q(x) = \frac{1}{\pi} \int_0^{\pi/2} \exp \left(-\frac{x^2}{2\sin^2\theta} \right) d\theta,$$

$$Q^2(x) = \frac{1}{\pi} \int_0^{\pi/4} \exp \left(-\frac{x^2}{2\sin^2\theta} \right) d\theta.$$

Based on above equations, we have

$$I_1 := \int_0^\infty Q \left(\frac{hd}{\sigma_0} \right) p_h(h) dh$$

$$= \int_0^\infty \frac{1}{\pi} \int_0^{\pi/2} \exp \left(-\frac{h^2 d^2}{2\sigma_0^2 \sin^2\theta} \right) d\theta p_h(h) dh$$

$$= \int_0^\infty \frac{1}{\pi} \int_0^{\pi/2} \exp\left(-\frac{ud^2}{2\sigma_0^2\sin^2\theta}\right) d\theta p_U(u) du$$

$$= \frac{1}{\pi} \int_0^{\pi/2} \int_0^\infty \exp\left(-\frac{ud^2}{2\sigma_0^2\sin^2\theta}\right) \int_0^\infty \frac{\lambda^2}{v} \exp\left[-\lambda\left(v+\frac{u}{v}\right)\right] dv du d\theta$$

$$= \frac{1}{\pi} \int_0^{\pi/2} \int_0^\infty \frac{\lambda^2}{v} e^{-\lambda v} \int_0^\infty \exp\left(-\frac{ud^2}{2\sigma_0^2\sin^2\theta}\right) \exp\left(-\lambda\frac{u}{v}\right) du dv d\theta$$

$$= \frac{1}{\pi} \int_0^{\pi/2} \int_0^\infty \frac{\lambda^2}{v} e^{-\lambda v} \frac{1}{\frac{d^2}{2\sigma_0^2\sin^2\theta} + \frac{\lambda}{v}} dv d\theta, \tag{4.44}$$

where we have used the fact that

$$\int_0^\infty \exp\left(-\frac{ud^2}{2\sigma_0^2\sin^2\theta}\right) \exp\left(-\lambda\frac{u}{v}\right) du = \frac{1}{\frac{d^2}{2\sigma_0^2\sin^2\theta} + \frac{\lambda}{v}}.$$

Define function $f(t) := \frac{1}{1+at}$ with $a > 0$. The Laplace transform of $f(t)$ is given by $F(s) = -\frac{1}{a} e^{\frac{s}{a}} \mathrm{Ei}(-\frac{s}{a})$, where Ei denotes the exponential integral defined by equation (4.27). Using this result, we have

$$\int_0^\infty \frac{1}{v} \frac{1}{\frac{d^2}{2\sigma_0^2\sin^2\theta} + \frac{\lambda}{v}} e^{-\lambda v} dv$$

$$= \frac{1}{\lambda} \int_0^\infty \frac{1}{1 + \frac{d^2}{2\lambda\sigma_0^2\sin^2\theta} v} e^{-\lambda v} dv$$

$$= -\frac{1}{\lambda} \left(\frac{d^2}{2\lambda\sigma_0^2\sin^2\theta}\right)^{-1} \exp\left[\lambda\left(\frac{d^2}{2\lambda\sigma_0^2\sin^2\theta}\right)^{-1}\right] \mathrm{Ei}\left[-\lambda\left(\frac{d^2}{2\lambda\sigma_0^2\sin^2\theta}\right)^{-1}\right]$$

$$= -\frac{2\sigma_0^2\sin^2\theta}{d^2} \exp\left(\frac{2\lambda^2\sigma_0^2\sin^2\theta}{d^2}\right) \mathrm{Ei}\left(-\frac{2\lambda^2\sigma_0^2\sin^2\theta}{d^2}\right). \tag{4.45}$$

Substituting equation (4.45) into (4.44), we have

$$I_1 = -\frac{1}{\pi} \int_0^{\pi/2} \frac{2\lambda^2\sigma_0^2\sin^2\theta}{d^2} \exp\left(\frac{2\lambda^2\sigma_0^2\sin^2\theta}{d^2}\right) \mathrm{Ei}\left(-\frac{2\lambda^2\sigma_0^2\sin^2\theta}{d^2}\right) d\theta$$

$$= -\frac{1}{\pi} \int_0^{\pi/2} \eta\sin^2\theta \exp\left(\eta\sin^2\theta\right) \mathrm{Ei}\left(-\eta\sin^2\theta\right) d\theta, \tag{4.46}$$

where

$$\eta := \frac{2\lambda^2\sigma_0^2}{d^2}.$$

Similarly we have

$$I_2 := \int_0^\infty \left[Q\left(\frac{hd}{\sigma_0}\right) \right]^2 p_h(h)\mathrm{d}h$$

$$= -\frac{1}{\pi} \int_0^{\pi/4} \eta \sin^2\theta \exp\left(\eta \sin^2\theta\right) \mathtt{Ei}\left(-\eta\sin^2\theta\right)\mathrm{d}\theta. \tag{4.47}$$

Substituting equations (4.46) and (4.47) into (4.8), we get the result (4.24).

Appendix 4.B Derivation of SER Formula (4.40)

Define a new random vector

$$\mathbf{z} = \begin{bmatrix} z_1 \\ z_2 \\ \vdots \\ z_{M-1} \end{bmatrix} := \begin{bmatrix} n_1 - n_2 \\ n_1 - n_3 \\ \vdots \\ n_1 - n_M \end{bmatrix}.$$

It is easy to obtain

$$\mathbb{E}(\mathbf{z}) = 0,$$

$$\mathbb{E}(\mathbf{z}\mathbf{z}^T) = \sigma_0^2 \mathbf{C}.$$

From the following matrix inverse lemma [4]:

$$(\mathbf{A} + \mathbf{BFD})^{-1} = \mathbf{A}^{-1} - \mathbf{A}^{-1}\mathbf{B}(\mathbf{DA}^{-1}\mathbf{B} + \mathbf{F}^{-1})^{-1}\mathbf{DA}^{-1}, \tag{4.48}$$

where $\mathbf{A}$ (invertible), $\mathbf{B}$, $\mathbf{D}$ and $\mathbf{F}$ (invertible) are matrices with appropriate dimensions, we have

$$\mathbf{C}^{-1} = (\mathbf{I}_{M-1} + \mathbf{cc}^T)^{-1} = \mathbf{I} - \mathbf{c}(\mathbf{c}^T\mathbf{c} + 1)^{-1}\mathbf{c}^T = \mathbf{I} - \frac{1}{M}\mathbf{cc}^T.$$

Note that $\mathbf{z}$ is Gaussian since it is a linear transform of Gaussian random variables. Based on above results, the pdf of $\mathbf{z}$ is given by

$$p_{\mathbf{z}}(\mathbf{z}) = \frac{1}{\sqrt{(2\pi)^{M-1}\det(\sigma_0^2\mathbf{C})}} \exp\left(-\frac{1}{2}\mathbf{z}^T(\sigma_0^2\mathbf{C})^{-1}\mathbf{z}\right)$$

$$= \frac{1}{\sqrt{(2\pi)^{M-1}\sigma_0^{2(M-1)}\det(\mathbf{C})}} \exp\left\{-\frac{1}{2\sigma_0^2}\left[\mathbf{z}^T\mathbf{z} - \frac{1}{M}\left(\sum_{i=1}^{M-1} z_i\right)^2\right]\right\}.$$

Equation (4.39) can be re-expressed as

$$P_{\mathrm{S,MFSK}|h}$$

$$= 1 - P\{(n_1 - n_2 > -hA) \cap (n_1 - n_3 > -hA) \cap \cdots \cap (n_1 - n_M > -hA)\}$$

$$= 1 - P\{(z_1 > -hA) \cap (z_2 > -hA) \cap \cdots \cap (z_{M-1} > -hA)\}$$

$$= \int_{-hA}^{\infty} \int_{-hA}^{\infty} \cdots \int_{-hA}^{\infty} p_{\mathbf{z}}(\mathbf{z}) d\mathbf{z}$$

$$= \frac{1}{\sqrt{(2\pi)^{M-1} \sigma_0^{2(M-1)} \det(\mathbf{C})}} \int_{-hA}^{\infty} \int_{-hA}^{\infty} \cdots \int_{-hA}^{\infty}$$

$$\exp\left\{-\frac{1}{2\sigma_0^2}\left[\mathbf{z}^T\mathbf{z} - \frac{1}{M}\left(\sum_{i=1}^{M-1} z_i\right)^2\right]\right\} dz_1 dz_2 \cdots dz_{M-1}.$$

This is equation (4.40).

References

[1] G. R. Cooper and C. D. McGillem. *Modern Communications and Spread Spectrum*. McGraw-Hill, Mexico, 1986.

[2] A. Goldsmith. *Wireless Communications*. Cambridge University Press, Cambridge, 2005.

[3] S. Haykin. *Communication Systems*. John Wiley & Sons, Inc., New York, 4th ed, 2001.

[4] S. M. Kay. *Fundamentals of Statistical Signal Processing, Vol. 2: Detection Theory*. Prentice Hall PTR, Upper Saddle River, New Jersey, 1998.

[5] J. G. Proakis. *Digital Communications*. McGraw-Hill, Singapore, 4th ed, 2001.

[6] J. G. Proakis and M. Salehi. *Fundamentals of Communication Systems*. Prentice-Hall, Upper Saddle River, New Jersey, 2005.

[7] T. S. Rappaport. *Wireless Communications: Principles and Practice,* 2nd ed. Prentice Hall PTR, Upper Saddle River, New Jersey, 2002.

[8] M. K. Simon and M.-S. Alouini. *Digital Communication over Fading Channels*. Wiley, New Jersey, 2nd ed, 2005.

[9] H. Taub and D. L. Schilling. *Principle of Communication Systems*. McGraw-Hill, New York, 1971.

[10] S. G. Wilson. *Digital Modulation and Coding*. Prentice Hall, Upper Saddle River, New Jersey, 1996.

5

MIMO for RFID

5.1 Introduction

Currently, an RFID system in the market is usually equipped with a single antenna at its reader or tag. However, the RFID research community has recently started to pay attention to using multiple antennas at either reader side or tag side [8, 21]. The reason is that using multiple antennas is an efficient approach to increasing the coverage of RFID, solving the NLoS problem, improving the reliability of data communications between the reader and tag and thus further extending the information-carrying ability of RFID. Besides, some advanced technology in multiple transmit and receive antennas (MIMO) can be used to solve the problem of detecting multiple objects simultaneously, see e.g. reference [29].

There have been several studies about RFID-MIMO. In general, these studies are somehow scattered in different topics. It is difficult to find the logical relationship among these studies. Therefore, the state of the art of the studies will be reviewed to a great extent in chronological order. The work [18] first showed the idea of using multiple antennas at the reader for both transmission and reception. In [8], the authors first proposed to use multiple antennas at the tag and showed the performance gain by equipping multiple antennas at the reader (for both transmission and reception) and the tag. In [9], the multipath fading for both single-antenna based RFID channel and RFID-MIMO channel was measured and compared. The improvement on the fading depth by using MIMO can be clearly seen from the measured power distribution (see, e.g. Figure 10 therein). In [12], the authors first proposed to apply the Alamouti space-time coding (STC) technique, which is now popularly used in wireless communication systems, to RFID systems. The reference [12] presented a closed-form expression for the bit error rate (BER) of RFID systems with non-coherent frequency-shift keying modulation and multiple transmit antennas at the tag and single transmit/receive antenna at the reader, where double-Rayleigh fading is assumed for the forward and backward links. Systematic studies for the application of STC to RFID-MIMO systems were reported in [11] and [36] almost simultaneously. The study in [36] focused on the space-time coding method for RFID-MIMO systems, where the real orthogonal design (ROD) was applied to the systems, while the study in [11] focused on theoretical analysis of the benefit of RFID-MIMO systems, where BER

Digital Signal Processing for RFID, First Edition. Feng Zheng and Thomas Kaiser.
© 2016 John Wiley & Sons, Ltd. Published 2016 by John Wiley & Sons, Ltd.

performance and diversity order of space-time coded RFID-MIMO systems were presented. The results in both [11] and [36] reveal similar findings about the performance improvement by using MIMO to RFID. The channel fading in [11, 36] is assumed to be double Rayleigh. When the RFID channel fading is characterized by double-Nakagami distribution, BER analysis for space-time coded RFID systems was reported in [26]. In [23], the interrogation range of ultrahigh-frequency-band (UHF-band) RFID with multiple transmit/receive antennas at the reader and single antenna at the tag was analysed, where the forward and backward channels were assumed to take Nakagami distribution. In [29], a blind source separation technique in antenna array was used to solve the multiple-tag identification problem, where the reader is equipped with multiple antennas. The work [2] applied a maximal ratio combining technique to an RFID receiver, where the channel of the whole chain, including a forward link, backscattering coefficient and backward link, was estimated and used as the weighting coefficient for the combining branches. In [24], a prototype for an RFID-MIMO system in UHF band was reported. In [3], both MIMO-based zero-forcing and MMSE receivers were used to deal with the multiple-tag identification problem where the channel of the whole chain was estimated using the approach similar to the one in [2].

The report [6] presented a technique for the measurement of multiple RF tag antennas simultaneously in a backscatter radio link, with which genuine mutual coupling effects among the antennas can be measured. It was reported in [28] that four antennas were fabricated in a given fixed surface at the tag. The measurement results showed that an increase of 83% in area gave a 300% increase in available power to turn on a given tag load and the operational distance of the powered device was increased to 100 cm by the four-antenna setup from roughly 40 cm for the single-antenna setup. In [34], it was shown how four patch antennas could be fabricated in a passive tag with a thickness of 62 mil, length of 12.5 cm and a width of 7.5 cm at the frequency of 5.8 GHz. In [31], a distributed antenna system (DAS) for RFID was proposed. In this system, multiple transmit and receive antennas for a reader are distributed in different locations in a given area. The received signals at different antennas are fed to a centralized processor to process. It was shown that, in an open office area of 10 m by 8 m, the system with four transmit antennas and four receive antennas could read passive tag ID at a success rate of 100%, compared to the conventional commercial passive RFID system with a successful reading rate of less than 60%. In [30], the radio signals from the antenna elements in the DAS were fed to the RFID reader by multi-mode optic fiber. It was shown by experiments that, for a triple-antenna DAS in a 10 m range using fiber links of 30 m, the backscattered signal power was increased by around 10 dB, compared with a conventional switched multi-antenna RFID system.

When multiple RFID tags are placed close to each other, each antenna of the tag behaves like a shielding and reflecting object to others and the radiation pattern and other EM properties will be detuned. In [25], the EM near-field coupling effect among the antennas of the stacked tags was investigated. It was shown there that both tag radiation pattern and other EM properties are greatly altered by neighbouring tags.

The results in [11, 28, 31, 36] suggest that the MIMO technique can be very promising for RFID technology.

In the aforementioned reports, the Alamouti STC technique has been shown to be extendable to RFID-MIMO systems. However, it can only apply to the case where the tag has two antennas. Since implementing four antennas at the tag has been shown to be possible in experiments, it is necessary to investigate the possibility of applying other STC techniques to RFID-MIMO systems so that the scenario of more than two antennas can be dealt with. In this chapter, we will study how to apply the ROD technique, proposed by Tarokh et al. in [33], to RFID-MIMO systems. This technique is suitable for the case where the tag is equipped up to eight antennas, which should be sufficient for RFID technology in the near future. In the ROD-based STC technique, the channel state information (CSI) should be available at the receiver of the reader. This is challenging. Therefore, differential STC will be further discussed in this chapter. The CSI is not required at either the reader side or the tag side for differential STC.

This chapter is organized as follows. Some main benefits of MIMO technique will be briefly discussed in Section 5.2. A modified RFID-MIMO channel model will be developed in Section 5.3. An optimal transmit signal design for the transmitter of the reader will be addressed in Section 5.4. The ROD-based space-time coding and decoding schemes for RFID-MIMO will be discussed in Section 5.5. Section 5.6 deals with differential STC for RFID-MIMO and Section 5.7 concludes the chapter.

5.2 MIMO Principle

To get an intuitive insight into the benefit of MIMO, let us investigate the SNR gain of MIMO wireless communication systems compared to the single-transmit and single-receive antenna (SISO) case. This makes sense since many kinds of system performance, such as channel capacity, BER, communication coverage and so on, are directly determined by SNR. To make the comparison fair, we maintain the constraint that the transmit power of the MIMO is the same as that of the SISO. It is assumed that the CSI is available at the receiver for the single-transmit and multiple-receive antenna (SIMO) case or at the transmitter for the multiple-transmit and single-receive antenna (MISO) case. Let N_T and N_R be the number of transmit and receive antennas, respectively.

First consider the SIMO case. The input-output relationship can be expressed as

$$\mathbf{y}_i(t) = h_i X(t) + \mathbf{n}_i(t), \tag{5.1}$$

where $X(t)$ and $\mathbf{y}_i(t)$, $i = 1, \ldots, N_R$, are the transmit and receive signals, respectively, h_i, $i = 1, \ldots, N_R$, are the channel fading from the transmit antenna to the ith receive antenna, and $\mathbf{n}_i(t)$, $i = 1, \ldots, N_R$, are the receiver noises with zero mean and variance σ_N^2. For the SISO case, the input-output relationship is expressed as

$$Y(t) = hX(t) + N(t),$$

where the symbols have the same meaning as those in (5.1) and the noise N is also of zero mean and variance σ_N^2. Suppose that all h and h_i, $i = 1, \ldots, N_R$, are complex Gaussian with zero mean and variance σ_h^2 and $\sigma_{h_i}^2$, respectively. Suppose that X, h_i, h, $\mathbf{n}_i$ and N are mutually independent.

At the SIMO receiver, we combine the received signals at different antennas using maximal ratio combining (MRC) as follows:

$$Y_{\text{SIMO}}(t) = \sum_{i=1}^{N_R} h_i^* \mathbf{y}_i(t) = \sum_{i=1}^{N_R} |h_i|^2 X(t) + \sum_{i=1}^{N_R} h_i^* \mathbf{n}_i(t).$$

Then the SNR gained is given by

$$\text{SNR}_{\text{SIMO}} = \frac{\mathbb{E}\left\{\left[\sum_{i=1}^{N_R} |h_i|^2 X(t)\right]\left[\sum_{i=1}^{N_R} |h_i|^2 X(t)\right]^*\right\}}{\mathbb{E}\left\{\left[\sum_{i=1}^{N_R} h_i^* \mathbf{n}_i(t)\right]\left[\sum_{i=1}^{N_R} h_i^* \mathbf{n}_i(t)\right]^*\right\}}. \tag{5.2}$$

From equation (5.2) and using the statistic properties of relevant signals we can obtain

SNR_{SIMO}

$$= \frac{\mathbb{E}\left[\sum_{i=1}^{N_R} |h_i|^4\right]\mathbb{E}[|X(t)|^2] + \mathbb{E}\left[\sum_{i_1=1}^{N_R} |h_{i_1}|^2 \sum_{i_2=1,i_2\neq i_1}^{N_R} |h_{i_2}|^2\right]\mathbb{E}[|X(t)|^2]}{\mathbb{E}\left[\sum_{i_1=1}^{N_R} |h_{i_1}|^2\right]\sigma_N^2}$$

$$= \frac{2\sum_{i=1}^{N_R} \sigma_{h_i}^4 + \sum_{i_1=1}^{N_R}\sum_{i_2=1,i_2\neq i_1}^{N_R} \sigma_{h_{i_1}}^2 \sigma_{h_{i_2}}^2}{\sum_{i=1}^{N_R} \sigma_{h_i}^2}\frac{\mathbb{E}[|X(t)|^2]}{\sigma_N^2}$$

$$= \left[\frac{\sum_{i_1=1}^{N_R} \sigma_{h_i}^4}{\sum_{i_1=1}^{N_R} \sigma_{h_i}^2} + \sum_{i_1=1}^{N_R} \sigma_{h_i}^2\right]\text{SNR}_T, \tag{5.3}$$

where SNR_T denotes the SNR at the transmitter side. In the second equality, we have used the property $\mathbb{E}[|h_i|^4] = 2\sigma_{h_i}^4$ for complex Gaussian random variables [10, p. 91], [22]. From equation (5.3) we can see that the SNR is increased exactly N_R-fold (i.e. the SNR of SIMO is $(N_R + 1)$-times of that of the SISO) by using MRC if all the links in the SIMO have the same fading power as that in the SISO.

Next, consider the MISO case. The input-output relationship can be expressed as

$$Y(t) = \sum_{i=1}^{N_T} h_i \mathbf{x}_i(t) + N(t), \tag{5.4}$$

where $Y(t)$ is the receive signal and $\mathbf{x}_i(t)$, $i = 1, \ldots, N_T$, are the transmit signals, and h_i, $i = 1, \ldots, N_T$, are the channel fading from the ith transmit antenna to the receive antenna. For the statistical properties of model (5.4), we make the similar assumptions as those for model (5.1). At the transmitter side, we pre-process the transmitted signals at different antenna branches as follows: the transmitted signal at each branch is weighted by its channel fading: $\mathbf{x}_i(t) \longrightarrow \frac{h_i^*}{\sqrt{\sum_{i=1}^{N_T} \sigma_{h_i}^2}}\mathbf{x}_i(t)$. Note that $\mathbb{E}[|\mathbf{x}_i(t)|^2] = \mathbb{E}[|X(t)|^2]$ in this situation. Then the overall transmitted power across all the N_T transmit antennas will be the same as the SISO case. The received signal is given by

$$Y(t) = \frac{1}{\sqrt{\sum_{i=1}^{N_T} \sigma_{h_i}^2}}\sum_{i=1}^{N_T} h_i h_i^* \mathbf{x}_i(t) + N(t).$$

Thus the received SNR is given by

$$
\mathrm{SNR}_{\mathrm{MISO}}
$$

$$
= \frac{\frac{1}{\sum_{i=1}^{N_T} \sigma_{h_i}^2} \mathbb{E}\left\{ \left[\sum_{i=1}^{N_T} |h_i|^2 \mathbf{x}_i(t) \right] \left[\sum_{i=1}^{N_T} |h_i|^2 \mathbf{x}_i(t) \right]^* \right\}}{\mathbb{E}\{|N(t)|^2\}}
$$

$$
= \frac{\mathbb{E}\left[\sum_{i=1}^{N_T} |h_i|^4 |\mathbf{x}_i(t)|^2 \right] + \mathbb{E}\left[\sum_{i_1=1}^{N_T} |h_{i_1}|^2 \mathbf{x}_{i_1}(t) \sum_{i_2=1, i_2 \neq i_1}^{N_T} |h_{i_2}|^2 \mathbf{x}_{i_2}^*(t) \right]}{\sigma_N^2 \sum_{i=1}^{N_T} \sigma_{h_i}^2}.
$$

Consider the case that all the transmit antennas transmit the same symbol, that is, $\mathbf{x}_{i_1}(t) = \mathbf{x}_{i_2}(t)$ for $i_1 \neq i_2$, and further suppose that the transmitted symbol is of zero mean. In this case, we have

$$
\mathrm{SNR}_{\mathrm{MISO}}
$$

$$
= \frac{\mathbb{E}\left[\sum_{i=1}^{N_T} |h_i|^4 |\mathbf{x}_i(t)|^2 \right] + \mathbb{E}\left[\sum_{i_1=1}^{N_T} |h_{i_1}|^2 \mathbf{x}_{i_1}(t) \sum_{i_2=1, i_2 \neq i_1}^{N_T} |h_{i_2}|^2 \mathbf{x}_{i_2}^*(t) \right]}{\sigma_N^2 \sum_{i=1}^{N_T} \sigma_{h_i}^2}
$$

$$
= \frac{\mathbb{E}\left[\sum_{i=1}^{N_T} |h_i|^4 \right] \mathbb{E}[|\mathbf{x}_i(t)|^2] + \mathbb{E}\left[\sum_{i_1=1}^{N_T} |h_{i_1}|^2 \sum_{i_2=1, i_2 \neq i_1}^{N_T} |h_{i_2}|^2 \right] \mathbb{E}[|\mathbf{x}_i(t)|^2]}{\sigma_N^2 \sum_{i=1}^{N_T} \sigma_{h_i}^2}
$$

$$
= \frac{2 \sum_{i=1}^{N_T} \sigma_{h_i}^4 + \sum_{i_1=1}^{N_T} \sum_{i_2=1, i_2 \neq i_1}^{N_T} \sigma_{h_{i_1}}^2 \sigma_{h_{i_2}}^2}{\sum_{i=1}^{N_T} \sigma_{h_i}^2} \frac{\mathbb{E}[|X(t)|^2]}{\sigma_N^2}
$$

$$
= \left[\frac{\sum_{i_1=1}^{N_T} \sigma_{h_i}^4}{\sum_{i_1=1}^{N_T} \sigma_{h_i}^2} + \sum_{i_1=1}^{N_T} \sigma_{h_i}^2 \right] \mathrm{SNR}_T.
$$

We obtain the same result as the SIMO case, that is, the SNR is increased exactly N_T-fold by using the pre-processing technique at the transmitter side if all the links in the MISO have the same fading power as that in the SISO case.

For the MIMO case, it can be shown based on the singular value decomposition [15] of the channel that a MIMO channel is equivalent to N_{TR} independent channels and thus the data transmission rate can be increased N_{TR}-folds in average compared to the SISO channel, where N_{TR} is the rank of the MIMO channel, which equals $\min\{N_T, N_R\}$ in generic.

From these discussions, we can see that the MIMO technology can yield a considerable gain in the system performance compared to the SISO system. Using different versions of MRC can generally harvest the gain. MRC is fundamentally derived from the principle of matching filters.

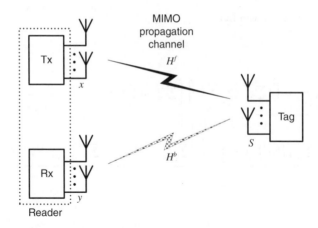

Figure 5.1   A block diagram of an RFID-MIMO system.

5.3   Channel Modelling of RFID-MIMO Wireless Systems

The block diagram of a general RFID-MIMO system is illustrated in Figure 5.1, where the reader and tag are equipped with N_{rd} and N_{tag} antennas, respectively. It is assumed in this chapter that the N_{rd} antennas at the reader are used for both reception and transmission. This assumption is just for brevity of the notation. It is straightforward to extend the approaches and results presented in this chapter to the case where the reader has different numbers of antennas for reception and transmission.

In terms of equation (1) of [8], the narrowband RFID-MIMO wireless channel can be expressed as

$$\mathbf{y}(t) = \mathbf{H}^b \mathbf{S}(t) \mathbf{H}^f \mathbf{x}(t) + \mathbf{n}(t), \tag{5.5}$$

where $\mathbf{x}$ (an $N_{rd} \times 1$ vector) is the transmitted signal at the reader, $\mathbf{y}$ (an $N_{rd} \times 1$ vector) is the received signal at the reader, $\mathbf{n}$ is the receiver noise, $\mathbf{H}^f$ (an $N_{tag} \times N_{rd}$ matrix) is the channel matrix from the reader to the tag, $\mathbf{H}^b$ (an $N_{rd} \times N_{tag}$ matrix) is the channel matrix from the tag to the reader, and $\mathbf{S}$ is the backscattering matrix, which is also called the signalling matrix.

The following assumption on RFID-MIMO channel is used often in this chapter.

Assumption 5.1   (i) The channels $\mathbf{H}^f$ and $\mathbf{H}^b$ are assumed to be complex Gaussian distributed; (ii) $\mathbf{H}^f$ and $\mathbf{H}^b$ are mutually independent and all the entries of $\mathbf{H}^f$ and $\mathbf{H}^b$ are independent of each other and (iii) Each entry of $\mathbf{H}^f$ and $\mathbf{H}^b$ is of zero mean and unit variance; (iv) $\mathrm{Re}(\mathbf{H}^f)$, $\mathrm{Im}(\mathbf{H}^f)$, $\mathrm{Re}(\mathbf{H}^b)$ and $\mathrm{Im}(\mathbf{H}^b)$ are mutually independent and have the same Gaussian distribution.

In most general case where the modulated backscatter signals at the tag are transferred between the antennas, the signalling matrix $\mathbf{S}$ is a full matrix [8]. However, no application of the full signalling matrix has been identified up to now [8]. Therefore, we will consider the situation where the RF tag antennas modulate backscatter with different signals and no signals are transferred between the antennas. In this case, the signalling matrix is a diagonal matrix [8]

$$\mathbf{S}(t) = \mathtt{diag} \{\Gamma_1(t), \Gamma_2(t) \ \ldots \ , \Gamma_{N_{tag}}(t)\} \quad \text{with } |\Gamma_i(t)| \leq 1,$$

where $\Gamma_i(t)$ is the backscattering coefficient of the ith antenna at the tag. The ith tag identity (ID) is contained in the coefficient $\Gamma_i(t)$.

Note that in the RFID system, the transmitted signal $\mathbf{x}$ is mainly used to carry the transmit power, while the information data (i.e. tag ID) is carried by $\mathbf{S}$. Therefore, the central issue for the RFID is to decode $\Gamma_1, \cdots, \Gamma_{N_{\text{tag}}}$ from the received signal $\mathbf{y}$. Next we transform equation (5.5) to the conventional form in signal processing. Let us define

$$\boldsymbol{\gamma}(t) = \begin{bmatrix} \Gamma_1(t) \\ \Gamma_2(t) \\ \vdots \\ \Gamma_{N_{\text{tag}}}(t) \end{bmatrix}, \quad \mathbf{H}^{\text{f}} = \begin{bmatrix} \mathbf{H}^{\text{f}}_1 \\ \mathbf{H}^{\text{f}}_2 \\ \vdots \\ \mathbf{H}^{\text{f}}_{N_{\text{tag}}} \end{bmatrix}. \tag{5.6}$$

Then equation (5.5) can be rewritten as

$$\begin{aligned}
\mathbf{y}(t) &= \mathbf{H}^{\text{b}} \texttt{diag}\, \{\Gamma_1(t), \Gamma_2(t) \ \dots \ , \Gamma_{N_{\text{tag}}}(t)\} \mathbf{H}^{\text{f}} \mathbf{x}(t) + \mathbf{n}(t) \\[4pt]
&= \mathbf{H}^{\text{b}} \texttt{diag}\, \{1, 0, \dots, 0\} \mathbf{H}^{\text{f}} \mathbf{x}(t) \Gamma_1(t) \\[4pt]
&\quad + \mathbf{H}^{\text{b}} \texttt{diag}\, \{0, 1, \dots, 0\} \mathbf{H}^{\text{f}} \mathbf{x}(t) \Gamma_2(t) + \cdots \\[4pt]
&\quad + \mathbf{H}^{\text{b}} \texttt{diag}\, \{0, 0, \dots, 1\} \mathbf{H}^{\text{f}} \mathbf{x}(t) \Gamma_{N_{\text{tag}}}(t) + \mathbf{n}(t) \\[4pt]
&= \mathbf{H}^{\text{b}} \begin{bmatrix} \mathbf{H}^{\text{f}}_1 \\ \mathbf{0} \\ \vdots \\ \mathbf{0} \end{bmatrix} \mathbf{x}(t) \Gamma_1(t) + \mathbf{H}^{\text{b}} \begin{bmatrix} \mathbf{0} \\ \mathbf{H}^{\text{f}}_2 \\ \vdots \\ \mathbf{0} \end{bmatrix} \mathbf{x}(t) \Gamma_2(t) + \cdots \\[4pt]
&\quad + \mathbf{H}^{\text{b}} \begin{bmatrix} \mathbf{0} \\ \mathbf{0} \\ \vdots \\ \mathbf{H}^{\text{f}}_{N_{\text{tag}}} \end{bmatrix} \mathbf{x}(t) \Gamma_{N_{\text{tag}}}(t) + \mathbf{n}(t),
\end{aligned}$$

which can be further rewritten as

$$\begin{aligned}
\mathbf{y}(t) &= \mathbf{H}^{\text{b}} \begin{bmatrix} \mathbf{H}^{\text{f}}_1 \mathbf{x}(t) & 0 & \cdots & 0 \\ 0 & \mathbf{H}^{\text{f}}_2 \mathbf{x}(t) & \cdots & 0 \\ \vdots & \vdots & \ddots & \vdots \\ 0 & 0 & \cdots & \mathbf{H}^{\text{f}}_{N_{\text{tag}}} \mathbf{x}(t) \end{bmatrix} \begin{bmatrix} \Gamma_1(t) \\ \Gamma_2(t) \\ \vdots \\ \Gamma_{N_{\text{tag}}}(t) \end{bmatrix} + \mathbf{n}(t) \\[4pt]
&= \mathbf{H}^{\text{b}} \breve{\mathbf{H}}(t) \boldsymbol{\gamma}(t) + \mathbf{n}(t), \tag{5.7}
\end{aligned}$$

where

$$\breve{\mathbf{H}}(t) := \begin{bmatrix} \mathbf{H}^{\text{f}}_1 \mathbf{x}(t) & 0 & \cdots & 0 \\ 0 & \mathbf{H}^{\text{f}}_2 \mathbf{x}(t) & \cdots & 0 \\ \vdots & \vdots & \ddots & \vdots \\ 0 & 0 & \cdots & \mathbf{H}^{\text{f}}_{N_{\text{tag}}} \mathbf{x}(t) \end{bmatrix}.$$

Equation (5.7) converts the original system model (5.5) to the conventional form in signal processing: the signal to be estimated or decoded is packed in a vector, whose entries are independent of each other.

5.4   Design of Reader Transmit Signals

5.4.1   Signal Design

In this section, it is assumed that the total transmit power at the reader across all the antennas is normalized to unity, that is, $\|\mathbf{x}\|^2 = 1$, where $\| \cdot \|$ denotes the 2-norm in the corresponding vector space. We will investigate how to distribute the transmit power at the reader to achieve good performance. According to the matching filter principle addressed in Section 3.2 and from equation (5.7) it can be seen that one way to maximize the received power for some tag is to let $\mathbf{x}$ match to some spatial channels, say $\mathbf{H}_i^{\mathrm{f}}$. This leads to

$$\mathbf{x} = \frac{(\mathbf{H}_{i_0}^{\mathrm{f}})^\dagger}{\|\mathbf{H}_{i_0}^{\mathrm{f}}\|}, \quad i_0 \in \{1, \ldots, N_{\mathrm{tag}}\}. \tag{5.8}$$

Equation (5.8) says that we can only match the transmit signal to some specific branch of the multiple channels. We can further choose i_0 such that $\mathbf{H}_{i_0}^{\mathrm{f}}$ is of maximal power among all $\mathbf{H}_i^{\mathrm{f}}$'s, that is,

$$i_0 = \arg\max_i \ \|\mathbf{H}_i^{\mathrm{f}}\|. \tag{5.9}$$

The combination of (5.8) and (5.9) maximally exploits the allowed transmit power of the reader.

At the receiver of the reader, the MMSE receiver is used to decode the information data $\boldsymbol{\gamma}$. Its estimate can be written as[1]

$$\hat{\boldsymbol{\gamma}} = (\mathbf{H}^{\mathrm{b}}\breve{\mathbf{H}})^\dagger \boldsymbol{\Phi}\mathbf{y} = (\mathbf{H}^{\mathrm{b}}\breve{\mathbf{H}})^\dagger \boldsymbol{\Phi}(\mathbf{H}^{\mathrm{b}}\breve{\mathbf{H}}\boldsymbol{\gamma} + \mathbf{n}),$$

$$\boldsymbol{\Phi} = [\mathbf{H}^{\mathrm{b}}\breve{\mathbf{H}}(\mathbf{H}^{\mathrm{b}}\breve{\mathbf{H}})^\dagger + \eta\mathbf{I}]^{-1}$$

$$= \{\mathbf{H}^{\mathrm{b}}D(\mathbf{H}^{\mathrm{f}})\|\mathbf{x}\|^2(\mathbf{H}^{\mathrm{b}})^\dagger + \eta\mathbf{I}\}^{-1}, \tag{5.10}$$

where

$$D(\mathbf{H}^{\mathrm{f}}) := \begin{bmatrix} \|\mathbf{H}_1^{\mathrm{f}}\|^2 & 0 & \cdots & 0 \\ 0 & \|\mathbf{H}_2^{\mathrm{f}}\|^2 & \cdots & 0 \\ \vdots & \vdots & \ddots & \vdots \\ 0 & 0 & \cdots & \|\mathbf{H}_{N_{\mathrm{tag}}}^{\mathrm{f}}\|^2 \end{bmatrix}$$

and η is a real positive parameter related with the power ratio between the signal $\boldsymbol{\gamma}$ and noise $\mathbf{n}$.

It can be seen that how to allocate the power of $\mathbf{x}$ does not affect the value of matrix $\boldsymbol{\Phi}$. It only affects the value of matrix $\breve{\mathbf{H}}$. From equation (5.10) we can see that the policy (5.8)–(5.9) maximizes the power ratio between the desired signal $\mathbf{H}^{\mathrm{b}}\breve{\mathbf{H}}\boldsymbol{\gamma}$ (or $\hat{\boldsymbol{\gamma}}$) and the noise $\mathbf{n}$. Because of the diagonal structure of matrix $\breve{\mathbf{H}}$, the power ratio maximization is applied to the selected i_0th tag, i.e., the signal branch for the i_0th tag identity will receive a maximal SNR, and hence will have the best BER performance.

---

[1] Here we have used the fact that $\mathbf{A}^\dagger(\mathbf{A}\mathbf{A}^\dagger + \rho\mathbf{I})^{-1} = (\mathbf{A}^\dagger\mathbf{A} + \rho\mathbf{I})^{-1}\mathbf{A}^\dagger$, which holds true when ρ is sufficiently large so that both $\mathbf{A}\mathbf{A}^\dagger + \rho\mathbf{I}$ and $\mathbf{A}^\dagger\mathbf{A} + \rho\mathbf{I}$ are invertible for matrix $\mathbf{A}$.

5.4.2 Simulation Results

When multiple antennas are equipped at the tag, there are two basic ways to modulate the backscattering coefficients γ among different antennas. The first way is that the backscatter circuits at different antennas use different data symbols (i.e. tag IDs) to modulate the backscattering coefficients, i.e., each tag antenna transmits its own data symbol. In this scenario, the signal vector γ is of the general form as shown in equation (5.6). The second way is that all the backscatter circuits at different antennas use the same data symbol (i.e. tag ID) to modulate the backscattering coefficients and then transmit the reflected signal among different antennas. In this case, the signal vector γ is of the following form

$$\gamma(t) = \Gamma_0(t) \begin{bmatrix} c_1 \\ c_2 \\ \vdots \\ c_{N_{\text{tag}}} \end{bmatrix}, \tag{5.11}$$

where $\Gamma_0(t)$ corresponds to the tag ID information, which is to be estimated, and c_1, $c_2, \ldots, c_{N_{\text{tag}}}$ are constants, which is determined by the backscatter modulation circuits. For simplicity, we assume that $c_1 = \cdots = c_{N_{\text{tag}}} = 1$.

In the first scenario, the multiple antennas at the tag are mainly used to increase the data rate of RFID. In the second scenario, the multiple antennas at the tag are mainly used to increase the diversity of the wireless channel and hence to improve the BER performance of RFID.

First we investigate the performance of RFID-MIMO systems for the first scenario. In the simulations throughout this section, the forward channel $\mathbf{H}^{\text{f}}$ and backward channel $\mathbf{H}^{\text{b}}$ follow Assumption 5.1. To compare, we also plot the BER curves for the case where $\mathbf{x}$ is a random vector whose entry is uniformly distributed among $\pm \frac{1}{\sqrt{N_{\text{rd}}}}$. It is seen that $\mathbf{x}$ is also of power unity.

In Figures 5.2–5.5, the SNR is defined as the ratio between the total power carried by signal vector γ and noise power at each receive antenna. The parameter η is chosen to be 1/SNR.

In Figures 5.2–5.4, the curve marked with 'match to best tag ant'. means the BER for the data symbol transmitted at the i_0th antenna at the tag, where the spatial matching policy (5.8)–(5.9) is applied; while the curve marked with 'match to 1st tag ant'. means the BER for the data symbol transmitted at the first antenna at the tag (i.e., the tag antenna with subscript 1), where the spatial matching policy (5.8) is applied to a fixed antenna (here this means the first antenna) at the tag. The curve marked with 'uniform' means the BER for the data symbols transmitted through all the tag antennas where the aforementioned uniformly distributed random vector (thereafter denoted as uniform policy) is transmitted at the reader, i.e., no spatial matching policy is applied at the reader.

Figure 5.2 shows the BERs for the case of $N_{\text{rd}} = 4$ and $N_{\text{tag}} = 2$. The BERs for two different symbols are plotted here. Figure 5.2(a) illustrates the BER for the optimally matched tag ID (i.e. the BER for the data symbol transmitted through the i_0th antenna at the tag), and Figure 5.2(b) illustrates the BER for all the tag IDs (i.e. the BER for all the data symbols transmitted through the multiple antennas at the tag). It can be seen from Figure 5.2(a) that the spatial matching policy at the reader yields a great SNR gain for the optimally matched tag ID compared to the uniform policy. Comparing the two curves marked with 'match to best tag ant'. and 'match to 1st tag ant', we can find that a further diversity gain is obtained by selecting the best channel among the two available spatial channels (two 4×1 channels). From Figure 5.2 (b) we can see that the BERs for all data symbols are roughly on the same order for different

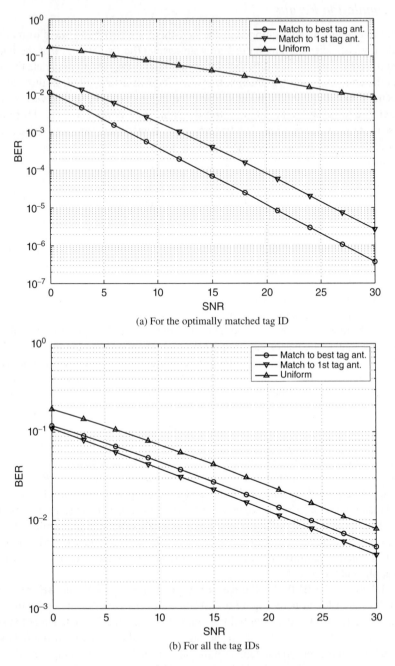

(a) For the optimally matched tag ID

(b) For all the tag IDs

Figure 5.2   BER of the RFID-MIMO system: $N_{rd} = 4$, $N_{tag} = 2$.

Table 5.1 BER comparison for a fixed SNR and different system configuration. SNR = 18 dB, and $N_{rd} = 4$.

N_{tag}	uniform	match to 1st tag	match to best tag
1	2.5×10^{-2}	4.2×10^{-5}	4.2×10^{-5}
2	1.9×10^{-3}	1.2×10^{-5}	2.1×10^{-6}
3	2.3×10^{-4}	5.7×10^{-6}	9.5×10^{-7}
4	4.2×10^{-5}	2.8×10^{-6}	7.0×10^{-7}

polices, even though the 'uniform' policy yields a little better BER than the other two policies for the unmatched tag ID symbols. The reason for this phenomenon is that the good channel is used to transmit the selected tag ID, while other unfavoured channels are used to transmit other tags' IDs. This is a sacrifice to the identification of other unfavoured tags' IDs.

Figure 5.3 shows the BERs for the case of $N_{rd} = 4$ and $N_{tag} = 4$. It sheds a similar light on the usage of the spatial matching policy versus the uniform policy as Figure 5.2 does. The major difference that is worth mentioning is that the SNR gain obtained by the spatial matching policy for the case of $N_{tag} = 4$ is much greater than that for the case of $N_{tag} = 2$.

Figure 5.4 shows the BERs for the case of $N_{rd} = 4$ and $N_{tag} = 1$. As it can be imagined, the case of 'match to best tag ant' collapses to the case of 'match to 1st tag ant'. This is indeed the case. We also see that a great SNR gain is obtained by using the spatial matching policy, compared to the case of using the uniform policy.

Figure 5.5 shows the BER performance of the RFID-MIMO system for the second scenario, where only one data symbol is transmitted across the multiple antennas at the tag in one time slot. In the figure, the legend 'N ant' means that $N (= 1, 2, 3, 4)$ antennas are equipped at the tag. The legend 'match 1st' means that the transmit signal at the reader is spatially matched to the channel corresponding to the first antenna at the tag, while the legend 'match best' means that the transmit signal at the reader is spatially matched to the best channel corresponding to the selected antenna at the tag. It can be seen that for a given number of antennas at the tag, to design the transmit signal at the reader with spatial matching to the best antennas at the tag gives about 2.5 dB gain in the SNR, compared to the case of being spatially matched to a fixed antenna at the tag. The more the antennas at the tag, the better the BER performance of the system. In the case of a single antenna at the tag, both policies produce the same BER performance.

Table 5.1 shows the BER comparison for a fixed SNR (18 dB) and different system configuration. It can be seen that, compared to the uniform transmit policy, the spatial matching transmit policy improves considerably the BER of the system.

5.5 Space-Time Coding for RFID-MIMO Systems[2]

5.5.1 A Review of Real Orthogonal Design

Let us first review the real orthogonal design proposed by Tarokh et al. in [33].

[2] A part of this section is reproduced from [36]. (Reproduced with permission from F. Zheng and T. Kaiser. A space-time coding approach for RFID MIMO systems. EURASIP J. Embedded Systems, 2012. doi:10.1186/1687-3963-2012-9, Available: http://jes.eurasipjournals.com/content/pdf/1687-3963-2012-9.pdf.)

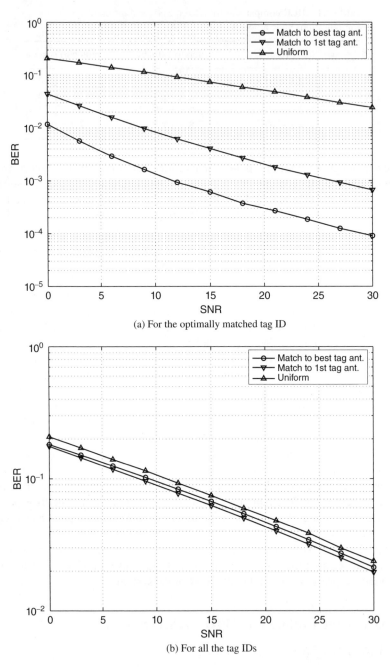

(a) For the optimally matched tag ID

(b) For all the tag IDs

Figure 5.3   BER of the RFID-MIMO system: $N_{rd} = 4$, $N_{tag} = 4$.

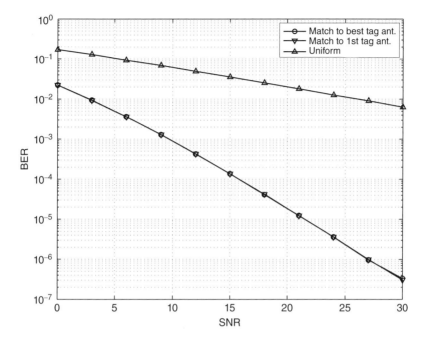

Figure 5.4   BER of the RFID-MIMO system: $N_{rd} = 4, N_{tag} = 1$.

Definition 5.1 [33] A real orthogonal design $\mathcal{G}$ of size m is an $m \times k$ matrix with entries 0, $\pm S_1, \pm S_2, \ldots, \pm S_k$ such that $\mathcal{G}\mathcal{G}^T = \mathbf{D}$, where $\mathbf{D}$ is a diagonal matrix with diagonal entries being $\mathbf{D}_{ii} = l_{i1}S_1^2 + l_{i2}S_2^2 + \cdots + l_{ik}S_k^2, i = 1, 2, \ldots, m$, and the coefficients $l_{i1}, l_{i2}, \ldots, l_{ik}$ are strictly positive integers.

In some cases, we need to explicitly specify the arguments of $\mathcal{G}$. In these cases, the ROD will be denoted by $\mathcal{G}(S_1, S_2, \ldots, S_k)$, where $S_1, S_2, \ldots, S_k$ are the arguments of $\mathcal{G}$.

The construction of general RODs can be found in [33]. For completeness, the RODs for the cases of $m = 2, 3, 4$, denoted by $\mathcal{G}^{(2)}, \mathcal{G}^{(3)}, \mathcal{G}^{(4)}$, respectively, are listed as follows:

$$\mathcal{G}^{(2)} = \begin{bmatrix} S_1 & -S_2 \\ S_2 & S_1 \end{bmatrix}, \tag{5.12}$$

$$\mathcal{G}^{(3)} = \begin{bmatrix} S_1 & -S_2 & -S_3 & -S_4 \\ S_2 & S_1 & S_4 & -S_3 \\ S_3 & -S_4 & S_1 & S_2 \end{bmatrix}, \tag{5.13}$$

$$\mathcal{G}^{(4)} = \begin{bmatrix} S_1 & -S_2 & -S_3 & -S_4 \\ S_2 & S_1 & S_4 & -S_3 \\ S_3 & -S_4 & S_1 & S_2 \\ S_4 & S_3 & -S_2 & S_1 \end{bmatrix}. \tag{5.14}$$

For the construction of $\mathcal{G}^{(5)}, \ldots, \mathcal{G}^{(8)}$, readers are referred to [33].

To develop the decoding algorithm for the ROD, let us define the companion of the ROD (CROD) as follows.

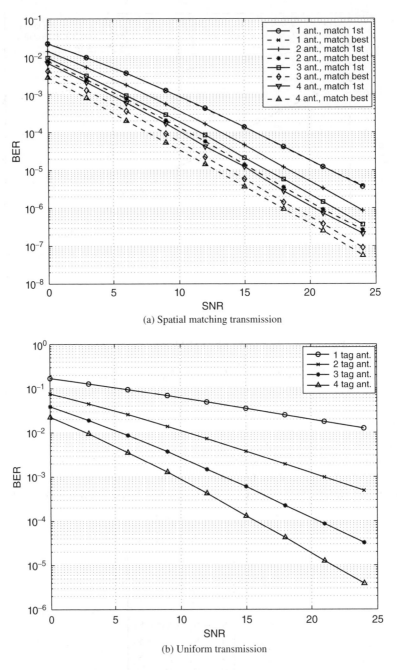

(a) Spatial matching transmission

(b) Uniform transmission

Figure 5.5 BER of the RFID-MIMO system. A single data symbol is transmitted across the multiple antennas at the tag in one time slot. $N_{rd} = 4$, $N_{tag} = 1, 2, 3,$ or 4.

Definition 5.2 A companion of a real orthogonal design $\mathcal{G}(S_1, S_2, \ldots, S_k)$, denoted by $\mathcal{G}_c(\alpha_1, \alpha_2, \ldots, \alpha_m)$, is a matrix satisfying the following equation

$$\begin{bmatrix} \alpha_1 & \alpha_2 & \cdots & \alpha_m \end{bmatrix} \mathcal{G}(S_1, S_2, \ldots, S_k) = \begin{bmatrix} S_1 & S_2 & \cdots & S_k \end{bmatrix} \mathcal{G}_c(\alpha_1, \alpha_2, \ldots, \alpha_m).$$

For the RODs as shown in equations (5.12)–(5.14), their CRODs are

$$\mathcal{G}_c^{(2)} = \begin{bmatrix} \alpha_1 & \alpha_2 \\ \alpha_2 & -\alpha_1 \end{bmatrix}, \tag{5.15}$$

$$\mathcal{G}_c^{(3)} = \begin{bmatrix} \alpha_1 & \alpha_2 & \alpha_3 & 0 \\ \alpha_2 & -\alpha_1 & 0 & \alpha_3 \\ \alpha_3 & 0 & -\alpha_1 & -\alpha_2 \\ 0 & -\alpha_3 & \alpha_2 & -\alpha_1 \end{bmatrix}, \tag{5.16}$$

$$\mathcal{G}_c^{(4)} = \begin{bmatrix} \alpha_1 & \alpha_2 & \alpha_3 & \alpha_4 \\ \alpha_2 & -\alpha_1 & -\alpha_4 & \alpha_3 \\ \alpha_3 & \alpha_4 & -\alpha_1 & -\alpha_2 \\ \alpha_4 & -\alpha_3 & \alpha_2 & -\alpha_1 \end{bmatrix}. \tag{5.17}$$

For a given ROD, the calculation of its CROD is presented in [19].

For the CRODs as defined in equations (5.15)–(5.17), it can be easily shown that the following equality

$$\mathcal{G}_c[\mathcal{G}_c]^T = \sum_{j=1}^{m} \alpha_j^2 \cdot \mathbf{I} \tag{5.18}$$

holds true, where the superscript $(\cdot)^T$ stands for the transpose (*without conjugate!*) of a matrix or vector. As can be seen from the discussion in Section 5.5.3, one can remove the inter-symbol interference (ISI) by using the property of CROD, but the diversity gain thus obtained from the multiple channels is limited when the channel is complex instead of real.

To find the decoding scheme, let us consider the property of $\mathcal{G}_c[\mathcal{G}_c]^\dagger$, where the superscript $(\cdot)^\dagger$ stands for the *conjugate* transpose of a matrix or vector. We have

$$\mathcal{G}_c^{(2)}[\mathcal{G}_c^{(2)}]^\dagger = \begin{bmatrix} \sum_{i=1}^{2} |\alpha_i|^2 & \alpha_1 \alpha_2^\dagger - \alpha_1^\dagger \alpha_2 \\ \star & \sum_{i=1}^{2} |\alpha_i|^2 \end{bmatrix}, \tag{5.19}$$

$$\mathcal{G}_c^{(3)}[\mathcal{G}_c^{(3)}]^\dagger = \begin{bmatrix} \sum_{i=1}^{3} |\alpha_i|^2 & \alpha_1 \alpha_2^\dagger - \alpha_1^\dagger \alpha_2 & \alpha_1 \alpha_3^\dagger - \alpha_1^\dagger \alpha_3 & -\alpha_2 \alpha_3^\dagger + \alpha_2^\dagger \alpha_3 \\ \star & \sum_{i=1}^{3} |\alpha_i|^2 & \alpha_2 \alpha_3^\dagger - \alpha_2^\dagger \alpha_3 & \alpha_1 \alpha_3^\dagger - \alpha_1^\dagger \alpha_3 \\ \star & \star & \sum_{i=1}^{3} |\alpha_i|^2 & -\alpha_1 \alpha_2^\dagger + \alpha_1^\dagger \alpha_2 \\ \star & \star & \star & \sum_{i=1}^{3} |\alpha_i|^2 \end{bmatrix}, \tag{5.20}$$

$$
\mathcal{G}_c^{(4)}[\mathcal{G}_c^{(4)}]^\dagger =
\begin{bmatrix}
\sum_{i=1}^{4} |\alpha_i|^2 & \zeta_1 & \zeta_2 & \zeta_3 \\[2mm]
\star & \sum_{i=1}^{4} |\alpha_i|^2 & \zeta_4 & \zeta_5 \\[2mm]
\star & \star & \sum_{i=1}^{4} |\alpha_i|^2 & \zeta_6 \\[2mm]
\star & \star & \star & \sum_{i=1}^{4} |\alpha_i|^2
\end{bmatrix},
\tag{5.21}
$$

where the entry marked with $\star$ means that its value can be inferred from the value of its corresponding symmetric entry, and

$$
\zeta_1 := \alpha_1 \alpha_2^\dagger - \alpha_1^\dagger \alpha_2 - \alpha_3 \alpha_4^\dagger + \alpha_3^\dagger \alpha_4,
$$

$$
\zeta_2 := \alpha_1 \alpha_3^\dagger - \alpha_1^\dagger \alpha_3 + \alpha_2 \alpha_4^\dagger - \alpha_2^\dagger \alpha_4,
$$

$$
\zeta_3 := \alpha_1 \alpha_4^\dagger - \alpha_1^\dagger \alpha_4 - \alpha_2 \alpha_3^\dagger + \alpha_2^\dagger \alpha_3,
$$

$$
\zeta_4 := \alpha_2 \alpha_3^\dagger - \alpha_2^\dagger \alpha_3 - \alpha_1 \alpha_4^\dagger + \alpha_1^\dagger \alpha_4,
$$

$$
\zeta_5 := \alpha_2 \alpha_4^\dagger - \alpha_2^\dagger \alpha_4 + \alpha_1 \alpha_3^\dagger - \alpha_1^\dagger \alpha_3,
$$

$$
\zeta_6 := \alpha_3 \alpha_4^\dagger - \alpha_3^\dagger \alpha_4 - \alpha_1 \alpha_2^\dagger + \alpha_1^\dagger \alpha_2.
$$

It can be checked that the structural property as shown in equations (5.19)–(5.21) also holds true for higher dimensional CRODs.

From equations (5.19)–(5.21) we can see that the entries of $\mathcal{G}_c[\mathcal{G}_c]^\dagger$ on the main diagonal are real, while the off-diagonal entries are imaginary! We will use this property to design the decoding algorithm for ROD.

5.5.2 Space-Time Coding for RFID-MIMO Systems

Using RODs and the corresponding CRODs, a general space-time encoding scheme and two decoding approaches for RFID-MIMO systems can be developed as follows.

Consider the equivalent RFID-MIMO channel (5.7). We rewrite it here:

$$
\mathbf{y}(t) = \mathbf{H}^b \check{\mathbf{H}}(t)\boldsymbol{\gamma}(t) + \mathbf{n}(t).
$$

Let T_f (a positive number) and K (a positive integer) denote, respectively, the period of a symbol and the length of one space-time block code. As in general STC schemes, the following assumption on the RFID channel is needed.

Assumption 5.2 It is assumed that the channels of both forward and backward links $\mathbf{H}^f$ and $\mathbf{H}^b$ are quasi-static, i.e., they do not change with time during the period of a coding block KT_f. The transmit signal $\mathbf{x}$ at the reader is also fixed during one coding block period KT_f.

Based on Assumption 5.2, the equivalent composite channel $\mathbf{H}^b \check{\mathbf{H}}$ will not change with time when we only consider the signal processing for one coding block. Let us define

$$\mathbf{A} = \mathbf{H}^b \mathbf{\breve{H}}. \tag{5.22}$$

Let $\mathcal{G}$ (of dimension $N_{tag} \times K$) be a ROD in variables $S_1, S_2, \ldots, S_K$, where $S_1, S_2, \ldots, S_K$ are the symbols to be transmitted at the N_{tag} transmit antennas in one STC frame. Define

$$\mathbf{w}(t) = \begin{bmatrix} w(t) \\ w(t - T_f) \\ \vdots \\ w(t - (K-1)T_f) \end{bmatrix},$$

where $w(t)$ is the baseband waveform of the transmit signal at the tag. The transmitted signal across the N_{tag} transmit antennas at the tag can be expressed as

$$\gamma(t) = \sqrt{\frac{E_0}{N_{tag}}} \, \mathcal{G}(S_1, S_2, \ldots, S_K)\mathbf{w}(t), \tag{5.23}$$

where E_0 is the total transmit power at the tag at each time instant. The scaling coefficient $\sqrt{\frac{E_0}{N_{tag}}}$ is to normalize the overall energy consumption per time slot at the tag side to be E_0 no matter how many antennas are deployed at the tag.

5.5.3   Two Space-Time Decoding Approaches for RFID-MIMO Systems

The received signal after sampling can be expressed as

$$\vec{\mathbf{y}} = \sqrt{\frac{E_0}{N_{tag}}} \mathbf{A}\mathcal{G}(S_1, S_2, \ldots, S_K) + \vec{\mathbf{n}}, \tag{5.24}$$

where $\vec{\mathbf{n}}$ is the receiver noise (a matrix) at the corresponding time instants. Notice that $\vec{\mathbf{y}}$ is of dimension $N_{rd} \times K$, since one frame of the transmitted signal contains the pulses of K time slots.

Let $[\mathbf{M}]_j$ denote the jth row of a matrix $\mathbf{M}$. Consider the jth row of the matrix $\vec{\mathbf{y}}$, which is the received signal at the jth antenna of the reader for time instants $1, \ldots, K$, respectively. Let

$$[\mathbf{A}]_j = \begin{bmatrix} \alpha_{j1} & \alpha_{j2} & \cdots & \alpha_{jN_{tag}} \end{bmatrix}.$$

Since the transmitted signal is space-time coded, the entries in $[\mathbf{y}]_j$ should be related to each other somehow. Right-hand multiplying both sides of equation (5.24) with the matrix $[\mathcal{G}_c(\alpha_{j1}, \alpha_{j2}, \ldots, \alpha_{jN_{tag}})]^T$, we have

$$\mathbf{z}_j := [\vec{\mathbf{y}}]_j [\mathcal{G}_c(\alpha_{j1}, \alpha_{j2}, \ldots, \alpha_{jN_{tag}})]^T \tag{5.25}$$

$$= \left\{ \sqrt{\frac{E_0}{N_{tag}}} [\mathbf{A}]_j \mathcal{G}(S_1, S_2, \ldots, S_K) + [\mathbf{n}]_j \right\} [\mathcal{G}_c(\alpha_{j1}, \alpha_{j2}, \ldots, \alpha_{jN_{tag}})]^T$$

$$= \left\{ \sqrt{\frac{E_0}{N_{tag}}} [S_1 \ S_2 \ \cdots \ S_K] \mathcal{G}_c(\alpha_{j1}, \alpha_{j2}, \ldots, \alpha_{jN_{tag}}) + [\vec{\mathbf{n}}]_j \right\}$$

$$\times [\mathcal{G}_c(\alpha_{j1}, \alpha_{j2}, \dots, \alpha_{jN_{\text{tag}}})]^T$$

$$= \sqrt{\frac{E_0}{N_{\text{tag}}}} \sum_{k=1}^{N_{\text{tag}}} (\alpha_{jk})^2 \, [S_1 \quad S_2 \quad \cdots \quad S_K] + [\bar{\mathbf{n}}]_j [\mathcal{G}_c(\alpha_{j1}, \alpha_{j2}, \dots, \alpha_{jN_{\text{tag}}})]^T. \quad (5.26)$$

From equation (5.26) we can see that the transmitted symbols $S_1, S_2, \dots, S_K$ are decoupled from each other in the processed signal $\mathbf{z}_j$ through the processing algorithm (5.25). However, it is not efficient to decode the symbols $S_1, S_2, \dots, S_K$ directly from (5.26) since the complex channel makes the phase of $\sum_{k=1}^{N_{\text{tag}}} [\alpha_{jk}]^2$ randomly change over $[0, 2\pi]$. Define

$$\beta_j = \sum_{k=1}^{N_{\text{tag}}} (\alpha_{jk})^2. \quad (5.27)$$

Multiplying both sides of (5.26) by $\beta_j^\dagger$ will remove the phase ambiguity of the equivalent channel. This gives

$$\bar{\mathbf{z}}_j := \beta_j^\dagger \mathbf{z}_j = |\beta_j|^2 \, [S_1 \quad S_2 \quad \cdots \quad S_K] + \bar{\mathbf{n}}_j, \quad (5.28)$$

where

$$\bar{\mathbf{n}}_j = \beta_j^\dagger \, [\bar{\mathbf{n}}]_j [\mathcal{G}_c(\alpha_{j1}, \alpha_{j2}, \dots, \alpha_{jN_{\text{tag}}})]^T.$$

To collect all the diversities provided by multiple-receive antennas at the reader, we sum up all $\bar{\mathbf{z}}_j$s. This gives

$$\bar{\mathbf{z}} := \sum_{j=1}^{N_{\text{rd}}} \bar{\mathbf{z}}_j = \sum_{j=1}^{N_{\text{rd}}} |\beta_j|^2 \, [S_1 \quad S_2 \quad \cdots \quad S_K] + \sum_{j=1}^{N_{\text{rd}}} \bar{\mathbf{n}}_j. \quad (5.29)$$

The symbols $S_1, S_2, \dots, S_K$ can be easily decoded from equation (5.29).

For the convenience of exposition in the next subsection, we call the encoding and decoding scheme discussed above as Scheme I.

Another decoding scheme (hereafter it is referred to as Scheme II) is to exploit the property of the matrix $\mathcal{G}_c(\mathcal{G}_c)^\dagger$, as shown in equations (5.19)–(5.21). Right-hand multiplying both sides of equation (5.24) with the matrix $[\mathcal{G}_c(\alpha_{j1}, \alpha_{j2}, \dots, \alpha_{jN_{\text{tag}}})]^\dagger$, we have

$$\mathbf{u}_j := [\bar{\mathbf{y}}]_j [\mathcal{G}_c(\alpha_{j1}, \alpha_{j2}, \dots, \alpha_{jN_{\text{tag}}})]^\dagger$$

$$= \left\{ \sqrt{\frac{E_0}{N_{\text{tag}}}} [S_1 \quad S_2 \quad \cdots \quad S_K] \mathcal{G}_c(\alpha_{j1}, \alpha_{j2}, \dots, \alpha_{jN_{\text{tag}}}) + [\bar{\mathbf{n}}]_j \right\}$$

$$\times [\mathcal{G}_c(\alpha_{j1}, \alpha_{j2}, \dots, \alpha_{jN_{\text{tag}}})]^\dagger$$

$$= \sqrt{\frac{E_0}{N_{\text{tag}}}} [S_1 \quad S_2 \quad \cdots \quad S_K] \cdot [\mathcal{G}_c(\alpha_{j1}, \alpha_{j2}, \dots, \alpha_{jN_{\text{tag}}})] \cdot [\mathcal{G}_c(\alpha_{j1}, \alpha_{j2}, \dots, \alpha_{jN_{\text{tag}}})]^\dagger$$

$$+ [\bar{\mathbf{n}}]_j [\mathcal{G}_c(\alpha_{j1}, \alpha_{j2}, \dots, \alpha_{jN_{\text{tag}}})]^\dagger. \quad (5.30)$$

From equations (5.30) and (5.19)–(5.21) we can see that, if the symbols $S_1, S_2, \dots, S_K$ are real, the symbol to be decoded, say S_k for some k, and the ISI caused by other symbols are projected into different subspaces in the complex plane: the desired signal is in the real subspace, while the ISI is in the imaginary subspace. Therefore, a very simple decoding method for this case works in the following way: From kth entry of $\mathbf{u}_j$ (denoted by $\mathbf{u}_{j,k}$), get the real part of $\mathbf{u}_{j,k}$ (denoted by Re $(\mathbf{u}_{j,k})$), and then decode S_k in terms of Re $(\mathbf{u}_{j,k})$.

The diversities provided by multiple-receive antennas at the reader can be collected in the following way:

$$\bar{\mathbf{u}}_k := \sum_{j=1}^{N_{\mathrm{rd}}} \mathrm{Re}\ (\mathbf{u}_{j,k}). \tag{5.31}$$

Then S_k can be decoded in terms of $\bar{\mathbf{u}}_k$.

5.5.4   Simulation Results

In this subsection, we investigate the SER or BER performance of both Schemes I and II. In Scheme I, QPSK modulation is used and the constellation of transmitted symbols is $\frac{\pm 1 \pm j}{\sqrt{2}}$. In Scheme II, BPSK modulation is used and the constellation of transmitted symbols is ± 1. Therefore, the SER in Scheme II reduces to BER. At the transmitter of the reader, the signal $\mathbf{x}$ takes the form of a random vector whose entry is uniformly distributed among $\pm \frac{1}{\sqrt{N_{\mathrm{rd}}}}$. It is seen that $\mathbf{x}$ is of unit power. The RFID channel follows Assumptions 5.1 and 5.2.

In Figures 5.6–5.12, the SNR is defined as $\frac{E_0}{\sigma_{\bar{\mathbf{n}}}^2}$, where $\sigma_{\bar{\mathbf{n}}}^2$ is the variance of each entry of noise vector $\bar{\mathbf{n}}$.

Figures 5.6–5.7 show the SER of Scheme I for different cases: Figure 5.6 illustrates how the SER changes with N_{tag} for fixed N_{rd}, i.e., when $N_{\mathrm{rd}} = 1$ and 4, respectively; while Figure 5.7 demonstrates how the SER changes with N_{rd} for fixed N_{tag}, i.e., when $N_{\mathrm{tag}} = 1$ and 4, respectively.

Figures 5.8–5.9 shows the BER of Scheme II for different cases: Figure 5.8 illustrates how the BER changes with N_{tag} for fixed N_{rd}, i.e., when $N_{\mathrm{rd}} = 1$ and 4, respectively; while Figure 5.9 demonstrates how the BER changes with N_{rd} for fixed N_{tag}, i.e., when $N_{\mathrm{tag}} = 1$ and 4, respectively.

From Figures 5.6–5.9 the following phenomena can be observed:

Claim 5.1   Comparing the dashed curves, which correspond to the performance of the RFID system with a single antenna at both reader and tag sides (naturally without space-time coding) and the solid curves in Figures 5.6(b), 5.7(b), 5.8(b) and 5.9(b), we see that deploying multiple antennas at both reader and tag can greatly improve the SER/BER performance of RFID systems.

Claim 5.2   When N_{rd} is fixed to be one, increasing N_{tag} *considerably* decreases the BER of the system in Scheme II, but only *marginally* decreases the SER of the system in Scheme I. For example, when SNR $= 18$ dB and $N_{\mathrm{rd}} = 1$, the BER of Scheme II decreases from 1.6×10^{-2} at $N_{\mathrm{tag}} = 1$ to 2.0×10^{-3} at $N_{\mathrm{tag}} = 2$ and 8.8×10^{-5} at $N_{\mathrm{tag}} = 4$, respectively. For the same SNR and N_{rd}, the SER of Scheme I decreases from 4.7×10^{-2} at $N_{\mathrm{tag}} = 1$ to 2.9×10^{-2} at $N_{\mathrm{tag}} = 2$ and 3.0×10^{-2} at $N_{\mathrm{tag}} = 4$, respectively. The reason for this phenomenon is that the channel diversity provided by N_{tag} antennas at the tag side is harvested by Scheme II [as seen from equations (5.19)–(5.21)], but not harvested by Scheme I [as seen from equation (5.26)].

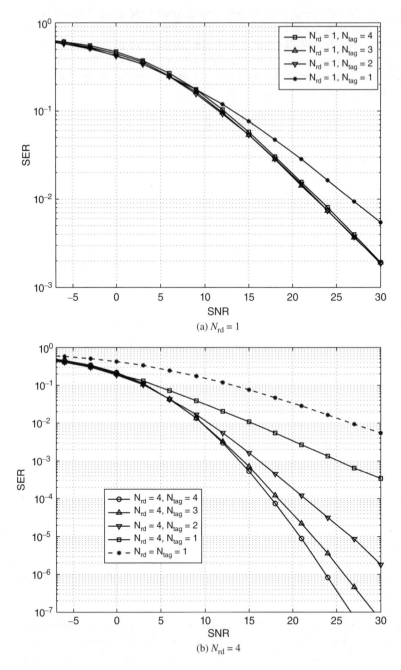

(a) $N_{rd} = 1$

(b) $N_{rd} = 4$

Figure 5.6 SER of RFID-MIMO systems for Scheme I with QPSK modulation: SER versus N_{tag}.

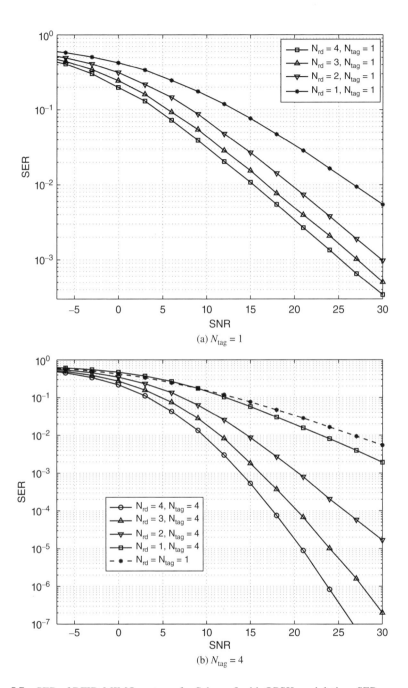

(a) $N_{\text{tag}} = 1$

(b) $N_{\text{tag}} = 4$

Figure 5.7   SER of RFID-MIMO systems for Scheme I with QPSK modulation: SER versus N_{rd}.

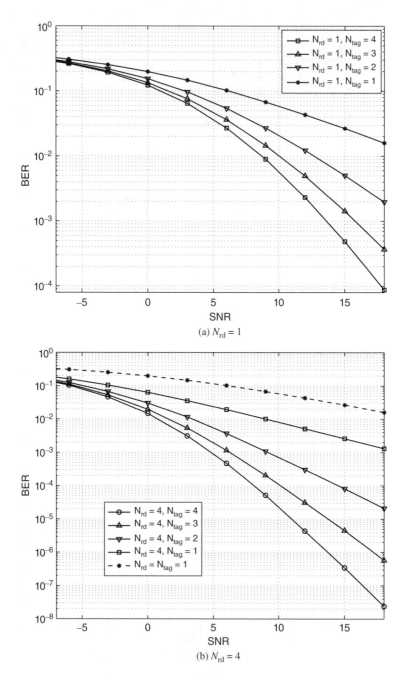

Figure 5.8 BER of RFID-MIMO systems for Scheme II with BPSK modulation: BER versus N_{tag}.

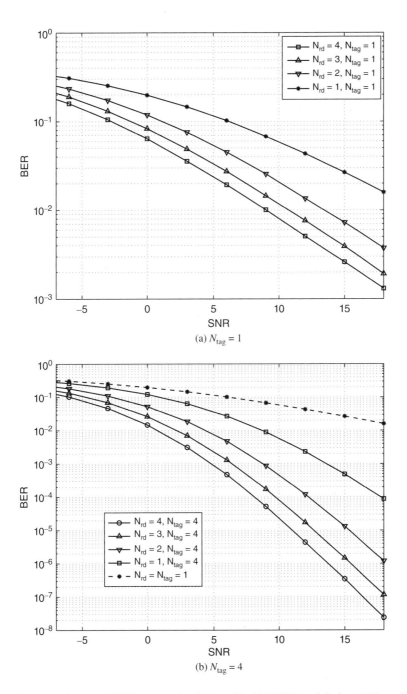

Figure 5.9   BER of RFID-MIMO systems for Scheme II with BPSK modulation: BER versus N_{rd}.

Claim 5.3 When N_{tag} is fixed to be one, increasing N_{rd} *noticeably* and *monotonically* decreases the SER or BER of the system. This phenomenon can be clearly seen from Figure 5.7(a) and Figure 5.9(a). The reason is that only the array gain is provided by the system when $N_{\text{tag}} = 1$ and is indeed collected by both Scheme I and Scheme II. Due to the double-Rayleigh fading channel, the system performance cannot be improved conspicuously by only exploiting this array gain.

Claim 5.4 When N_{rd} (or N_{tag}) is fixed and greater than one, increasing N_{tag} (or N_{rd}) *greatly* decreases the SER or BER of the system, especially for Scheme II. For example, when SNR = 18 dB and $N_{\text{tag}} = 4$, the SER of Scheme I decreases from 3.0×10^{-2} at $N_{\text{rd}} = 1$ to 2.7×10^{-3} at $N_{\text{rd}} = 2$ and 7.5×10^{-5} at $N_{\text{rd}} = 4$, respectively. For the same SNR and N_{tag}, the BER of Scheme II decreases from 8.8×10^{-5} at $N_{\text{rd}} = 1$ to 1.2×10^{-6} at $N_{\text{rd}} = 2$ and 2.4×10^{-8} at $N_{\text{rd}} = 4$, respectively. To achieve BER $= 8.8 \times 10^{-5}$ for the case of Scheme II and $N_{\text{tag}} = 4$, the SNR gain is about 7.5 dB and 10 dB, respectively, by deploying $N_{\text{rd}} = 2$ and $N_{\text{rd}} = 4$ antennas at the reader, compared to the single-antenna setup at the reader. On the other hand, to achieve BER $= 1.3 \times 10^{-3}$ for the case of Scheme II and $N_{\text{rd}} = 4$, the SNR gain is about 9 dB and 13.5 dB, respectively, by deploying $N_{\text{tag}} = 2$ and $N_{\text{tag}} = 4$ antennas at the tag, compared to the single-antenna setup at the tag. This is dramatic improvement for the system performance.

Claim 5.5 Scheme II yields much better SER performance than Scheme I. There are two reasons. The first reason, which is obvious, is that different symbol constellations are used in Schemes I and II. In these simulations, one symbol in Scheme I actually carries two bit information, while one symbol in Scheme II carries only one-bit information. The second reason, which is somewhat subtle to see, is that the diversity gain harvested by Scheme I is lower than that harvested by Scheme II, even though Scheme II throws away the signal in another half signal space. This observation can be seen by comparing equations (5.19)–(5.21) and (5.31) (for Scheme II) and equations (5.26), (5.27) and (5.29) (for Scheme I). For Scheme I, it is seen from (5.26) and (5.27) that the N_{tag} independent channels are not coherently summed. In (5.29), the N_{rd} independent summed channels are further summed. Thus Scheme I yields a diversity order of N_{rd} and the system-inherited diversity order N_{tag} is sacrificed. For Scheme II, it is seen from (5.19)–(5.21) that the N_{tag} independent channels are first coherently summed, yielding a diversity order of N_{tag}. From (5.31), the N_{rd} independent summed channels are further summed, yielding a diversity order of N_{rd}. Thus a total diversity order of $N_{\text{rd}} \times N_{\text{tag}}$ is obtained in Scheme II.

Claim 5.6 Comparing Figures 5.6–5.7 and Figures 5.8–5.9, we can conclude that it is better to deploy as many antennas as possible at the reader. At least the number of antennas at the reader side should be not less than the number of antennas at the tag side. In this way, the full channel diversity generated by multiple antennas at the tag can be maximally exploited.

It may be argued that it is not fair to compare the SER performance of Scheme I and Scheme II, since the former uses QPSK modulation while the latter uses BPSK modulation. To make the comparison complete, the BER performance of Scheme I with BPSK modulation is shown in Figures 5.10–5.11 for the corresponding cases. The results in Figures 5.6–5.11 show that the BER performance of Scheme I is much worse than that of Scheme II, even though the BER of Scheme I with BPSK modulation is lower than the SER of Scheme I with QPSK

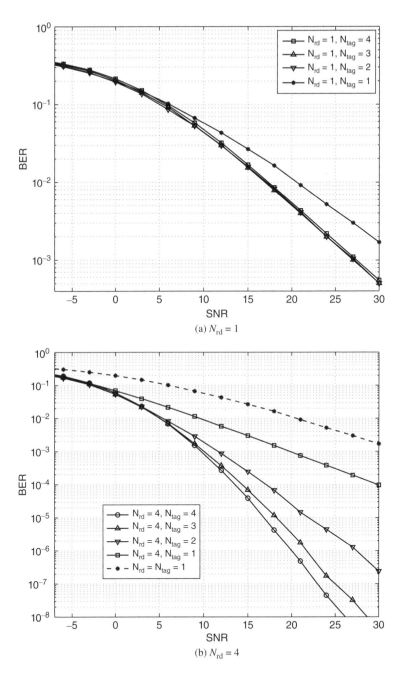

Figure 5.10   BER of RFID-MIMO systems for Scheme I with BPSK modulation: BER versus N_{tag}.

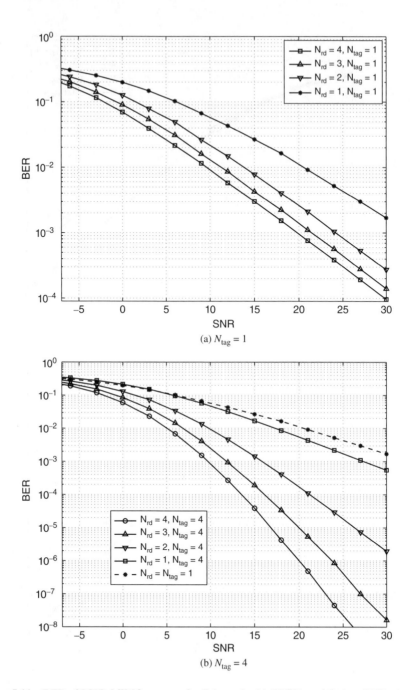

(a) $N_{tag} = 1$

(b) $N_{tag} = 4$

Figure 5.11   BER of RFID-MIMO systems for Scheme I with BPSK modulation: BER versus N_{rd}.

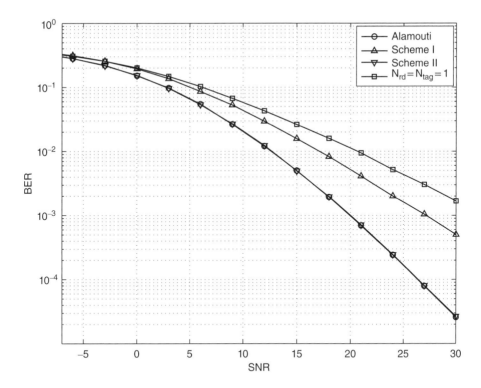

Figure 5.12 A comparison among Scheme I, Scheme II and the Alamouti STC. For the curves marked with 'Scheme I', 'Scheme II' and 'Alamouti', $N_{tag} = 2$ and $N_{rd} = 1$.

modulation for the same configuration of antenna numbers at the reader and tag. By comparing Figures 5.10–5.11 and Figures 5.8–5.9 we can see that Claims 5.1–5.6 obtained based on the comparison between Figures 5.6–5.7 and Figures 5.8–5.9 also hold true qualitatively.

From the aforementioned phenomena, the following conclusions can be drawn: if the required data rate is not high, it is better to use a real-symbol constellation for the transmitted symbols at the tag and correspondingly to use Scheme II decoding policy at the reader receiver; by keeping the cost of the system under constraint, it is better to deploy multiple tag antennas and reader antennas, and the number of reader antennas should be at least equal to the number of tag antennas.

It is interesting to compare the ROD based STC and Alamouti STC. The latter is described in Appendix 5.A. Figure 5.12 shows the comparison. It can be seen that Scheme II and Alamouti STC yield the same BER performance, both are better than Scheme I. This is due to the fact that both Scheme II and Alamouti STC collect all the available channel diversities, while Scheme I does not.

Finally, let us compare the complexity of Scheme I and Scheme II. Both Scheme I and Scheme II perform the same processing, as shown in equations (5.12)–(5.14), for the trans-mitted symbols at the tag. As seen from (5.12)–(5.14), the symbol processing at the tag is quite simple: only the sign of the symbols to be transmitted needs to be changed at some time slots for some antennas. For the processing of a block of space-time decoding at the reader, Scheme I needs $N_{rd}(K^2 + K + N_{tag})$ complex multiplications and $N_{rd}K(K - 1) + (N_{rd} - 1)K +$

$N_{rd}(N_{tag} - 1) = N_{rd}(K^2 + N_{tag} - 1) - K$ complex additions, and Scheme II needs $N_{rd}K^2$ complex multiplications, $N_{rd}K(K - 1)$ complex additions and $(N_{rd} - 1)K$ real additions. Therefore, the computational burden of Scheme II is a little less than that of Scheme I. With regard to the hardware cost of the proposed STC technique, the main increase in the cost arises from the deployment of multiple antennas. The cost increase for the involved signal processing unit is negligible at either tags or readers, since the space-time encoding is very simple, which can be easily dealt with by the embedded chip at tags, and the required computational burden for the space-time decoding at readers is also negligible compared to relatively strong computational power of readers.

5.6 Differential Space-Time Coding for RFID-MIMO Systems

The STC approach discussed in the preceding section needs to know the channel fading of RFID communication links, including both forward and backward links, or equivalent composite channel matrix as shown in equation (5.22). This is especially challenging to RFID systems. In this section, a differential space-time coding (DSTC) scheme for RFID-MIMO systems will be presented, where the CSI of RFID-MIMO links will not be required at either reader or tag side. For narrowband MIMO communication systems, several DSTC approaches [14, 16, 32] were proposed almost simultaneously. The approach in [32] is essentially an extension of Alamouti STC scheme and applies to the case of two transmit antennas. The approaches in [14, 16], essentially equivalent to each other, are designed based on group theory. Report [27] extended the approach in [32] to RFID-MIMO systems.

Since the approach in [16] is more general and some explicit designs for DSTC are provided, we will extend the DSTC scheme in [16] to RFID-MIMO systems and investigate its performance for RFID-MIMO systems in this section.

5.6.1 A Review of Unitary DSTC

In this subsection, we need some preliminary about group theory, complex matrix Gaussian distribution and ML decoder design for unitary space-time coding systems, which are presented in Appendices 5.B, 5.C and 5.D, respectively. The approach presented in this subsection is mainly from [16].

Let us consider a MIMO communication system equipped with N_T transmit antennas and N_R receive antennas. The system is modelled by

$$\mathbf{y} = \mathbf{Hx} + \mathbf{n}, \tag{5.32}$$

where $\mathbf{x}$ and $\mathbf{y}$ are transmit and receive signals, respectively, $\mathbf{H}$ is channel fading matrix, which is complex Gaussian distributed and $\mathbf{n}$ is receiver noise vector, which is also complex Gaussian distributed. Let C denote the constellation of the transmitted symbols. For example, for QPSK modulation, $C = \{1, j, -1, -j\}$. Suppose that the block length of the concerned STC is K. A transmitted space-time code is an element of $C^{N_T \times K}$. Let $\mathcal{G}$ be a group of $K \times K$ unitary matrices, and let $\mathbf{Q}$ be an $N_T \times K$ matrix such that $\mathbf{QG} \in C^{N_T \times K}$ for all $\mathbf{G} \in \mathcal{G}$. Then the collection of matrices

$$\mathbf{Q}\mathcal{G} := \{\mathbf{QG} : \mathbf{G} \in \mathcal{G}\}$$

is called a *group code* of length K over the constellation C [16].

For example, consider the case of $N_T = K = 2$. Take [16]

$$\mathcal{G} = \left\{ \pm \begin{bmatrix} 1 & 0 \\ 0 & 1 \end{bmatrix}, \pm \begin{bmatrix} J & 0 \\ 0 & -J \end{bmatrix}, \pm \begin{bmatrix} 0 & 1 \\ -1 & 0 \end{bmatrix}, \pm \begin{bmatrix} 0 & J \\ J & 0 \end{bmatrix} \right\}, \tag{5.33}$$

$$\mathbf{Q} = \begin{bmatrix} 1 & 1 \\ -1 & 1 \end{bmatrix}. \tag{5.34}$$

Then $\mathbf{Q}\mathcal{G}$ is a group code over the QPSK constellation $C = \{1, J, -1, -J\}$.

For any number of transmit antennas and any constellation, a group code can be constructed in the following way: choose $\mathbf{Q}$ to be an $N_T \times K$ matrix whose elements are taken from C, and $\mathcal{G}$ to be a group of $K \times K$ permutation matrices. Of course, the performance of so constructed group codes needs to be further investigated.

The unitary DSTC scheme works in the following way:

step 0: At the initial time, the transmitter sends block code $S_0 = \mathbf{Q}$.
step i: At time slot i, the transmitter sends block code S_i according to

$$S_i = S_{i-1} \mathbf{G}_i, \quad \mathbf{G}_i \in \mathcal{G}, \ i = 1, 2, \dots . \tag{5.35}$$

The closure property of the group ensures that $S_i \in \mathbf{Q}\mathcal{G}$ for all i.

From the encoding scheme (5.35) it can be seen that, to the receiver, only $\mathbf{G}_i$ is uncertain. Therefore, the information between the receiver and transmitter is carried by $\mathbf{G}_i$ instead of the symbol in the constellation C. Hence, the information rate between the receiver and transmitter is given by

$$R = \frac{1}{K} \log_2 |\mathcal{G}| \ (\text{bits/s/Hz}),$$

where $|\mathcal{G}|$ stands for the cardinality of $\mathcal{G}$.

For the decoding at the receiver, let us start with the ML decoder.

Based on equation (5.32) and considering the transmitted and received signal in a space-time coding block, i.e. during K time slots, we can write the received signal in a space-time coding block as follows:

$$\mathbf{Y}_i = \mathbf{H}\mathbf{X}_i + \mathbf{N}_i = \sqrt{\rho}\,\mathbf{H}S_i + \mathbf{N}_i, \tag{5.36}$$

where ρ stands for the SNR (so that $\mathbf{X}_i = \sqrt{\rho}\,S_i$ and the power of each symbol in a space-time block code S_i is normalized to be unity), and

$$\mathbf{Y}_i = [\mathbf{y}(t) \quad \mathbf{y}(t+1) \quad \cdots \quad \mathbf{y}(t+K-1)],$$

$$\mathbf{X}_i = [\mathbf{x}(t) \quad \mathbf{x}(t+1) \quad \cdots \quad \mathbf{x}(t+K-1)],$$

$$\mathbf{N}_i = [\mathbf{n}(t) \quad \mathbf{n}(t+1) \quad \cdots \quad \mathbf{n}(t+K-1)].$$

Here the subscript i is related with discrete time t.

To find an optimal decoder, we need some assumptions on the fading channel and receiver noise. In the following, we will assume that the fading channel and receiver noise satisfy the following assumption.

Assumption 5.3 Suppose that (i) all the entries of $\mathbf{H}$ are complex Gaussian distributed, independent of each other, of mean zero and variance unity, (ii) all the entries of $\mathbf{N}$ are complex

Gaussian distributed, independent of each other, of mean zero and variance unity, and (iii) $\mathbf{H}$ and $\mathbf{N}$ are mutually independent.

To optimally decode the transmitted symbol sequence, we should consider the received signals $\mathbf{Y}_i$, $i = 1, 2, \cdots$, jointly. The decoder designed in this way is too complicated. Instead, we consider only two consecutive received signal blocks $\mathbf{Y}_{i-1}$ and $\mathbf{Y}_i$. Let

$$\vec{\mathbf{Y}}_i = [\mathbf{Y}_{i-1} \ \mathbf{Y}_i], \quad \vec{\mathbf{N}}_i = [\mathbf{N}_{i-1} \ \mathbf{N}_i]. \tag{5.37}$$

Let us use $\mathbf{G}$ to denote the selected group element $\mathbf{G}_i$ for transmission at time instant i, that is, $S_i = S_{i-1}\mathbf{G}$. Our objective is to find $\mathbf{G}$ such that the ML function $p(\vec{\mathbf{Y}}_i|\mathbf{G})$ is maximized. From equations (5.36) and (5.37) we have

$$\begin{aligned}
\vec{\mathbf{Y}}_i &= [\sqrt{\rho}\, \mathbf{H}S_{i-1} + \mathbf{N}_{i-1} \quad \sqrt{\rho}\, \mathbf{H}S_{i-1}\mathbf{G} + \mathbf{N}_i] \\
&= \sqrt{\rho}\, \mathbf{H} [S_{i-1} \quad S_{i-1}\mathbf{G}] + [\mathbf{N}_{i-1} \quad \mathbf{N}_i] \\
&= \sqrt{\rho}\, \mathbf{H}\, \vec{S}_i + \vec{\mathbf{N}}_i,
\end{aligned} \tag{5.38}$$

where

$$\vec{S}_i := [S_{i-1} \quad S_{i-1}\mathbf{G}].$$

By induction and the unitary property of group $\mathcal{G}$, it can be easily shown that all the transmitted block codes satisfy the condition

$$S_i S_i^\dagger = \frac{1}{2}\alpha\mathbf{I} \ \text{ for all } \ i = 1, 2, \cdots \tag{5.39}$$

if

$$S_0 S_0^\dagger = \frac{1}{2}\alpha\mathbf{I}, \tag{5.40}$$

where α is a constant. Since S_0 is under the designer's choice, we can always choose it such that condition (5.40) holds true. Therefore, we can always assume that condition (5.39) holds true. Thus we have

$$\vec{S}_i(\vec{S}_i)^\dagger = S_{i-1}S_{i-1}^\dagger + S_{i-1}\mathbf{G}\mathbf{G}^\dagger S_{i-1}^\dagger = \alpha\mathbf{I}. \tag{5.41}$$

The STC satisfying equation (5.41) is called unitary STC [13, 16].

From these facts we can see that the system model (5.38) is of the form (5.52) with the transmitted block codes satisfying condition (5.51). Thus the ML decoder (5.62) is applicable to system (5.38).

In the following it is assumed that $K = N_T$. Due to this assumption and equation (5.39) we have $S_i^\dagger S_i = \frac{1}{2}\alpha\mathbf{I}$ for all i.

Since

$$(\vec{S}_i)^\dagger \vec{S}_i = \begin{bmatrix} \frac{1}{2}\alpha\mathbf{I} & \frac{1}{2}\alpha\mathbf{G} \\ \frac{1}{2}\alpha\mathbf{G}^\dagger & \frac{1}{2}\alpha\mathbf{I} \end{bmatrix},$$

we get the ML estimate of $\mathbf{G}$ as follows:

$$\{\hat{\mathbf{G}}, \hat{S}_{i-1}\} = \arg\max_{\mathbf{G}\in\mathcal{G}, S_{i-1}} \mathtt{tr}(\vec{\mathbf{Y}}_i(\vec{S}_i)^\dagger \vec{S}_i \vec{\mathbf{Y}}_i^\dagger)$$

$$= \arg\max_{\mathbf{G}\in\mathcal{G}} \mathtt{tr}(\mathbf{Y}_i \mathbf{G}^\dagger \mathbf{Y}_{i-1}^\dagger + \mathbf{Y}_{i-1}\mathbf{G}\mathbf{Y}_i^\dagger)$$

$$= \arg\max_{\mathbf{G}\in\mathcal{G}} \mathfrak{R}[\mathtt{tr}(\mathbf{Y}_{i-1}\mathbf{G}\mathbf{Y}_i^\dagger)]$$

$$= \arg\max_{\mathbf{G}\in\mathcal{G}} \mathfrak{R}[\mathtt{tr}(\mathbf{G}\mathbf{Y}_i^\dagger \mathbf{Y}_{i-1})]. \qquad (5.42)$$

In the second equality, the subscript S_{i-1} in the operation arg max is dropped since the argument of the function $\mathtt{tr}$ does not depend on S_{i-1}.

The decoder (5.42) is a sub-optimal decoder for unitary DSTC scheme (5.35), since only two consecutive transmit block codes are considered.

5.6.2 Application of Unitary DTSC to RFID

Consider the equivalent RFID-MIMO system (5.7). For clarity, we rewrite it here

$$\mathbf{y}(t) = \mathbf{A}\boldsymbol{\gamma}(t) + \mathbf{n}(t),$$

$$\mathbf{A} = \mathbf{H}^b \breve{\mathbf{H}}, \qquad (5.43)$$

where $\mathbf{y}$ and $\mathbf{n}$ are the received signal and noise, respectively, at the reader's receiver, $\boldsymbol{\gamma}$ is the backscattering signal at the tag, $\breve{\mathbf{H}}$ and $\mathbf{H}^b$ are the fading channels of forward (including the effect of transmitted signal at the reader's transmitter) and backward links, respectively, as defined in Section 5.3.

Let $\mathbf{Q}\mathcal{G}$ be a group code of length K over a constellation $\mathcal{C}$ for N_{tag} antennas at the tag. Suppose that the channels of both forward and backward links do not change with time during a two-block-coding period $2KT_f$, where T_f is a symbol period. The transmit signal $\mathbf{x}$ at the reader is also fixed during a two-block-coding period $2KT_f$. Therefore, the equivalent composite channel $\mathbf{A}$ will not change with time when we only consider the signal processing for a two-block-coding period.

The unitary DSTC scheme for RFID-MIMO systems works in the following way:

step 0: At the initial time, the tag transmits block code $\boldsymbol{\Gamma}_0 = \sqrt{\rho}S_0$ with $S_0 = \mathbf{Q}$ and ρ being the transmit power.

step i: At time slot i, the tag sends block code $\boldsymbol{\Gamma}_i$ according to

$$\boldsymbol{\Gamma}_i = \sqrt{\rho}S_i, \quad \text{with}$$

$$S_i = S_{i-1}\mathbf{G}_i, \quad \mathbf{G}_i \in \mathcal{G}, \; i = 1, 2, \cdots$$

where

$$\boldsymbol{\Gamma}_i = [\boldsymbol{\gamma}(t) \quad \boldsymbol{\gamma}(t+1) \quad \cdots \quad \boldsymbol{\gamma}(t+K-1)]$$

with the subscript i being corresponding to time t.

Decoder (5.42) is used to decode the transmitted code $\mathbf{G}_i$, that is,

$$\hat{\mathbf{G}}_i = \arg \max_{\mathbf{G} \in \mathcal{G}} \Re[\mathtt{tr}(\mathbf{G}\mathbf{Y}_i^{\dagger}\mathbf{Y}_{i-1})], \qquad (5.44)$$

where $\mathbf{Y}_i = [\mathbf{y}(t) \quad \mathbf{y}(t+1) \quad \cdots \quad \mathbf{y}(t+K-1)]$.

Note that for RFID-MIMO system (5.43), the equivalent composite fading matrix $\mathbf{A}$ generally does not satisfy condition (i) in Assumption 5.3 (by replacing $\mathbf{H}$ with $\mathbf{A}$). Therefore, decoder (5.44) is not a maximum likelihood receiver in general. However, since it is difficult, if not impossible, to find the optimal receiver, we still use (5.44) to decode the transmitted code $\mathbf{G}$ and resort to simulations to investigate the performance of the receiver.

5.6.3 Simulation Results

In this subsection, simulation results for two examples of unitary DSTC will be shown. The first example is based on the unitary DSTC $\{\mathcal{G}, \mathbf{Q}\}$ as given by equations (5.33)–(5.34). The second example is based on the unitary DSTC $\{\mathcal{G}, \mathbf{Q}\}$ as given by the following equations [17]:

$$\mathcal{G} = \{\mathbf{I}, \mathbf{G}_0, \mathbf{G}_0^2, \ldots, \mathbf{G}_0^{15}\}, \quad \mathbf{G}_0 = \begin{bmatrix} 0 & 0 & 0 & j \\ 1 & 0 & 0 & 0 \\ 0 & 1 & 0 & 0 \\ 0 & 0 & 1 & 0 \end{bmatrix}, \qquad (5.45)$$

$$\mathbf{Q} = \begin{bmatrix} 1 & -1 & -1 & 1 \\ 1 & 1 & -1 & -1 \\ 1 & -1 & 1 & -1 \\ 1 & 1 & 1 & 1 \end{bmatrix}. \qquad (5.46)$$

The symbol constellation is also based on QPSK.

For the first example, the results of Alamouti STC scheme with QPSK modulation will be used as a comparison. To make the comparison fair, we choose

$$\rho = \frac{E_0}{N_{\text{tag}}}$$

for DSTC schemes (5.33)–(5.34) and (5.45)–(5.46), where E_0 is the total transmit power at the tag at each time instant, and define SNR as $\frac{E_0}{\sigma_{\mathbf{n}}^2}$, where $\sigma_{\mathbf{n}}^2$ is the variance of each entry of the noise vector $\mathbf{n}$. The scaling coefficient $\sqrt{\frac{E_0}{N_{\text{tag}}}}$ is to normalize the overall energy consumption per time slot at the tag side to be E_0 no matter how many antennas are deployed at the tag.

The simulation results for unitary DSTC (5.33)–(5.34) for the case of two tag antennas are shown in Figure 5.13, where the simulation results for Alamouti STC with QPSK modulation are also illustrated. It is seen that, to achieve the same SER, unitary DSTC (5.33)–(5.34) needs about $2 \sim 3$ dB more SNR than the Alamouti STC does for the same number of reader antennas. Note that the data rates for DSTC (5.33)–(5.34) and Alamouti STC systems are 1.5 bits/s/Hz and 2 bits/s/Hz, respectively. The benefit obtained by the DSTC scheme, as a compromise to the performance sacrifice in data rate and SER (or required more SNR), is that both reader and tag do *not* need the knowledge about CSI.

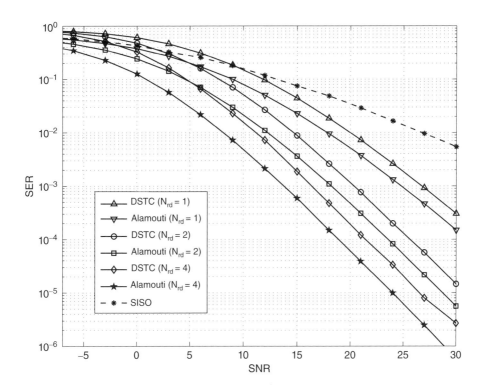

Figure 5.13   Code error rate for unitary DSTC (5.33)–(5.34) for the case of two tag antennas.

The simulation results for unitary DSTC (5.45)–(5.46) for the case of four tag antennas are shown in Figure 5.14. It is seen that the SER performance of DSTC scheme (5.45)–(5.46) is greatly improved, compared to DSTC scheme (5.33)–(5.34). Note that the data rate for DSTC (5.45)–(5.46) is 1 bit/s/Hz. This is a compromise to the aforementioned performance gain.

The comparison among DSTC (5.33)–(5.34) (named as DSTC I), DSTC (5.45)–(5.46) (named as DSTC II) and Alamouti STC (named as ASTC) is detailed in Table 5.2.

From Table 5.2 we can see that, to achieve an SER of 10^{-3}, the system DSTC II (8,4) offers an SNR gain of 17.0 dB and 19.5 dB, respectively, compared to the systems ASTC (1,2) and DSTC I (1,2), and to achieve an SER of 10^{-5} the system DSTC II (8,4) offers an SNR gain of 16.0 dB and 18.5 dB, respectively, compared to the systems ASTC (2,2) and DSTC I (2,2). This SNR gain is significant to passive RFID systems.

5.7   Summary

In this chapter, we have discussed the space-time encoding and decoding problem for RFID-MIMO systems. A mathematical model for this kind of system is developed from the viewpoint of signal processing, which makes it easy to design the transmit signals for both readers and tags and to decode the tag's ID. Based on this model, the signal design for both readers and tags has been addressed.

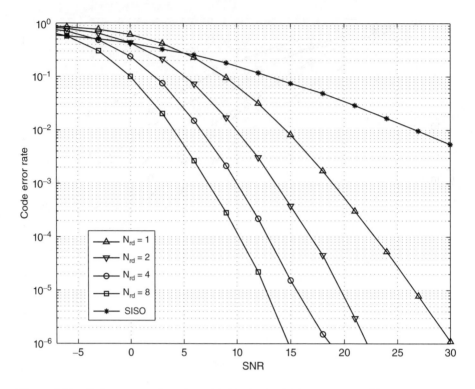

Figure 5.14   Code error rate for unitary DSTC (5.45)–(5.46) for the case of four tag antennas.

Table 5.2   A comparison of the required SNRs to achieve the same SER for different RFID-STC systems, where the notation DSTC I (m_1, m_2) means that the system uses DSTC (5.33)–(5.34) and is equipped with m_1 antennas at its reader and m_2 antennas at its tag. Similar notations apply to ASTC and DSTC II schemes too.

system	data rate (bits/s/Hz)	CSI required?	required SNR SER $= 10^{-3}$	SER $= 10^{-5}$
ASTC (1,2)	2.0	yes	24.5 dB	$\gg 30$ dB
DSTC I (1,2)	1.5	no	27.0 dB	$\gg 30$ dB
DSTC II (1,4)	1.0	no	19.0 dB	26.5 dB
ASTC (2,2)	2.0	yes	18.0 dB	28.5 dB
DSTC I (2,2)	1.5	no	20.5 dB	≈ 31.0 dB
DSTC II (2,4)	1.0	no	13.5 dB	19.5 dB
ASTC (4,2)	2.0	yes	13.5 dB	23.5 dB
DSTC I (4,2)	1.5	no	16.5 dB	26.5 dB
DSTC II (4,4)	1.0	no	10.0 dB	15.5 dB
DSTC II (8,4)	1.0	no	7.5 dB	12.5 dB

For the reader's transmitter design, two spatial matching transmission schemes (namely matched to a fixed antenna at the tag or matched to a selected antenna at the tag) for the reader's signals are proposed. Simulation results show that the proposed approaches can greatly improve the BER performance of RFID-MIMO systems, compared to the uniform transmit policy. In many application situations, the reader should recognize a large amount of tags in a short time. This problem is challenging. The approach developed in Section 5.4 provides a possible solution to this problem. If the signals coming from all the tags, no matter whether the tags are equipped with multiple antennas or a single antenna each, are considered simultaneously, then we have an equivalent RFID-MIMO system. It can be easily seen that the first transmission way for the tag investigated in Section 5.4 can be straightforwardly applied to this scenario.

For the tag's transmitter design, two STC schemes, namely Scheme I and Scheme II, are proposed. Simulation results illustrate that the proposed approaches can greatly improve the SER/BER performance of RFID systems, compared to non space-time encoded RFID systems. The SER/BER performance for Scheme I and Scheme II is thoroughly compared and it is found that Scheme II with the innate real-symbol constellation yields a better SER/BER performance than Scheme I.

As is commonly assumed in the STC technique, the channel state information is required to be available at the receiver side of the reader to use the techniques of Scheme I and Scheme II. The channel estimation problem for RFID systems has been discussed in [2, 3, 20], where a method for estimating the channel of the whole chain, including forward link, backscattering coefficient and backward link, is presented. However, to estimate the forward and backward channels $\mathbf{H}^f$ and $\mathbf{H}^b$ separately remains an open issue.

Another approach to solving the CSI problem is to use differential space-time coding and decoding technique. Using the unitary DSTC scheme, CSI is not required at either reader side or tag side. As shown in Figures 5.13–5.14 and Table 5.2, with some sacrifice in data rates, using 8 and 4 antennas at the reader and tag, respectively, and employing the corresponding unitary DSTC scheme can yield considerable SER performance improvement or SNR gains, compared to Alamouti STC scheme.

Appendix 5.A Alamouti Space-Time Coding for Narrowband Systems

Alamouti coding is one of the most popularly used STC schemes in narrowband MIMO systems. In this Appendix, a brief revisit for the Alamouti coding scheme will be presented to show its basic idea.

The basic structure for the transmitter and receiver of Alamouti STC scheme is illustrated in Figure 5.15. As shown in Figure 5.15, we consider a narrowband system with two transmit antennas and one receive antenna. The channel fading gains from the two transmit antennas to the receive antenna are denoted as h_1 and h_2, respectively, and the received signal at time t is denoted by $y(t)$.

When discussing the Alamouti STC, the following assumptions are required.

Assumption 5.4 The channel impulse responses are frequency-flat and independent of each other.

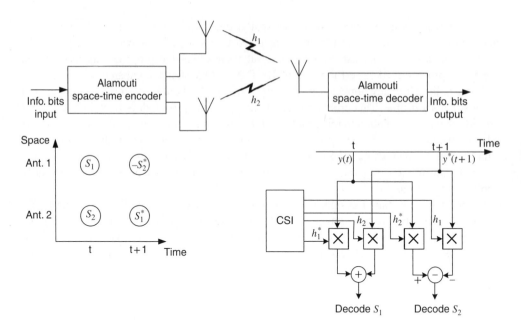

Figure 5.15   Alamouti STC scheme for 2×1 narrowband MIMO.

Assumption 5.5   The channel fading h_1 and h_2 do not change across two consecutive symbol transmissions.

Assumption 5.6   The channel state information h_1 and h_2 are available at the receiver.

According to the Alamouti STC scheme, two information symbols denoted as S_1 and S_2 are transmitted at two consecutive time instants in the following way [1]: at time t, S_1 and S_2 are transmitted from antenna 1 and antenna 2, respectively; at time $t + 1$, $-S_2^*$ and S_1^* are transmitted from antenna 1 and antenna 2, respectively. Let us define

$$S := \begin{bmatrix} S_1 & -S_2^* \\ S_2 & S_1^* \end{bmatrix},$$

$$\mathbf{w}(t) := \begin{bmatrix} w(t) \\ w(t - T_f) \end{bmatrix},$$

where $w(t)$ is the baseband waveform of the transmit signal and T_f is a symbol period, similar to the notations in Section 5.5. The transmitted signal across the two transmit antennas can be expressed as

$$\mathbf{x}(t) = \sqrt{\frac{E_0}{2}}\, S\mathbf{w}(t), \tag{5.47}$$

where E_0 is the total power used for the transmission of each symbol in a block code, which is also the total transmit power at each time instant.

Let $n(t)$ be the receiver noise at time instant t. Using the similar convention for notations as in Section 5.5, let us define

$$\vec{y} := [y(t) \quad y(t+1)],$$

$$\vec{n} := [n(t) \quad n(t+1)].$$

Let the channel matrix be denoted by

$$\mathbf{h} := [h_1 \quad h_2]$$

and define decoding matrix by

$$D := \begin{bmatrix} h_1^* & h_2 \\ h_2^* & -h_1 \end{bmatrix}.$$

Then the received signal after sampling can be expressed as

$$\vec{y} = \sqrt{\frac{E_0}{2}} \mathbf{h} S + \vec{n}.$$

To decode the transmit symbol, the received signal is processed as follows:

$$\mathbf{z} = D \begin{bmatrix} [\vec{y}]_1 \\ ([\vec{y}]_2)^* \end{bmatrix}.$$

It can be easily shown that

$$\begin{cases} \mathbf{z}_1 = (|h_1|^2 + |h_2|^2)S_1 + h_1^* n(t) + h_2 n^*(t+1), \\ \mathbf{z}_2 = (|h_1|^2 + |h_2|^2)S_2 + h_2^* n(t) - h_1 n^*(t+1). \end{cases} \tag{5.48}$$

Therefore, the information symbols S_1 and S_2 are decoupled from each other and can be easily decoded from the linearly processed receiver outputs $\mathbf{z}_1$ and $\mathbf{z}_2$, respectively.

The Alamouti STC idea for the 2×1 MISO system can be easily extended to the case of $2 \times N_R$ MIMO, as illustrated in Figure 5.16.

The information symbols S_1 and S_2 are also transmitted in the same way as in the case of 2×1 MISO. Now denote the received signals at receive antenna i as $y_i(t)$ and $y_i(t+1)$ ($i = 1, \ldots, N_R$), respectively, at time instants t and $t+1$. Similarly, the receiver noises at receive antenna i are denoted as $n_i(t)$ and $n_i(t+1)$ ($i = 1, \ldots, N_R$), respectively, at time instants t and $t+1$. Let the channel fading gain from transmit antenna i to receive antenna j be h_{ji} ($i = 1, 2$, $j = 1, 2, \ldots, N_R$). Define

$$\vec{y} := [\mathbf{y}(t) \quad \mathbf{y}(t+1)], \quad \mathbf{y}(t) = \begin{bmatrix} y_1(t) \\ y_2(t) \\ \vdots \\ y_{N_R}(t) \end{bmatrix},$$

$$\vec{n} := [\mathbf{n}(t) \quad \mathbf{n}(t+1)], \quad \mathbf{n}(t) = \begin{bmatrix} n_1(t) \\ n_2(t) \\ \vdots \\ n_{N_R}(t) \end{bmatrix},$$

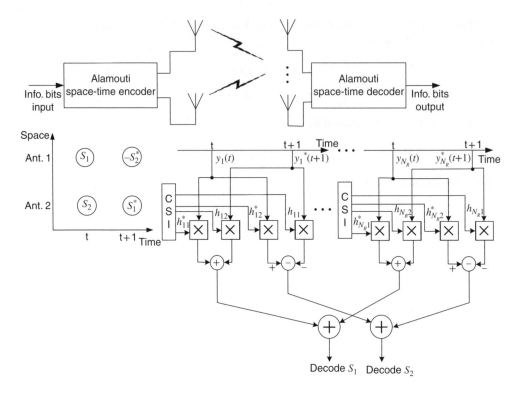

Figure 5.16   Alamouti STC scheme for $2 \times N_R$ narrowband MIMO.

$$\mathbf{H} := \begin{bmatrix} h_{11} & h_{12} \\ h_{21} & h_{22} \\ \vdots & \vdots \\ h_{N_R 1} & h_{N_R 2} \end{bmatrix},$$

$$D_j := \begin{bmatrix} h_{j1}^* & h_{j2} \\ h_{j2}^* & -h_{j1} \end{bmatrix}.$$

Then the received signal after sampling can be expressed as

$$\vec{\mathbf{y}} = \sqrt{\frac{E_0}{2}} \mathbf{H} S + \vec{\mathbf{n}}.$$

To decode the transmit symbol, the received signal is processed as follows:

$$\mathbf{z} = \sum_{j=1}^{N_R} D_j \begin{bmatrix} [\vec{\mathbf{y}}]_{j1} \\ ([\vec{\mathbf{y}}]_{j2})^* \end{bmatrix}.$$

It can be shown that

$$
\begin{cases}
\mathbf{z}_1 = \displaystyle\sum_{j=1}^{N_R} \sum_{i=1}^{2} |h_{ji}|^2 S_1 + \sum_{j=1}^{N_R} h_{j1}^* n_j(t) + h_{j2} n_j^*(t+1), \\[4mm]
\mathbf{z}_2 = \displaystyle\sum_{j=1}^{N_R} \sum_{i=1}^{2} |h_{ji}|^2 S_2 + \sum_{j=1}^{N_R} h_{j2}^* n_j(t) - h_{j1} n_j^*(t+1).
\end{cases}
\tag{5.49}
$$

Again, it is observed that the information symbols S_1 and S_2 are decoupled from each other and hence can be decoded from the combined receiver outputs $\mathbf{z}_1$ and $\mathbf{z}_2$ respectively.

Appendix 5.B Definition of Group

A group $\mathcal{G}$ is a set of elements $\{X, Y, \dots , \}$, together with an operation $\bullet$, called product, such that the following conditions hold [5]:

(i) *Closure property*: $\forall X, Y \in \mathcal{G} \;\Rightarrow\; X \bullet Y \in \mathcal{G}$.
(ii) *Associative law*: $\forall X, Y, Z \in \mathcal{G} \;\Rightarrow\; X \bullet (Y \bullet Z) = (X \bullet Y) \bullet Z$.
(iii) There exists a unique element $I \in \mathcal{G}$ with the property

$$ I \bullet X = X \bullet I = X, \forall X \in \mathcal{G}. $$

The element I is called the *identity element* of the group.
(iv) For any $X \in \mathcal{G}$, there exists an element $X^{-1} \in \mathcal{G}$ such that

$$ X^{-1} \bullet X = X \bullet X^{-1} = I. $$

The element X^{-1} is called the inverse of X.

It is apparent that all the unitary matrices of a given dimension, say $n \times n$ with n being a positive integer, under the matrix product, form a group. The identity matrix $\mathbf{I}$ is the identity element of this group, and the inverse of any element (matrix) in this group is the Hermitian transpose of this element.

Appendix 5.C Complex Matrix/Vector Gaussian Distribution

For the complex vector Gaussian distribution, several forms of definition exist in the literature: some are restrictive, some are loose. A direct idea is to expand the real and imagery parts of all the entries of the vector as separate variables and to stack them one after another, forming an augmented real vector. Then define the distribution based on real Gaussian distribution for this augmented real vector. This way will give a loosely defined complex vector Gaussian distribution. The drawback of this definition is that it is not convenient to deal with relevant signal processing topics based on this definition. Another idea for defining the complex Gaussian distribution is based directly on the complex vector. This way will lead to an elegant but more restrictive definition. However, this kind of definition will facilitate the elaboration of relevant signal processing.

For the complex matrix Gaussian distribution, a direct approach is to convert the matrix to a vector by stacking all the columns of the matrix one by one and then to apply complex vector Gaussian distribution to this augmented vector. This way will give a very cumbersome expression for the complex matrix Gaussian distribution. Similar to the vector case, a direct way to define the distribution is based on the complex matrix itself. In this appendix, we will adopt this approach by combining the definitions of real matrix Gaussian distribution [4, 7] and complex vector Gaussian distribution [35].

The probability density function for a complex random matrix $\mathbf{X}$ (with dimension $m \times n$) has the following form

$$p_{\mathbf{X}}(\mathbf{X}|\mathbf{M}, \mathbf{U}, \mathbf{V}) = \frac{1}{\pi^{mn}|\mathbf{U}|^n|\mathbf{V}|^m} \exp\{-\mathrm{tr}[\mathbf{V}^{-1}(\mathbf{X}-\mathbf{M})^\dagger\mathbf{U}^{-1}(\mathbf{X}-\mathbf{M})]\}, \tag{5.50}$$

where

$$\mathbf{M} = \mathbb{E}(\mathbf{X}),$$

$$\mathbf{U} = \mathbb{E}[(\mathbf{X}-\mathbf{M})(\mathbf{X}-\mathbf{M})^\dagger],$$

$$\mathbf{V} = \mathbb{E}[(\mathbf{X}-\mathbf{M})^\dagger(\mathbf{X}-\mathbf{M})]/c$$

with c being a constant which makes the integral of $p_{\mathbf{X}}(\mathbf{X}|\mathbf{M}, \mathbf{U}, \mathbf{V})$ over the definition domain of $\mathbf{X}$ be unity.

Note that, based on definition (5.50), the covariance between the real part and imaginary part of the entries of the concerned random vector/matrix should satisfy some constraints [35].

Appendix 5.D Maximum Likelihood Receiver for Unitary STC

Consider the MIMO system (5.32) and space-time block codes $S_l \in C^{N_T \times K}$, $l = 1, 2, \ldots, L$, where $K \,(\geq N_T)$ and L are two positive integers, and C denotes the constellation of the transmitted symbols. The STC $\{S_l\}$ is said to be unitary if [13, 16]

$$S_l S_l^\dagger = \alpha \mathbf{I} \quad \text{for } all \ l, \tag{5.51}$$

where α is a constant.

Denote with $\mathbf{X}$ the transmitted signal for a space-time block. Let ρ be the transmitted power per time instant. Then $\mathbf{X}$ can be written as

$$\mathbf{X} = \sqrt{\rho}\, S, \quad S \in \{S_l\}.$$

The received signal for the space-time block can be written as

$$\mathbf{Y} = \mathbf{H}\mathbf{X} + \mathbf{N} = \sqrt{\rho}\,\mathbf{H}S + \mathbf{N}, \tag{5.52}$$

where $\mathbf{N}$ denotes the noise matrix during the period of a concerned space-time block.

Suppose that the channel matrix $\mathbf{H}$ and noise $\mathbf{N}$ satisfy Assumption 5.3. From Assumption 5.3 we have

$$\mathbb{E}(\mathbf{H}) = \mathbf{0}, \quad \mathbb{E}(\mathbf{N}) = \mathbf{0}, \tag{5.53}$$

$$\mathbb{E}(\mathbf{H}^\dagger\mathbf{H}) = N_R\mathbf{I}_{N_T}, \quad \mathbb{E}(\mathbf{H}\mathbf{H}^\dagger) = N_T\mathbf{I}_{N_R}, \tag{5.54}$$

$$\mathbb{E}(\mathbf{N}^\dagger\mathbf{N}) = N_R\mathbf{I}_K, \quad \mathbb{E}(\mathbf{N}\mathbf{N}^\dagger) = K\mathbf{I}_{N_R}. \tag{5.55}$$

Based on equations (5.51)–(5.55), we have

$$\mathbf{M_Y} := \mathbb{E}(\mathbf{Y}) = \mathbf{0}, \tag{5.56}$$

$$\mathbf{U_Y} := \mathbb{E}[\mathbf{YY}^\dagger] = \rho\,\mathbb{E}(\mathbf{H}SS^\dagger\mathbf{H}^\dagger) + K\mathbf{I}_{N_R} = (\rho\alpha N_T + K)\mathbf{I}_{N_R}, \tag{5.57}$$

$$\mathbf{V_Y} := \mathbb{E}[\mathbf{Y}^\dagger\mathbf{Y}]/c = (\rho\,S^\dagger N_R\mathbf{I}_{N_T}S + N_R\mathbf{I}_K)/c = (\rho N_R S^\dagger S + N_R\mathbf{I}_K)/c. \tag{5.58}$$

Substituting equations (5.56)–(5.58) into (5.50), we can obtain the likelihood function of $\mathbf{Y}$ for a given S as follows:

$$p_{\mathbf{Y}}(\mathbf{Y}|S) = \frac{1}{\pi^{N_R K}(\rho\alpha N_T + K)^{N_R K}|(\rho N_R S^\dagger S + N_R\mathbf{I})/c|^{N_T}}$$
$$\times \exp\{-\mathrm{tr}[c(\rho N_R S^\dagger S + N_R\mathbf{I})^{-1}\mathbf{Y}^\dagger(\rho\alpha N_T + K)^{-1}\mathbf{Y}]\}. \tag{5.59}$$

Notice that for any two compatible matrices $\mathbf{A}$ and $\mathbf{B}$, we have the following identity

$$|\mathbf{I} + \mathbf{AB}| = |\mathbf{I} + \mathbf{BA}|.$$

In terms of this identity, we obtain

$$|\rho N_R S^\dagger S + N_R\mathbf{I}| = N_R^K|\mathbf{I} + \rho SS^\dagger| = N_R^K(\alpha\rho + 1)^{N_T}.$$

Therefore, we have

$$p_{\mathbf{Y}}(\mathbf{Y}|S) = \tilde{c}\exp\left\{-\mathrm{tr}\left[\frac{c}{N_R}(\rho S^\dagger S + \mathbf{I})^{-1}\mathbf{Y}^\dagger(\rho\alpha N_T + K)^{-1}\mathbf{Y}\right]\right\}$$
$$= \tilde{c}\exp\left\{-\frac{c}{N_R(\rho\alpha N_T + K)}\mathrm{tr}\left[(\rho\,S^\dagger S + \mathbf{I})^{-1}\mathbf{Y}^\dagger\mathbf{Y}\right]\right\}, \tag{5.60}$$

where

$$\tilde{c} = \frac{c^{KN_T}}{\pi^{N_R K}(\rho\alpha N_T + K)^{N_R K}N_R^{KN_T}(\alpha\rho + 1)^{N_T^2}}.$$

Using the matrix inverse lemma (4.48), we get

$$(\rho\,S^\dagger S + \mathbf{I})^{-1} = \mathbf{I} - \rho\,S^\dagger(\mathbf{I} + S\rho\,S^\dagger)^{-1}S = \mathbf{I} - \frac{\rho}{1 + \alpha\rho}S^\dagger S. \tag{5.61}$$

Substituting equation (5.61) into (5.60) yields

$$p_{\mathbf{Y}}(\mathbf{Y}|S) = \tilde{c}\exp\left\{-\frac{c}{N_R(\rho\alpha N_T + K)}\mathrm{tr}\left[\mathbf{Y}^\dagger\mathbf{Y} - \frac{\rho}{1 + \alpha\rho}S^\dagger S\mathbf{Y}^\dagger\mathbf{Y}\right]\right\}.$$

Since the first item in the function tr is irrelevant with the test symbol S, the ML decoder for S is given by

$$\hat{S} = \arg\max_{S\in\{S_1,\ldots,S_L\}} p_{\mathbf{Y}}(\mathbf{Y}|S)$$
$$= \arg\max_{S\in\{S_1,\ldots,S_L\}} \mathrm{tr}(S^\dagger S\mathbf{Y}^\dagger\mathbf{Y}) = \arg\max_{S\in\{S_1,\ldots,S_L\}}\mathrm{tr}(\mathbf{Y}S^\dagger S\mathbf{Y}^\dagger). \tag{5.62}$$

It is nice to see that the ML decoder (5.62) is of a correlator structure.

References

[1] S. M. Alamouti. A simple transmit diversity technique for wireless communications. *IEEE J. Sel. Areas Commun.*, 16:1451–1458, 1998.

[2] C. Angerer, R. Langwieser, G. Maier, and M. Rupp. Maximal ratio combining receivers for dual antenna RFID readers. In *IEEE 2009 Int. Microwave Workshop on Wireless Sensing, Local Positioning, and RFID*, Cavtat, Croatia, 24–25 Sept. 2009.

[3] C. Angerer, R. Langwieser, and M. Rupp. RFID reader receivers for physical layer collision recovery. *IEEE Trans. Commun.*, 58:3526–3537, 2010.

[4] S. F. Arnold. *The theory of linear models and multivariate analysis*. John Wiley & Sons, Inc., New York, 1981.

[5] D. S. Dummit and R. M. Foote. *Abstract Algebra*, 3rd. edition, John Wiley & Sons, Inc., Hoboken, New Jersey, 2003.

[6] G. D. Durgin and A. Rohatgi. Multi-antenna RF tag measurement system using back-scattered spread spectrum. In *2008 IEEE Int. Conf. on RFID*, pages 1–7, Las Vegas, USA, 16–17 Apr. 2008.

[7] P. Dutilleul. The MLE algorithm for the matrix normal distribution. *J. Statist. Comput. Simul.*, 64:105–123, 1999.

[8] J. D. Griffin and G. D. Durgin. Gains for RF tags using multiple antennas. *IEEE Trans. Antennas Propag.*, 56:563–570, 2008.

[9] J. D. Griffin and G. D. Durgin. Multipath fading measurements for multi-antenna backscatter RFID at 5.8 GHz. In *Proc. 2009 IEEE Int. Conf. RFID*, pages 322–329, Orlando, USA, 27–28 Apr. 2009.

[10] S. Haykin. *Adaptive Filter Theory*. Prentice Hall, Upper Saddle River, NJ, 4th edition, 2002.

[11] C. He, X. Chen, Z. J. Wang, and W. Su. On the performance of MIMO RFID backscattering channels. *EURASIP J. Wireless Communications and Networking*, 2012. doi:10.1186/1687-1499-2012-357, Available: http://jwcn .eurasipjournals.com/content/pdf/1687-1499-2012-357.pdf.

[12] C. He and Z. J. Wang. Gains by a space-time-code based signaling scheme for multiple-antenna RFID tags. In *Proc. 23rd Canadian Conf. Electrical and Computer Engineering*, Calgary, Canada, 2–5 May 2010.

[13] B. M. Hochwald and T. L. Marzetta. Unitary space-time modulation for multiple-antenna communications in Rayleigh flat fading. *IEEE Trans. Inform. Theory*, 46:543–564, 2000.

[14] B. M. Hochwald and W. Sweldens. Differential unitary space-time modulation. *IEEE Trans. Commun.*, 48:2041–2052, 2000.

[15] R. A. Horn and C. R. Johnson. *Matrix Analysis*. Cambridge University Press, Cambridge, 1986.

[16] B. L. Hughes. Differential space-time modulation. *IEEE Trans. Inform. Theory*, 46:2567–2578, 2000.

[17] B. L. Hughes. Optimal space-time constellations from groups. *IEEE Trans. Inform. Theory*, 49:401–410, 2003.

[18] M. A. Ingram, M. F. Demirkol, and D. Kim. Transmit diversity and spatial multiplexing for RF links using modulated backscatter. In *Int. Symp. Signals, Systems, and Electronics*, Tokyo, Japan, 24–27 July 2001.

[19] T. Kaiser and F. Zheng. *Ultra Wideband Systems with MIMO*. John Wiley & Sons, Ltd., Chichester, 2010.

[20] J. Kaitovic, M. Simko, R. Langwieser, and M. Rupp. Channel estimation in tag collision scenarios. In *2012 IEEE Int. Conf. on RFID*, pages 74–80, Orlando, Florida, USA, 3–5 Apr. 2012.

[21] N. C. Karmakar (ed.). *Handbook of Smart Antennas for RFID Systems*. John Wiley & Sons, Inc., Hoboken, New Jersey, 2010.

[22] S. M. Kay. *Fundamentals of Statistical Signal Processing: Vol. 1: Estimation Theory*. Prentice-Hall, Upper Saddle River, New Jersey, 1993.

[23] D.-Y. Kim, H.-S. Jo, H. Yoon, C. Mun, B.-J. Jang, and J.-G. Yook. Reverse-link interrogation range of a UHF MIMO-RFID system in Nakagami-M fading channels. *IEEE Trans. Industrial Electronics*, 57:1468–1477, 2010.

[24] R. Langwieser, C. Angerer, and A. L. Scholtz. A UHF frontend for MIMO applications in RFID. In *Proc. IEEE 2010 Radio and Wireless Symp.*, pages 124–127, New Orleans, USA, 10–14 Jan. 2010.

[25] F. Lu, X. Chen, and T. T. Ye. Performance analysis of stacked RFID tags. In *2009 IEEE Int. Conf. on RFID*, pages 330–337, Orlando, Florida, USA, 27–28 Apr. 2009.

[26] K. Maichalernnukul, F. Zheng, and T. Kaiser. BER analysis of space-time coded RFID system in Nakagami-M fading channels. *Electronics Letters*, 50(5):405–407, 2014.

[27] K. Maichalernnukul, F. Zheng, and T. Kaiser. A differential space-time coded RFID system. In *2014 6th Int. Congress on Ultra Modern Telecommunications and Control Systems and Workshops*, pages 338–340, St. Petersburg, Russia, 6–8 Oct. 2014.

[28] M. Mi, M. H. Mickle, C. Capelli, and H. Switf. RF energy harvesting with multiple antennas in the same space. *IEEE Antennas Propagation Mag.*, 47(5):100–106, 2005.

[29] A. F. Mindikoglu and A.-J. van der Veen. Separation of overlapping RFID signals by antenna arrays. In *Proc. IEEE Int. Conf. Acoustics, Speech and Signal Processing 2008*, pages 2737–2740, Las Vegas, USA, 31 Mar. – 4 Apr. 2008.

[30] S. Sabesan, M. Crisp, R. V. Penty, and I. H. White. Demonstration of improved passive UHF RFID coverage using optically-fed distributed multi-antenna system. In *2010 IEEE Int. Conf. on RFID*, pages 102–109, Orlando, Florida, USA, 14–16 Apr. 2010.

[31] S. Sabesan, M. Crisp, R. V. Penty, and I. H. White. An error free passive UHF RFID system using a new form of wireless signal distribution. In *2012 IEEE Int. Conf. on RFID*, pages 58–65, Orlando, Florida, USA, 3–5 Apr. 2012.

[32] V. Tarokh and H. Jafarkhani. A differential detection scheme for transmit diversity. *IEEE J. Sel. Areas Commun.*, 18:1169–1174, 2000.

[33] V. Tarokh, H. Jafarkhani, and A. R. Calderbank. Space-time block codes from orthogonal designs. *IEEE Trans. Inform. Theory*, 45:1456–1467, 1999.

[34] M. S. Trotter, C. R. Valenta, G. A. Koo, B. R. Marshall, and G. D. Durgin. Multi-antenna techniques for enabling passive RFID tags and sensors at microwave frequencies. In *2012 IEEE Int. Conf. on RFID*, pages 1–7, Orlando, Florida, USA, 3–5 Apr. 2012.

[35] A. van den Bos. The multivariate complex normal distribution–a generalization. *IEEE Trans. Inf. Theory*, 41:537–539, 1995.

[36] F. Zheng and T. Kaiser. A space-time coding approach for RFID MIMO systems. *EURASIP J. Embedded Systems*, 2012. doi:10.1186/1687-3963-2012-9, Available: http://jes.eurasipjournals.com/content/pdf/1687-3963-2012-9.pdf.

6

Blind Signal Processing for RFID

6.1 Introduction

In RFID practice, we often meet the situation where several transponders are present in the reading zone of a single reader at the same time. For example, think of the case of a shopping trolley carrying several goods passing through the checkout in a supermarket. A great benefit in saving labor for the supermarket and in reducing the time for the customer would be gained if the reader can identify all the goods in the shopping trolley simultaneously. Therefore, it is important to study the techniques to identify the identities of multiple tags simultaneously. In principle, two approaches can be used to do this job. The first one is to use collision avoidance techniques such as tree-splitting or Aloha algorithm from a networking viewpoint. The second one is to use source separation techniques from a signal processing viewpoint. In this chapter, we will investigate the second approach.

Since few RFID tags are equipped with training symbols as the case of wireless communications, it is often difficult to estimate RFID channels. Therefore, blind source (or signal) separation (BSS) techniques are ideal tools to deal with multiple-tag ID identification problem. Report [10] first showed an approach on how to combine MIMO technique with BSS techniques to solve the problem of detecting multiple objects simultaneously.

A schematic diagram of a blind signal processing (BSP) problem is illustrated in Figure 6.1, where $\mathbf{s}$, $\mathbf{y}$ and $\mathbf{n}$ denote the source signal, measured or observed signal and measurement noise, respectively. $\mathbf{H}$ is the channel and k denotes discrete time. Note that $\mathbf{s}$, $\mathbf{y}$ and $\mathbf{n}$ can be either vectors or scalars and $\mathbf{H}$ can be either a matrix or a scalar. In the BSP problem, neither $\mathbf{s}$ nor $\mathbf{H}$ is known, and our goal is to extract or estimate either $\mathbf{s}$ or $\mathbf{H}$ from $\mathbf{y}$. The former problem (to estimate $\mathbf{s}$) is called blind source separation. In communication systems, it is also called blind equalization. The latter problem (to estimate $\mathbf{H}$) is called blind channel estimation or blind identification.

The input-output relationship for the system as illustrated in Figure 6.1 can be written as

$$\mathbf{y}(k) = \mathbf{H}(k) \otimes \mathbf{s}(k) + \mathbf{n}(k), \tag{6.1}$$

Digital Signal Processing for RFID, First Edition. Feng Zheng and Thomas Kaiser.
© 2016 John Wiley & Sons, Ltd. Published 2016 by John Wiley & Sons, Ltd.

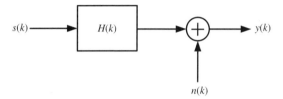

Figure 6.1   An illustration for the schema of blind signal processing problem.

where $\otimes$ stands for the convolution operation. We can have a brief look for the challenge of the BSP problem by considering the following two simple cases.

The first case is that both **s** and **y** are scalars, so that both **H** and **n** are also scalars. Further, assume that the channel is frequency-flat. Then equation (6.1) reduces to

$$y(k) = h(k)s(k) + n(k), \tag{6.2}$$

where y, s, h and n have the same meaning as that of **y**, **s**, **H** and **n**, respectively. From equation (6.2) we can see that, for any a claimed estimate of s, say $\hat{s}_0$, $\hat{s}_1 := c\hat{s}_0$ can be also taken as an estimate of s, if there is no any knowledge about channel h or signal s. Here, c is a constant.

The second case is that **s** is a vector, i.e., the source signals come from several sources. Also assume that the channel is frequency-flat. Let us write

$$\mathbf{s} = \begin{bmatrix} \mathbf{s}_1 \\ \mathbf{s}_2 \end{bmatrix}.$$

Correspondingly, the channel matrix is decomposed as

$$\mathbf{H} = [\mathbf{H}_1 \ \ \mathbf{H}_2].$$

Therefore, equation (6.1) reduces to

$$\mathbf{y}(k) = \mathbf{H}_1(k)\mathbf{s}_1(k) + \mathbf{H}_2(k)\mathbf{s}_2(k) + \mathbf{n}(k). \tag{6.3}$$

For any a claimed estimate of $\mathbf{s}_1$ and $\mathbf{s}_2$ (denoted by $\hat{\mathbf{s}}_1$ and $\hat{\mathbf{s}}_2$, respectively), define $\hat{\mathbf{s}}_1' := \hat{\mathbf{s}}_1 + \mathbf{c}_1$ and $\hat{\mathbf{s}}_2' := \hat{\mathbf{s}}_2 + \mathbf{c}_2$, where $\mathbf{c}_1$ and $\mathbf{c}_2$ are two constant vectors satisfying

$$\mathbf{H}_1\mathbf{c}_1 + \mathbf{H}_2\mathbf{c}_2 = 0 \quad \Leftrightarrow \quad \mathbf{H}\begin{bmatrix} \mathbf{c}_1 \\ \mathbf{c}_2 \end{bmatrix} = 0. \tag{6.4}$$

From equation (6.3) it is clear that $\hat{\mathbf{s}}_1'$ and $\hat{\mathbf{s}}_2'$ are also an estimate for $\mathbf{s}_1$ and $\mathbf{s}_2$, respectively. Note that equation (6.4) says that the vector $\mathbf{c} := [\mathbf{c}_1^T \ \ \mathbf{c}_2^T]^T$ lies in the null space of the channel matrix **H**. Model (6.3) characterizes a typical BSS problem.

From the above discussions, it can be seen that some additional conditions (or a prior knowledge) about the channel and/or source signals must be applied (or exploited) to make the BSP problem be properly defined (or have a proper solution). This kind of *a priori* knowledge can be broadly classified as statistical knowledge and deterministic knowledge, and hence the BSS approaches can be broadly categorized as a statistical method and deterministic method. General statistical *a priori* knowledge about source signals include the following: the source signals are independent of each other, or the mean value or even higher-order moments of the source signals are known. General deterministic *a priori* knowledge about source signals include the following: the source signals are often of finite alphabet, or the amplitudes of source signals are known. For the latter case, the widely used property is that the amplitude of source signals is a constant, i.e. a constant modulus (CM), such as in the case of PSK modulation.

General methods in developing blind estimations of overlapped source signals is first to select a proper penalty function as an optimization criterion by exploiting the aforementioned properties of the source signals and then to utilize a proper optimization tool to process the measured signals so that the penalty function is minimized, thus giving an estimate of the source signals.

The study of the BSS problem started from the pioneer work of Sato [12], Benveniste *et al.* [2] and Godard [9], where recursive algorithms were used to adaptively adjust the tap gains of the corresponding equalizers, so that the transmitted signals could be estimated or decoded. The term *blind equalization* was officially suggested in 1984 in [1]. The study of BSP was booming during 1980s and 1990s when various kinds of BSP algorithms were proposed and wide applications of these algorithms were found in communication systems, acoustic signal processing, oceanic engineering and so on. Among these results, two of them are especially remarkable. The first is about the convergence property of Godard's constant modulus algorithm (CMA), which was analytically established in 1985 by Foschini [8] and in 1991 by Ding *et al.* [6] for different cases. Before their works, the effectiveness of the corresponding blind algorithms was most demonstrated through simulations. The second is about the blind identifiability of single-input multiple-output (SIMO) linear channels by using only the second-order statistics, discovered by Tong, Xu and Kailath in [14]. Basically, the result in [14] says that using only the second-order statistics can identify linear discrete channels when the number of the output signals is greater than the number of the input signals.

The development of BSP studies was well documented in the survey papers [4, 13, 11] and monograph [7].

Blind signal separation based on statistical *a priori* knowledge of the source signals often needs a lot of measurements of the observed signals [4]. It does not suit the application of RFID well. On the other hand, BSP based on deterministic *a priori* knowledge of the source signals does not need to conduct comprehensive observations/measurements to the source signals. Therefore, this kind of BSS algorithms is preferred in RFID applications. Especially, an analytical constant modulus algorithm (ACMA) was proposed by van der Veen in [17] (see also [16]), which needs only moderate measurements for the signals to be estimated. Blind identification of tags' IDs provides a niche for the application of ACMA. In this chapter, we will discuss how to apply ACMA to RFID.

This chapter is organized as follows. A channel model for multiple-tag RFID-MIMO systems is presented in Section 6.2. The analytical constant modulus algorithm in [16] for blind source separation is reviewed in Section 6.3. The application of ACMA to multiple-tag RFID systems is investigated in Section 6.4. Section 6.5 concludes this chapter.

Notations: In this chapter, we use the following notational convention in addition to the general notations pointed out in Chapter 1. The notation rank stands for the rank of a matrix, row($\mathbf{X}$) the subspace spanned by the rows of matrix $\mathbf{X}$, ker the kernel space of a matrix, and dim the dimension of a matrix or subspace. For a vector or a matrix (say $\mathbf{A}$), we use $\mathbf{A}_{ij}$ or $\mathbf{A}_i$ to denote its relevant entry. When confusion might arise from the usage of this convention (e.g. when $\mathbf{A}_i$ and $\mathbf{A}_j$ themselves are different vectors or matrices), we use $[\mathbf{A}]_{ij}$ or $[\mathbf{A}]_i$ to denote its relevant entry.

6.2 Channel Model of Multiple-Tag RFID-MIMO Systems

6.2.1 Channel Model of Single-Tag RFID-MIMO Systems

Let us first consider the channel model of a single-tag RFID-MIMO system. The block diagram of a single-tag RFID-MIMO system is illustrated in Figure 5.1 in the previous chapter, where both the reader and tag are equipped with multiple antennas. Suppose that the reader and tag are equipped with N_{rd} and N_{tag} antennas, respectively, and the N_{rd} antennas at the reader are used for both reception and transmission. Let $\mathbf{x}$ (an $N_{rd} \times 1$ vector) be the transmitted signal at the reader, $\mathbf{y}$ (an $N_{rd} \times 1$ vector) the received signal at the reader, $\mathbf{n}$ the receiver noise, $\mathbf{H}^f$ (an $N_{tag} \times N_{rd}$ matrix) the channel matrix from the reader to the tag, $\mathbf{H}^b$ (an $N_{rd} \times N_{tag}$ matrix) the channel matrix from the tag to the reader, and $\mathbf{S}$ the backscattering matrix. Assume that the RF tag antennas modulate backscatter with different signals and no signals are transferred between the antennas. In this case $\mathbf{S}$ can be expressed as

$$\mathbf{S}(t) = \mathtt{diag}\ \{\Gamma_1(t), \Gamma_2(t)\ \cdots, \Gamma_{N_{tag}}(t)\},$$

where $\Gamma_i(t)$ is the backscattering coefficient of the ith antenna at the tag. Define

$$\boldsymbol{\gamma}(t) = \begin{bmatrix} \Gamma_1(t) \\ \Gamma_2(t) \\ \vdots \\ \Gamma_{N_{tag}}(t) \end{bmatrix}.$$

Based on these notations and in terms of the result of Section 5.3 of Chapter 5, the single-tag RFID-MIMO system can be modelled as follows:

$$\mathbf{y}(t) = \overset{\circ}{\mathbf{H}}(t)\boldsymbol{\gamma}(t) + \mathbf{n}(t), \tag{6.5}$$

where

$$\overset{\circ}{\mathbf{H}}(t) := \mathbf{H}^b \check{\mathbf{H}}(t),$$

$$\check{\mathbf{H}}(t) := \begin{bmatrix} [\mathbf{H}^f]_1 \mathbf{x}(t) & 0 & \cdots & 0 \\ 0 & [\mathbf{H}^f]_2 \mathbf{x}(t) & \cdots & 0 \\ \vdots & \vdots & \ddots & \vdots \\ 0 & 0 & \cdots & [\mathbf{H}^f]_{N_{tag}} \mathbf{x}(t) \end{bmatrix},$$

$$\mathbf{H}^f = \begin{bmatrix} [\mathbf{H}^f]_1 \\ [\mathbf{H}^f]_2 \\ \vdots \\ [\mathbf{H}^f]_{N_{tag}} \end{bmatrix}.$$

6.2.2 Channel Model of Multiple-Tag RFID-MIMO Systems

Now let us consider the channel model of a multiple-tag RFID-MIMO system. Suppose that N_T tags are placed in the reading zone of a reader. The reader is equipped with N_{rd} antennas and the ith tag is equipped with $N_{tag,i}$ antennas. The system is illustrated in Figure 6.2.

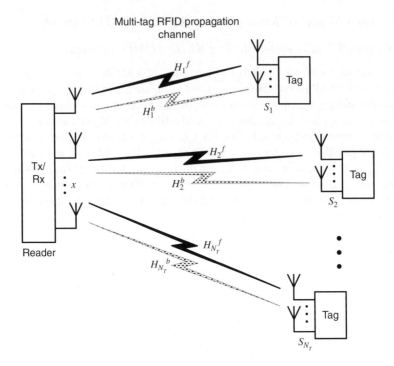

Figure 6.2   An illustration of multiple-tag RFID-MIMO systems.

From equation (6.5) we can see that the channel from the reader to the ith tag, and back to the reader again can be described by matrix $\overset{\circ}{\mathbf{H}}_i(t)$, where

$$\overset{\circ}{\mathbf{H}}_i(t) := \mathbf{H}_i^b \check{\mathbf{H}}_i(t),$$

$$\check{\mathbf{H}}_i(t) := \begin{bmatrix} [\mathbf{H}_i^f]_1 \mathbf{x}(t) & 0 & \cdots & 0 \\ 0 & [\mathbf{H}_i^f]_2 \mathbf{x}(t) & \cdots & 0 \\ \vdots & \vdots & \ddots & \vdots \\ 0 & 0 & \cdots & [\mathbf{H}_i^f]_{N_{\text{tag},i}} \mathbf{x}(t) \end{bmatrix},$$

$\mathbf{H}_i^b$ is the backward channel matrix from the ith tag to the reader, and $\mathbf{H}_i^f$ is the forward channel matrix from the reader to the ith tag. Based on equation (6.5), the received signal of the reader at time t_k can be written as

$$\mathbf{y}(t_k) = \sum_{i=1}^{N_T} \overset{\circ}{\mathbf{H}}_i(t_k)\boldsymbol{\gamma}_i(t_k) + \mathbf{n}(t_k) = \vec{\mathbf{H}}(t_k)\vec{\boldsymbol{\gamma}}(t_k) + \mathbf{n}(t_k),$$

where

$$\vec{\mathbf{H}}(t_k) := [\overset{\circ}{\mathbf{H}}_1(t_k) \ \overset{\circ}{\mathbf{H}}_2(t_k) \ \cdots \ \overset{\circ}{\mathbf{H}}_{N_T}(t_k)], \quad \vec{\boldsymbol{\gamma}}(t_k) := \begin{bmatrix} \gamma_1(t_k) \\ \gamma_2(t_k) \\ \vdots \\ \gamma_{N_T}(t_k) \end{bmatrix}.$$

Grouping all the received signals of the reader and the transmitted signals of the tags at different time instants $t_1, ..., t_K$ as

$$\mathbf{Y} = [\mathbf{y}(t_1)\ \mathbf{y}(t_2)\ \cdots\ \mathbf{y}(t_K)],\ \ \mathbf{S} = [\vec{\gamma}(t_1)\ \vec{\gamma}(t_2)\ \cdots\ \vec{\gamma}(t_K)],$$

and supposing that the channel does not change within the considered time frame, i.e.,

$$\vec{\mathbf{H}}(t_1) = \vec{\mathbf{H}}(t_2) = \cdots = \vec{\mathbf{H}}(t_K) := \vec{\mathbf{H}},$$

the input-output relationship from the tags to the reader can be re-expressed as

$$\mathbf{Y} = \mathbf{HS} + \vec{\mathbf{n}}, \qquad\qquad\qquad (6.6)$$

where

$$\vec{\mathbf{n}} = [\mathbf{n}(t_1)\ \mathbf{n}(t_2)\ \cdots\ \mathbf{n}(t_K)].$$

Equation (6.6) is the basis for applying ACMA to multiple-tag ID identification problem.

6.3   An Analytical Constant Modulus Algorithm[1]

Consider the system illustrated in Figure 6.3, where N_T sources, namely $s_1, s_2, ..., s_{N_T}$, transmit their signals to N_R receive antennas. The received signal at each antenna is the superposition of all the N_T source signals. The received signals from all the receive antennas are fed to the BSP block. The goal of the BSP block is to separate and then to recover the signals transmitted by the N_T sources. In BSP, it is generally assumed that the task of signal recovery is completed once the signals from different sources are separated.

Note that in Figure 6.3 neither the sources nor the receive antennas are assumed of any specific structures, which makes the approach described here applicable to a wide range of scenarios. Denote with $\mathbf{H}$ (an $N_R \times N_T$ matrix) the channel from the N_T sources to the N_R

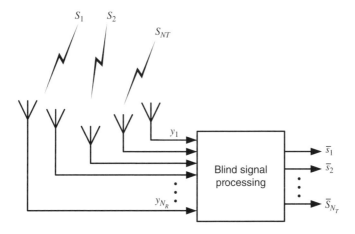

Figure 6.3   An illustration of blind signal processing technique.

---

[1] The main material in this section is adapted from reference [17].

receive antennas. Denote with $\mathbf{s}(t_k)$ (an N_T vector) and $\mathbf{y}(t_k)$ (an N_R vector) the vectors of transmit signals and receive signals, respectively, at time instants $t_k, k = 1, \ldots, K$. Assume that the channel $\mathbf{H}$ does not change during the considered measurements. The measurement noise is not considered temporarily for ease of exposition on the algorithm. Then the input-output relationship of the channel can be expressed by the following equation

$$\mathbf{y}(t_k) = \mathbf{H}\mathbf{s}(t_k), \quad k = 1, \ldots, K.$$

In BSP, it is assumed that the channel $\mathbf{H}$ is unknown to the receiver. The only knowledge is the received data and some properties of the transmitted symbols (such as constant modulus) and the channel (such as not changing for a period of time). Therefore, one generally needs to consider collectively the received signals for a period of time to estimate the hidden signal. Grouping the measurements and the transmitted signals at different time instants as

$$\mathbf{Y} = [\mathbf{y}(t_1) \ \mathbf{y}(t_2) \ \cdots \ \mathbf{y}(t_K)], \quad \mathbf{S} = [\mathbf{s}(t_1) \ \mathbf{s}(t_2) \ \cdots \ \mathbf{s}(t_K)],$$

the input-output relationship of the channel, or the data model, can be re-expressed as

$$\mathbf{Y} = \mathbf{H}\mathbf{S}. \tag{6.7}$$

For convenience, we call the measurements at time instants $t_1, t_2, \ldots, t_K$ the measurements of one snapshot.

In the data model, it is assumed that the transmitted signals are of constant modulus, that is,

$$|\mathbf{S}_{ij}| = 1, \quad i = 1, \ldots, N_T, \quad j = 1, \ldots, K.$$

Besides, it is further assumed that $\mathbf{H}$ and $\mathbf{S}$ are of full rank.

The CM blind source separation problem is stated as follows.

Problem 6.1 For a given data matrix $\mathbf{Y}$, find an integer δ and a $\delta \times N_R$ matrix $\mathbf{W}$ such that

$$\mathbf{W}\mathbf{Y} = \overline{\mathbf{S}} \quad \text{with } |\overline{\mathbf{S}}_{ij}| = 1, \quad i = 1, \ldots, \delta, \quad j = 1, \ldots, K, \tag{6.8}$$

where $\overline{\mathbf{S}}$ is of full rank and δ is as large as possible under the constraint $\delta \leq N_T$.

In Problem 6.1, if $\delta = N_T$, then all the signals from N_T sources are separated. On the other hand, if $\delta < N_T$, then only a part of the source signals are separated.

Clearly, if $\mathbf{W}$ is a solution to Problem 6.1, any reordering and any phase rotation of the row vectors of $\mathbf{W}$ is also a solution to Problem 6.1. This kind of possible ambiguity in the solutions cannot be solved by using only the CM property of the transmitted signals. In the sequel, we consider the solution to Problem 6.1 as unique in the sense that the aforementioned ambiguity is neglected.

Note that the goal of the source separation problem is perfectly achieved if

$$\overline{\mathbf{S}} = \mathbf{S}. \tag{6.9}$$

When equation (6.9) is satisfied, the transmitted signals are fully reconstructed. However, the restriction implied by equation (6.9) is too strong since it even requires that the reconstructed signals are in the same ordering as that of the transmitted signals, which is clearly unnecessary and often difficult to realize. The requirement imposed by equation (6.9) makes Problem 6.1 unnecessarily complicated.

Condition (6.8) says that the (a part of, i.e. δ) source signals are considered to be separated and reconstructed up to a reordering and/or phase rotation. From the simulations in Section 6.4, we will see that phase rotations in the reconstructed signals often happen.

The claim that Problem 6.1 admits a unique solution is equivalent to the claim that the data matrix $\mathbf{Y}$ admits a unique factorization $\{\mathbf{Y} = \mathbf{H}\overline{\mathbf{S}}, |\overline{\mathbf{S}}_{ij}| = 1\}$.

It is known (see e.g. [15, 18]) that, for the case of $\delta = N_T$, when $\mathbf{H}$ and $\mathbf{S}$ are of full rank and the sources generating matrix $\mathbf{S}$ are 'sufficiently independent' and have 'sufficient phase richness', the uniqueness of the solution to Problem 6.1 is guaranteed with a probability of 1 if the number of measurements K is 'sufficiently large'. It is proved in [17] that the uniqueness of the solution is guaranteed when $K \geq 2N_T$.

To ease exposition, let us introduce some notations. For two matrices $\mathbf{A}$ and $\mathbf{B}$, define their Kronecker product as

$$\mathbf{A} \otimes \mathbf{B} := [\mathbf{A}_{ij}\mathbf{B}].$$

For a matrix $\mathbf{A} = [\mathbf{A}_{ij}] \in \mathbb{C}^{m_1 \times m_2}$, define the vec operation as

$$\mathrm{vec}(\mathbf{A}) = \begin{bmatrix} \mathbf{A}_{11} \\ \mathbf{A}_{12} \\ \vdots \\ \mathbf{A}_{1m_2} \\ \mathbf{A}_{21} \\ \vdots \\ \mathbf{A}_{2m_2} \\ \vdots \\ \mathbf{A}_{m_1 1} \\ \vdots \\ \mathbf{A}_{m_1 m_2} \end{bmatrix}.$$

For a vector $\mathbf{v} = [\mathbf{v}_i] \in \mathbb{C}^{m^2}$, define the mat operation as

$$\mathrm{mat}(\mathbf{v}) = \begin{bmatrix} \mathbf{v}_1 & \mathbf{v}_2 & \cdots & \mathbf{v}_m \\ \mathbf{v}_{m+1} & \mathbf{v}_{m+2} & \cdots & \mathbf{v}_{2m} \\ \vdots & \vdots & \ddots & \vdots \\ \mathbf{v}_{m^2-m+1} & \mathbf{v}_{m^2-m+2} & \cdots & \mathbf{v}_{m^2} \end{bmatrix}.$$

Based on the above definitions, it can be proven that an equation in vector form

$$\eta = \psi \otimes \phi = \mathrm{vec}(\psi\phi^T) \tag{6.10}$$

is equivalent to the equation in matrix form

$$\mathrm{mat}(\eta) = \psi\phi^T \tag{6.11}$$

for any three vectors $\eta \in \mathbb{C}^{m^2 \times 1}$, $\psi \in \mathbb{C}^{m \times 1}$, and $\phi \in \mathbb{C}^{m \times 1}$.

For two given positive integers m and n, define the set of CM signals as

$$S_{\mathrm{cm}} = \{s \in \mathbb{C}^{m \times n} : |s_{ij}| = 1, \ \forall i \in 1, \ldots, m, \ j \in 1, \ldots, n\}.$$

Note that in the definition of S_{cm}, the dimension of an element of S_{cm}, implied by the problem under discussion, is not strengthened. The key point is that the amplitude of each entry of an element of S_{cm} is equal to the unity.

It is unwise to solve Problem 6.1 using brute force if all the N_T source signals are to be obtained since $\mathbf{W}$ contains $N_T \times N_R$ unknowns. However, noticing the fact that the data matrix $\mathbf{Y}$ is well structured, we can parameterize $\mathbf{W}$ to an $N_T \times N_T$-unknown problem, or essentially an N_T-unknown problem (see equation (6.16)). To achieve this goal, we need to analyze the geometric structure of data matrix $\mathbf{Y}$.

First note that Problem 6.1 is equivalent to the problem that find all linearly independent signals $\bar{\mathbf{s}}$ such that

$$\bar{\mathbf{s}} \in \text{row}(\mathbf{Y}), \tag{6.12}$$

$$\bar{\mathbf{s}} \in S_{cm}. \tag{6.13}$$

From condition (6.12) one can see that an immediate procedure for simplifying Problem 6.1 is to use a singular value decomposition (SVD) of $\mathbf{Y}$. Let $\mathbf{Y} = \mathbf{U}\mathbf{\Sigma}\mathbf{V}^\dagger$ be an SVD of $\mathbf{Y}$, where $\mathbf{U} \in \mathbb{C}^{N_R \times N_R}$, $\mathbf{\Sigma} \in \mathbb{R}^{N_R \times K}$, $\mathbf{V} \in \mathbb{C}^{K \times K}$, $\mathbf{U}$ and $\mathbf{V}$ are unitary matrices, and $\mathbf{\Sigma} = [\sigma_{ij}]$ is a diagonal matrix in the sense of $\sigma_{ij} = 0$ for all $i \neq j$. The diagonal entries σ_{ii} are of the property $\sigma_{11} \geq \sigma_{22} \geq \cdots \geq \sigma_{N_{TR}N_{TR}} \geq 0$, where $N_{TR} := \min\{N_T, N_R\}$. The values of σ_{ii}, $i = 1, \ldots, N_{TR}$, are proportional to the powers of the source signals. Suppose that m_0 be the integer such that $\sigma_{m_0 m_0} > 0$ and $\sigma_{m_0+1, m_0+1} = 0$ [2] and partition $\mathbf{U}$ and $\mathbf{V}$ as

$$\mathbf{U} = [\mathbf{u}_1 \quad \cdots \quad \mathbf{u}_{m_0} \quad \mathbf{u}_{m_0+1} \quad \cdots \quad \mathbf{u}_{N_R}],$$

$$\mathbf{V} = [\mathbf{v}_1 \quad \cdots \quad \mathbf{v}_{m_0} \quad \mathbf{v}_{m_0+1} \quad \cdots \quad \mathbf{v}_K].$$

A well-known property of the SVD of a matrix, say $\mathbf{Y}$, is that the column space of $\mathbf{Y}$ is spanned by column vectors $\{\mathbf{u}_1, \mathbf{u}_2, \ldots, \mathbf{u}_{m_0}\}$ and the row space of $\mathbf{Y}$ is spanned by row vectors $\{\mathbf{v}_1^\dagger, \mathbf{v}_2^\dagger, \ldots, \mathbf{v}_{m_0}^\dagger\}$, i.e.,

$$\text{row}(\mathbf{Y}) = \text{span}(\mathbf{v}_1^\dagger, \mathbf{v}_2^\dagger, \ldots, \mathbf{v}_{m_0}^\dagger).$$

Therefore, condition (6.12) is equivalent to

$$\bar{\mathbf{s}} \in \text{span}(\mathbf{v}_1^\dagger, \mathbf{v}_2^\dagger, \ldots, \mathbf{v}_{m_0}^\dagger) \iff \exists \mathbf{w} \in \mathbb{C}^{1 \times m_0} \text{ such that } \bar{\mathbf{s}} = \mathbf{w} \begin{bmatrix} \mathbf{v}_1^\dagger \\ \mathbf{v}_2^\dagger \\ \vdots \\ \mathbf{v}_{m_0}^\dagger \end{bmatrix}.$$

Condition (6.13) is essentially a quadratic constraint problem. In [17], this problem is converted to a generalized matrix pencil problem. To describe this problem conversion, we define

$$\mathbf{V}_1 := \begin{bmatrix} \mathbf{v}_1^\dagger \\ \mathbf{v}_2^\dagger \\ \vdots \\ \mathbf{v}_{m_0}^\dagger \end{bmatrix} := [\bar{\mathbf{v}}_1 \quad \bar{\mathbf{v}}_2 \quad \cdots \quad \bar{\mathbf{v}}_K],$$

---

[2] In general $m_0 = N_T$ if $N_R \geq N_T$.

where $\mathbf{V}_1$ is the matrix of the first m_0 rows of $\mathbf{V}^\dagger$, and $\overline{\mathbf{v}}_k$ is the kth column of matrix $\mathbf{V}_1$. Define matrix $\mathbf{P}$ by

$$\mathbf{P} := \begin{bmatrix} (\overline{\mathbf{v}}_1 \otimes \overline{\mathbf{v}}_1^*)^T \\ (\mathbf{v}_2 \otimes \mathbf{v}_2^*)^T \\ \vdots \\ (\overline{\mathbf{v}}_K \otimes \overline{\mathbf{v}}_K^*)^T \end{bmatrix} = \begin{bmatrix} (\mathrm{vec}(\overline{\mathbf{v}}_1 \overline{\mathbf{v}}_1^\dagger))^T \\ (\mathrm{vec}(\overline{\mathbf{v}}_2 \overline{\mathbf{v}}_2^\dagger))^T \\ \vdots \\ (\mathrm{vec}(\overline{\mathbf{v}}_K \overline{\mathbf{v}}_K^\dagger))^T \end{bmatrix}$$

and $K \times K$ unitary matrix $\mathbf{Q}$ by

$$\mathbf{Q} = \mathbf{I} - 2\frac{\mathbf{q}\mathbf{q}^T}{\mathbf{q}^T\mathbf{q}}, \quad \mathbf{q} = \begin{bmatrix} 1 \\ 1 \\ \vdots \\ 1 \end{bmatrix} - \begin{bmatrix} \sqrt{K} \\ 0 \\ \vdots \\ 0 \end{bmatrix}.$$

Partition the product $\mathbf{QP}$ as

$$\mathbf{QP} =: \begin{bmatrix} \overline{\mathbf{p}} \\ \overline{\mathbf{P}} \end{bmatrix},$$

$$\overline{\mathbf{p}} \in \mathbb{C}^{1 \times m_0^2}, \quad \overline{\mathbf{P}} \in \mathbb{C}^{(K-1) \times m_0^2},$$

i.e., $\overline{\mathbf{P}}$ is simply the remaining rows by removing the first row of the matrix $\mathbf{QP}$.

Using the above definitions, it is shown in [17] that Problem 6.1 is equivalent to finding all linearly independent nonzero vectors $\alpha \in \mathbb{C}^{m_0^2}$ satisfying

$$\overline{\mathbf{P}}\alpha = \mathbf{0}, \tag{6.14}$$

$$\alpha = \mathbf{w}^T \otimes \mathbf{w}^\dagger. \tag{6.15}$$

For each solution $\mathbf{w}$, scaled such that $\|\mathbf{w}\| = \sqrt{K}$, the corresponding CM signal is given by $\overline{\mathbf{s}} = \mathbf{w}\mathbf{V}_1$.

Let $\{\mathbf{z}_1, \mathbf{z}_2, \ldots, \mathbf{z}_\delta\}$ be a basis of $\ker(\overline{\mathbf{P}})$, which can be obtained from an SVD of $\overline{\mathbf{P}}$. From equation (6.14) we can parameterize the solution α as a linear combination of the basis of the kernel of $\overline{\mathbf{P}}$, i.e.,

$$\alpha = \alpha_1 \mathbf{z}_1 + \alpha_2 \mathbf{z}_2 + \cdots + \alpha_\delta \mathbf{z}_\delta.$$

Based on the equivalence property between (6.10) and (6.11), we see that equations (6.14)–(6.15) are equivalent to

$$\alpha_1 \mathbf{Z}_1 + \alpha_2 \mathbf{Z}_2 + \cdots + \alpha_\delta \mathbf{Z}_\delta = \mathbf{w}^T \mathbf{w}^* := \overline{\mathbf{w}}^\dagger \overline{\mathbf{w}}, \tag{6.16}$$

where $\mathbf{Z}_i := \mathrm{mat}(\mathbf{z}_i)$, $i = 1, \ldots, \delta$, and $\overline{\mathbf{w}} := \mathbf{w}^*$.

This procedure is formally stated as Problem 6.2 [17]:

Problem 6.2 Let $\mathbf{Y}$ be the given data matrix, from which the matrices $\mathbf{Z}_i$, $i = 1, \ldots, \delta$, are obtained as in the aforementioned procedure. Then Problem 6.1 is equivalent to the following problem: find all linearly independent nonzero vectors $[\alpha_1 \ \alpha_2 \ \cdots \ \alpha_\delta]$ such that equation (6.16) is satisfied. For each solution $\overline{\mathbf{w}}$, scaled such that $\|\overline{\mathbf{w}}\| = \sqrt{K}$, the corresponding CM signal is given by $\mathbf{s} = \overline{\mathbf{w}}^* \mathbf{V}_1$.

Suppose that m_0 CM signals are solved from the data matrices $\{\mathbf{Z}_1, \ldots, \mathbf{Z}_\delta\}$ and the corresponding solutions are denoted by the vectors $\boldsymbol{\alpha}_i := [\alpha_{i1}\ \alpha_{i2}\ \cdots\ \alpha_{i\delta}]$ and $\overline{\mathbf{w}}_i, i = 1, \ldots, m_0$, respectively. These solutions are characterized by the following equations

$$
\begin{cases}
\alpha_{11}\mathbf{Z}_1 + \alpha_{12}\mathbf{Z}_2 + \cdots + \alpha_{1\delta}\mathbf{Z}_\delta = \overline{\mathbf{w}}_1^\dagger \overline{\mathbf{w}}_1, \\
\alpha_{21}\mathbf{Z}_1 + \alpha_{22}\mathbf{Z}_2 + \cdots + \alpha_{2\delta}\mathbf{Z}_\delta = \overline{\mathbf{w}}_2^\dagger \overline{\mathbf{w}}_2, \\
\qquad\qquad\qquad \vdots \\
\alpha_{m_01}\mathbf{Z}_1 + \alpha_{m_02}\mathbf{Z}_2 + \cdots + \alpha_{m_0\delta}\mathbf{Z}_\delta = \overline{\mathbf{w}}_{m_0}^\dagger \overline{\mathbf{w}}_{m_0}.
\end{cases}
\tag{6.17}
$$

Since vectors $\boldsymbol{\alpha}_i, i = 1, \ldots, m_0$, are independent of each other, one can solve $\mathbf{Z}_1, \ldots, \mathbf{Z}_\delta$ from equation (6.17), which can be expressed as the following

$$
\begin{cases}
\mathbf{Z}_1 = \lambda_{11}\overline{\mathbf{w}}_1^\dagger \overline{\mathbf{w}}_1 + \lambda_{12}\overline{\mathbf{w}}_2^\dagger \overline{\mathbf{w}}_2 + \cdots + \lambda_{1m_0}\overline{\mathbf{w}}_{m_0}^\dagger \overline{\mathbf{w}}_{m_0} = \overline{\mathbf{W}}^\dagger \boldsymbol{\Lambda}_1 \overline{\mathbf{W}}, \\
\mathbf{Z}_2 = \lambda_{21}\overline{\mathbf{w}}_1^\dagger \overline{\mathbf{w}}_1 + \lambda_{22}\overline{\mathbf{w}}_2^\dagger \overline{\mathbf{w}}_2 + \cdots + \lambda_{2m_0}\overline{\mathbf{w}}_{m_0}^\dagger \overline{\mathbf{w}}_{m_0} = \overline{\mathbf{W}}^\dagger \boldsymbol{\Lambda}_2 \overline{\mathbf{W}}, \\
\qquad\qquad\qquad \vdots \\
\mathbf{Z}_\delta = \lambda_{\delta 1}\overline{\mathbf{w}}_1^\dagger \overline{\mathbf{w}}_1 + \lambda_{\delta 2}\overline{\mathbf{w}}_2^\dagger \overline{\mathbf{w}}_2 + \cdots + \lambda_{\delta m_0}\overline{\mathbf{w}}_{m_0}^\dagger \overline{\mathbf{w}}_{m_0} = \overline{\mathbf{W}}^\dagger \boldsymbol{\Lambda}_\delta \overline{\mathbf{W}},
\end{cases}
$$

where

$$
\overline{\mathbf{W}} = \begin{bmatrix} \overline{\mathbf{w}}_1 \\ \overline{\mathbf{w}}_2 \\ \vdots \\ \overline{\mathbf{w}}_{m_0} \end{bmatrix},
$$

$$
\boldsymbol{\Lambda}_i = \texttt{diag}\ \{\lambda_{i1}, \lambda_{i2}, \ldots, \lambda_{im_0}\}, i = 1, 2, \ldots, \delta.
$$

The above treatment is formally stated as Problem 6.3 [17]:

Problem 6.3 Suppose $K > N_T^2$ and dim ker $(\overline{\mathbf{P}}) = \delta$. Then the CM blind source separation in Problem 6.1 or Problem 6.2 is equivalent to the following simultaneous diagonalization problem: find a full rank matrix $\overline{\mathbf{W}} \in \mathbb{C}^{m_0 \times N_T}$ such that

$$
\begin{cases}
\mathbf{Z}_1 = \overline{\mathbf{W}}^\dagger \boldsymbol{\Lambda}_1 \overline{\mathbf{W}}, \\
\mathbf{Z}_2 = \overline{\mathbf{W}}^\dagger \boldsymbol{\Lambda}_2 \overline{\mathbf{W}}, \\
\qquad \vdots \\
\mathbf{Z}_\delta = \overline{\mathbf{W}}^\dagger \boldsymbol{\Lambda}_\delta \overline{\mathbf{W}},
\end{cases}
\tag{6.18}
$$

where $\boldsymbol{\Lambda}_1, \ldots, \boldsymbol{\Lambda}_\delta \in \mathbb{C}^{m_0 \times m_0}$ are diagonal matrices.

If matrices $\mathbf{Z}_i, i = 1, \ldots, \delta$, are Hermitian, one can use the approach developed in [3, 5] to find the solution of matrix $\overline{\mathbf{W}}$. In general, one can use the approach developed in [17] to find the solution of matrix $\overline{\mathbf{W}}$ or vector $\boldsymbol{\alpha}$. In the following, the approach in [17] is outlined.

In the approach of [17], there is no need to strictly diagonalize the matrices $\mathbf{Z}_i$, $i = 1, \ldots, \delta$. Instead, what we need to do is to simultaneously transform them into upper triangular matrices. Suppose that there are unitary matrices $\mathbf{U}_L$ and $\mathbf{U}_R$ such that

$$
\begin{cases}
\mathbf{U}_L \mathbf{Z}_1 \mathbf{U}_R = \mathbf{R}_1, \\
\mathbf{U}_L \mathbf{Z}_2 \mathbf{U}_R = \mathbf{R}_2, \\
\quad \vdots \\
\mathbf{U}_L \mathbf{Z}_\delta \mathbf{U}_R = \mathbf{R}_\delta,
\end{cases}
\tag{6.19}
$$

where $\mathbf{R}_i$, $i = 1, \ldots, \delta$, are upper triangular. Then it can be proven that the vector $[\alpha_1\ \alpha_2\ \cdots\ \alpha_\delta]$ satisfies equation (6.16) if and only if

$$
\mathtt{rank}(\alpha_1 \mathbf{R}_1 + \alpha_2 \mathbf{R}_2 + \cdots + \alpha_\delta \mathbf{R}_\delta) = 1.
\tag{6.20}
$$

From equation (6.20) it can be shown that all the independent vectors $[\alpha_1\ \alpha_2\ \cdots\ \alpha_\delta]$ are given by the rows of matrix $\boldsymbol{\Psi}$ defined by

$$
\boldsymbol{\Psi} =
\begin{bmatrix}
[\mathbf{R}_1]_{11} & [\mathbf{R}_1]_{22} & \cdots & [\mathbf{R}_1]_{\delta\delta} \\
[\mathbf{R}_2]_{11} & [\mathbf{R}_2]_{22} & \cdots & [\mathbf{R}_2]_{\delta\delta} \\
\vdots & \vdots & \ddots & \vdots \\
[\mathbf{R}_\delta]_{11} & [\mathbf{R}_\delta]_{22} & \cdots & [\mathbf{R}_\delta]_{\delta\delta}
\end{bmatrix}^{-1}.
$$

In summary, we see that the original CM blind source separation Problem 6.1 for data model (6.7) is solved if equation (6.19) has a solution. In this case, the independent vectors α's are given by the rows of matrix $\boldsymbol{\Psi}$ and hence the corresponding vectors $\mathbf{w}$'s can be solved based on the SVD of matrix $\alpha_1 \mathbf{Z}_1 + \alpha_2 \mathbf{Z}_2 + \cdots + \alpha_\delta \mathbf{Z}_\delta$ in equation (6.16).

Taking measurement noise into account, the input-output relationship of the channel, or the data model, can be re-expressed as

$$
\mathbf{Y} = \mathbf{H}\mathbf{S} + \vec{\mathbf{n}},
\tag{6.21}
$$

where $\vec{\mathbf{n}}$ is a matrix, whose kth column is the measurement noise vector at time t_k. The approach described above can be directly applied to data model (6.21).

The above approach to BSS problem is summarized in the following algorithm. To easy the exposition in the sequel, we call it ACMA.

Algorithm 6.1 ACMA [17]. Initial data: given data matrix $\mathbf{Y}$, as described by (6.21).

Step 1 Calculate $\mathtt{row}(\mathbf{Y})$ by using SVD of $\mathbf{Y}$, forming matrix $\mathbf{V}_1$.
Step 2 Construct matrix $\mathbf{P}$ from $\mathbf{V}_1$ and matrix $\overline{\mathbf{P}}$ from $\mathbf{Q}$ and $\mathbf{P}$.
Step 3 Calculate $\{\mathbf{z}_1, \ldots, \mathbf{z}_\delta\}$ by using SVD of $\overline{\mathbf{P}}$, forming matrices $\{\mathbf{Z}_1, \ldots, \mathbf{Z}_\delta\}$.
Step 4 Solve the simultaneous diagonalization problem (6.18) or simultaneous triangularization problem (6.19), forming matrices $\{\mathbf{R}_1, \ldots, \mathbf{R}_\delta\}$.
Step 5 Calculate matrix $\boldsymbol{\Psi}$ from $\{\mathbf{R}_1, \ldots, \mathbf{R}_\delta\}$.
Step 6 Construct vectors α's from matrix $\boldsymbol{\Psi}$.
Step 7 Solve vectors $\mathbf{w}$'s from equation (6.16) based on the SVD of matrix $\alpha_1 \mathbf{Z}_1 + \alpha_2 \mathbf{Z}_2 + \cdots + \alpha_\delta \mathbf{Z}_\delta$.
Step 8 Recover the signals from $\mathbf{w}$'s and $\mathbf{V}_1$.

6.4   Application of ACMA to Multiple-Tag RFID Systems

In this section, we apply ACMA to multiple-tag RFID systems. In the simulation studies, the 8-ary phase-shift keying (8-PSK) modulation is used at the transmitter of each tag. The constellation of transmitted symbols is shown in Figure 6.4. The multiple-tag RFID system model is characterized by equation (6.6).

At the transmitter of the reader, the signal $\mathbf{x}$ takes the form of a random vector whose entry is uniformly distributed among $\pm \frac{1}{\sqrt{N_{rd}}}$. It is seen that $\mathbf{x}$ is of unity power. Each entry of the channels $\mathbf{H}^f$ and $\mathbf{H}^b$ is of mean zero and variance unity.

In the figures to be shown, the SNR is defined as $\frac{E_0}{\sigma_{\mathbf{n}}^2}$, where $\sigma_{\mathbf{n}}^2$ is the variance of each entry of noise vector $\mathbf{n}$ and E_0 is the power of each symbol transmitted at the tag.

The simulations are conducted in the following way. First, K measurement samples in a snapshot are generated based on multi-tag RFID channel model (6.6). Then these K measurement samples, as an ensemble, are fed to ACMA, producing an output matrix $\mathbf{WV}_1 \in \mathbb{C}^{m_0 \times K}$. Each column of matrix $\mathbf{WV}_1$ corresponds to the separated source signals from different tags at one measured instant. This process is repeated M times, i.e. total M runs of ACMA, to produce statistically reliable results.

The system performance can be observed qualitatively by the scatter plot of the processed signal. Quantitatively, it can be evaluated by the average modulus error (AME), which is

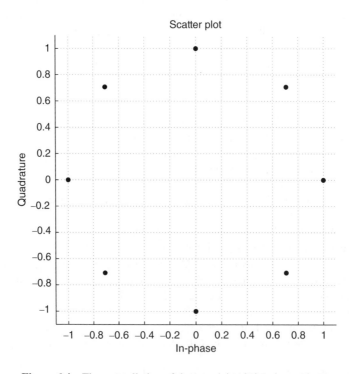

Figure 6.4   The constellation of the transmitted signal at each tag.

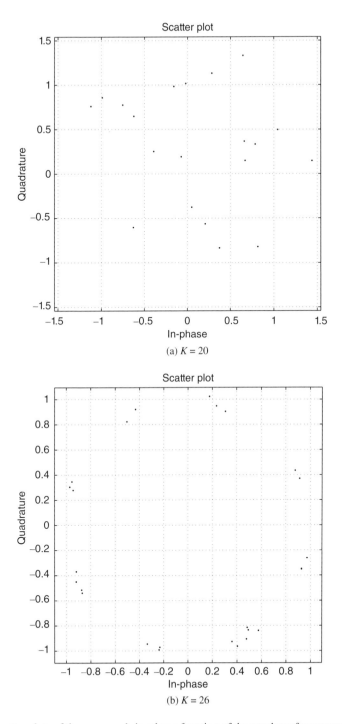

Figure 6.5 Scatter plots of the processed signal as a function of the number of measurements K, where $N_{rd} = 8$, $N_{tag} = 1$, $m_0 = 5$, SNR = 20 dB and $K = 20$ and 26, respectively.

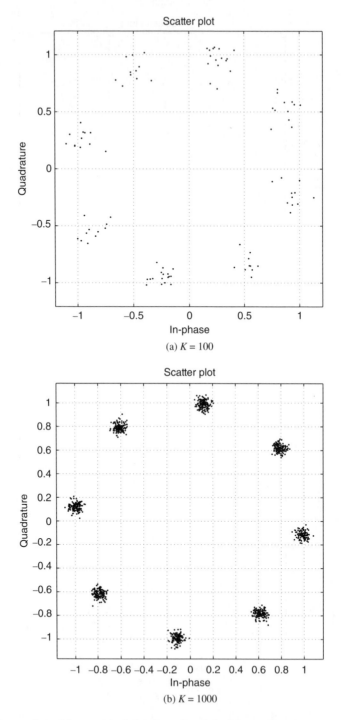

Figure 6.6 Scatter plots of the processed signal as a function of the number of measurements K, where $N_{rd} = 8$, $N_{tag} = 1$, $m_0 = 5$, SNR = 20 dB and $K = 100$ and 1000, respectively.

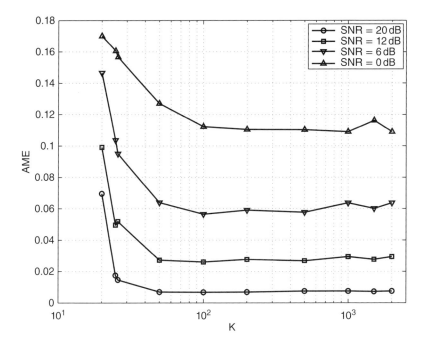

Figure 6.7 The average modulus error of the processed signal vs. K (the number of measurements) for different SNR, where $N_{rd} = 8$, $N_{tag} = 1$ and $m_0 = 5$.

defined as

$$\text{AME} = \sqrt{\frac{1}{Mm_0K} \sum_{m=1}^{M} \sum_{i=1}^{m_0} \sum_{j=1}^{K} (|[\mathbf{WV}_1]_{ij}| - 1)^2}. \tag{6.22}$$

Note that M is different from K, since all the K sampled data, as an ensemble, are processed jointly; while these M batched samples are processed independently using the same algorithm. Hence the signals $\mathbf{WV}_1$ change within different run indices. From the simulation results it is found that when K is small, for example, when $K = 20$, it is sufficient to choose $M = 100$ to get smooth and reliable results. When K is large, for example, when $K = 200$, a smaller M is also sufficient to give smooth and reliable results.

When printing the scatter plot, we should only print the processed signal for one ensemble, that is, K samples of measurements. If all the M runs of the processed signals are printed in the same scatter plot, then it might be difficult to observe the source separation because of phase rotation caused by ACMA. On the other hand, when plotting the AME curve, all the processed signals for the M runs should be taken into account, as shown by equation (6.22).

Figures 6.5–6.7 show how the performance of the system depends on K, the number of measurements in one snapshot. Figures 6.5 and 6.6 illustrate the scatter plots of the processed signal, and Figure 6.7 depicts the AME. It can be seen that when K is larger than 25 (the square of the number of the tags), the signals coming from different tags can be successfully separated, while when K is smaller than 25 (say 20), the received signals cannot be separated

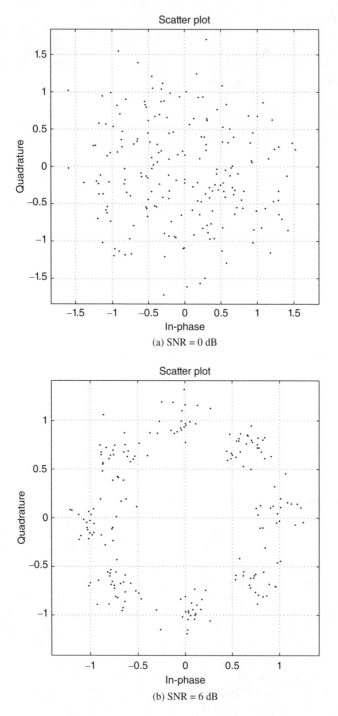

Figure 6.8 Scatter plots of the processed signal for different SNR, where $N_{rd} = 8$, $N_{tag} = 1$, $m_0 = 5$, $K = 200$ and SNR = 0 dB and 6 dB, respectively.

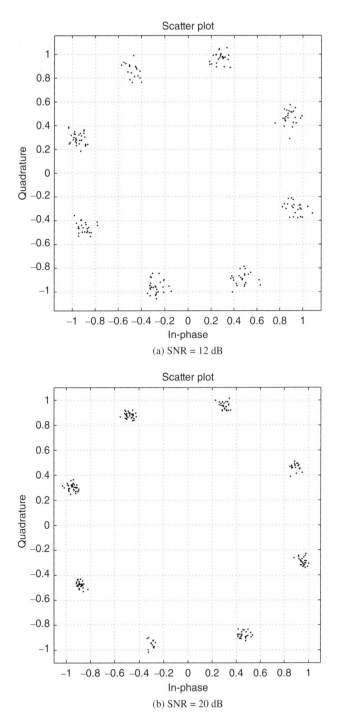

Figure 6.9 Scatter plots of the processed signal for different SNR, where $N_{rd} = 8$, $N_{tag} = 1$, $m_0 = 5$, $K = 200$ and SNR = 12 dB and 20 dB, respectively.

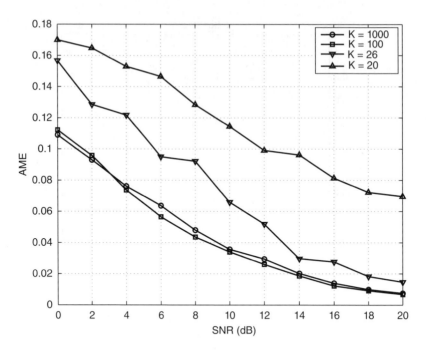

Figure 6.10 The average modulus error of the processed signal versus SNR for different K, where $N_{rd} = 8, N_{tag} = 1$ and $m_0 = 5$.

using ACMA (see Figure 6.5). The system performance, characterized by AME, is significantly improved by increasing K when K is smaller than 50. With K further increasing beyond 100, the system performance is saturated (see Figure 6.7). Therefore, in the sequent simulations, we fix K to be 200 to get reliable results while keeping the computational burden reasonably low.

Figures 6.8–6.10 show how the performance of the system depends on SNR. Figures 6.8 and 6.9 illustrate the scatter plots of the processed signal and Figure 6.10 depicts the AME. From Figures 6.8 and 6.9 it is seen that the overlapped signals can be successfully separated when SNR $\geq$ 6 dB for $K = 200$. Figure 6.10 shows that in this region the AME satisfies AME ≤ 0.06. Thus based on Figure 6.10 it can be expected that for the case of $K = 26$, ACMA can separate the overlapped signals when SNR > 12 dB and it fails to do so when SNR ≤ 8 dB. In the case of $K = 20$, ACMA cannot separate the overlapped signals even when SNR is as high as 20 dB. Figure 6.11 shows the scatter plots for some relevant cases, which, together with Figure 6.5(a), confirm the aforementioned claim.

Figures 6.12–6.14, illustrating the scatter plots and AME curves, respectively, of the processed signal, show how the performance of the system depends on N_{rd}, the number of antennas at the reader. Figures 6.12 and 6.13 show that, when $N_{rd} \geq 5 = m_0$ for the case of high SNR (20 dB), ACMA can separate the overlapped signals. Increasing N_{rd} can only moderately improve the AME performance. Based on Figure 6.14 it is expected that, for the case of low SNR (say 6 dB), the parameter N_{rd} plays a critical role for the separability of the received signals. For example, when SNR $= 6$ dB, ACMA with $N_{rd} = 10$ can separate the signals, while

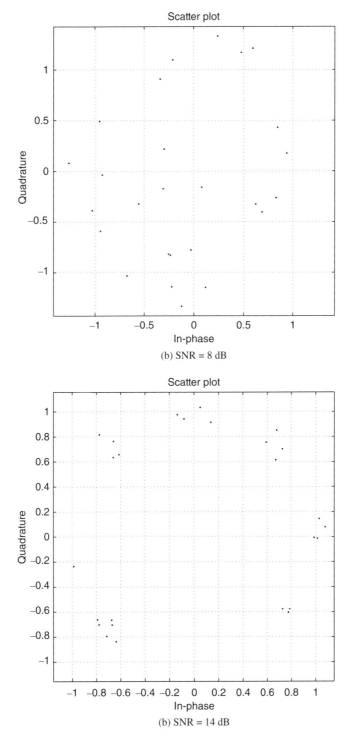

Figure 6.11 Scatter plots of the processed signal for different SNR, where $N_{rd} = 8$, $N_{tag} = 1$, $m_0 = 5$ and $K = 26$.

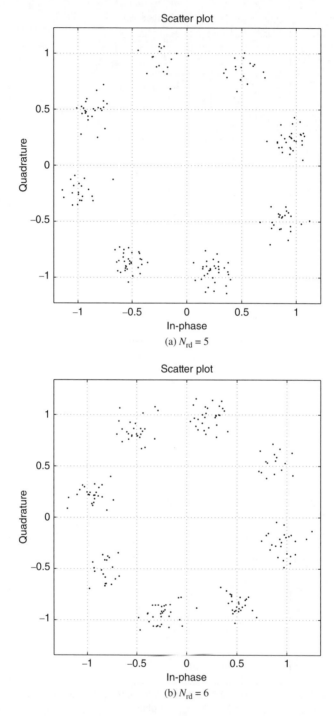

Figure 6.12 Scatter plots of the processed signal for different N_{rd}, where $N_{tag} = 1$, $m_0 = 5$, $K = 200$, SNR = 20 dB and $N_{rd} = 5$ and 6, respectively.

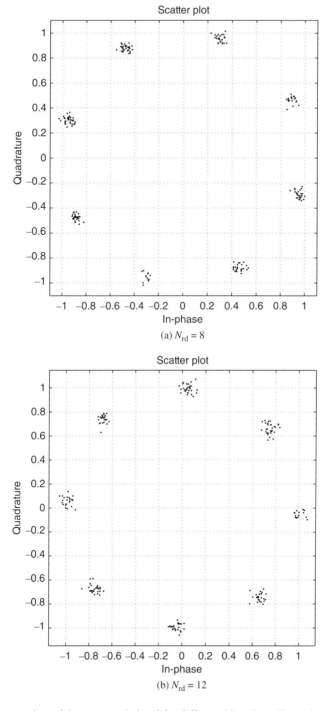

Figure 6.13 Scatter plots of the processed signal for different N_{rd}, where $N_{tag} = 1$, $m_0 = 5$, $K = 200$, SNR = 20 dB and $N_{rd} = 8$ and 12, respectively.

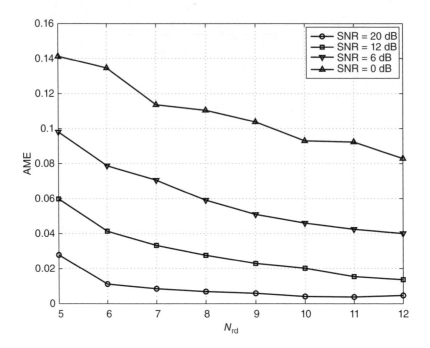

Figure 6.14 The average modulus error of the processed signal versus N_{rd} for different SNR, where $N_{tag} = 1$, $m_0 = 5$ and $K = 200$.

ACMA with $N_{rd} = 7$ cannot, as shown in Figure 6.15. Notice that a prerequisite for ACMA to be able to run is that $N_{rd} \geq m_0$.

Figures 6.16–6.18 show how the performance of the system depends on m_0, the number of tags. Figures 6.16 and 6.17 illustrate the scatter plots of the processed signal and Figure 6.18 depicts the AME curves. It can be seen that the AME of the system slowly increases with m_0 when the SNR is kept at a moderate level, for example, in the range between 6 and 20 dB. However, when m_0 approaches to the critical number N_{rd}, the system performance will dramatically deteriorate. When $m_0 = N_{rd}$, ACMA fails separating the overlapped signals.

From the scatter plots in this section, it can be seen that group phase rotation (which means that the reconstructed signals in one snapshot have roughly the same phase shift) in the reconstructed signals often happen. For this reason, the separation property of the recovered signals cannot be clearly seen if we put several scatter plots at different runs/snapshots together.

6.5 Summary

In this chapter, we have discussed the identification problem of multiple-tag RFID systems. First, a mathematical model for multi-tag RFID systems is presented based on the RFID-MIMO system model developed in Chapter 5. Then the analytical constant modulus algorithm in [17] is briefly reviewed. Finally, ACMA is applied to the multiple-tag identification problem. It is found that under moderate SNR and when the number of measurements to the multiple tags in one snapshot is sufficiently high, the overlapped signals coming from

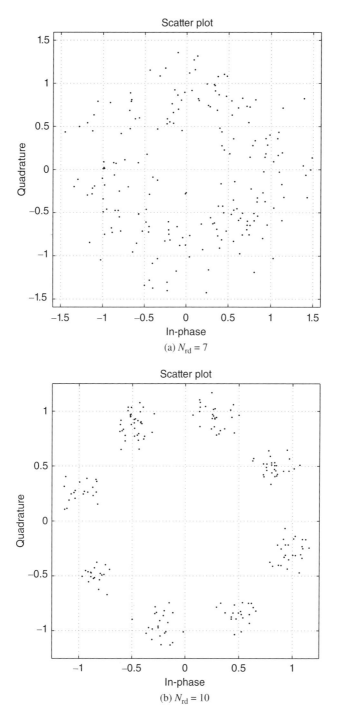

Figure 6.15   Scatter plots of the processed signal for two critical cases: $N_{\text{tag}} = 1$, $m_0 = 5$, $K = 200$ and SNR = 6 dB.

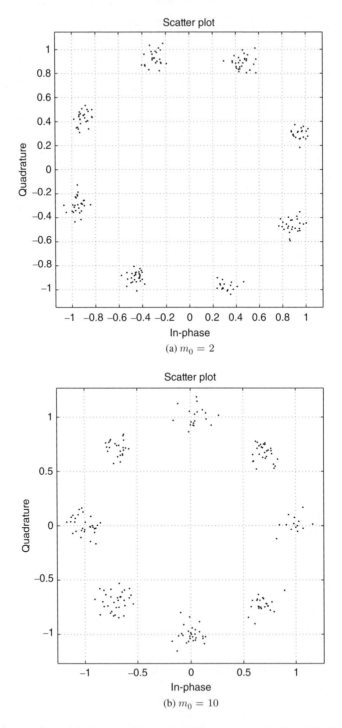

Figure 6.16   Scatter plots of the processed signal for different m_0, where $N_{rd} = 12$, $N_{tag} = 1$, $K = 200$, SNR = 12 dB and $m_0 = 2$ and 10, respectively.

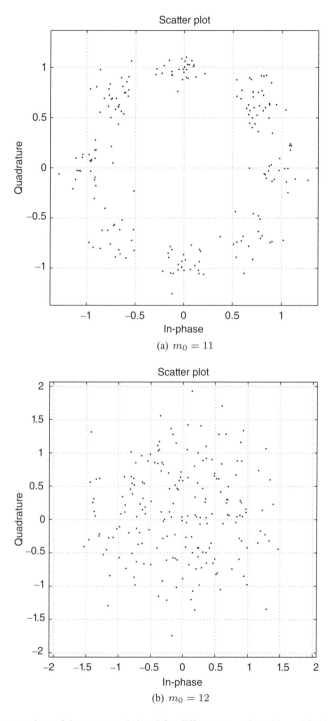

Figure 6.17 Scatter plots of the processed signal for different m_0, where $N_{rd} = 12$, $N_{tag} = 1$, $K = 200$, SNR = 12 dB and $m_0 = 11$ and 12, respectively.

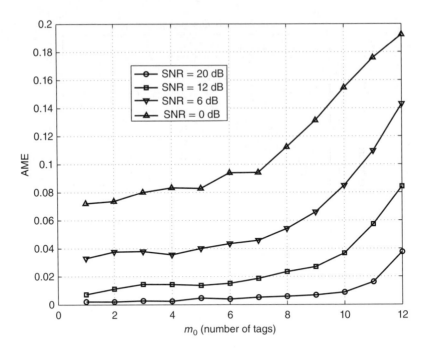

Figure 6.18   The average modulus error of the processed signal versus m_0 for different SNR, where $N_{rd} = 12$, $N_{tag} = 1$ and $K = 200$.

multiple tags can be successfully separated at the reader receiver if the number of the tags is less than the number of receiving antennas at the reader. Generally, the performance of ACMA is improved with increasing of the number of measurements to the multiple tags, the number of receiving antennas at the reader and the SNR. However, the average modulus error of the processed signals saturates when the number of the measurements is high, and it steadily decreases with the SNR and the number of receiving antennas at the reader in low SNR regime.

From the simulation results, some design guidelines for ACMA multiple-tag RFID systems can be drawn. First, the number of measurements to the tags in one snapshot should be sufficiently high, but not necessarily too high. In any case, it should be larger than the square of the number of tags to be identified simultaneously. Second, if the SNR for the application scenarios is low, increasing the number of antennas at the reader is helpful for increasing the separability of the overlapped signals; on the other hand, if the SNR is high, the effect of increasing the number of antennas at the reader on the system performance is marginal. In any case, the number of antennas at the reader should be greater than or equal to the number of tags to be identified simultaneously.

References

[1] A. Benveniste and M. Goursat. Blind equalizers. *IEEE Trans. Commun.*, 32:871–883, 1984.
[2] A. Benveniste, M. Goursat, and G. Ruget. Robust identification of a nonminimum phase system: blind adjustment of a linear equalizer in data communications. *IEEE Trans. Automatic Control*, 25:385–399, 1980.

[3] A. Bunse-Gerstner, R. Byers, and V. Mehrmann. Numerical methods for simultaneous diagonalization. *SIAM J. Mat. Anal. Appl.*, 14:927–949, 1993.

[4] J.-F. Cardoso. Blind signal separation: Statistical principles. *Proceedings of the IEEE*, 86:2009–2025, 1998.

[5] J.-F. Cardoso and A. Souloumiac. Jacobi angles for simultaneous diagonalization. *SIAM J. Mat. Anal. Appl.*, 17:161–164, 1996.

[6] Z. Ding, R. A. Kennedy, B. D. O. Anderson, and C. R. Johnson, Jr. Ill-convergence of Goddard blind equalizers in data communications. *IEEE Trans. Commun.*, 39:1313–1328, 1991.

[7] Z. Ding and Y. Li. *Blind Equalization and Identification*. Marcel Dekker, New York, 2001.

[8] G. J. Foschini. Equalizing without altering or detecting data. *AT&T Technical Journal*, 64:1885–1911, 1985.

[9] D. N. Godard. Self-recovering equalization and carrier tracking in two-dimensional data communication systems. *IEEE Trans. Commun.*, 28:1867–1875, 1980.

[10] A. F. Mindikoglu and A.-J. van der Veen. Separation of overlapping RFID signals by antenna arrays. In *Proc. IEEE Int. Conf. Acoustics, Speech and Signal Processing 2008*, pages 2737–2740, Las Vegas, USA, 31 Mar.–4 Apr. 2008.

[11] R. Johnson Jr., P. Schniter, T. J. Endres, J. D. Behm, D. R. Brown, and R. A. Casas. Blind equalization using the constant modulus criterion: A review. *Proceedings of the IEEE*, 86:1927–1950, 1998.

[12] Y. Sato. A method of self-recovering equalization for multilevel amplitude-modulation. *IEEE Trans. Commun.*, 23:679–682, 1975.

[13] L. Tong and S. Perreau. Multichannel blind identification: From subspace to maximum likelihood methods. *Proceedings of the IEEE*, 86:1951–1968, 1998.

[14] L. Tong, G. Xu, and T. Kailath. Blind identification and equalization based on secondorder statistics: A time domain approach. *IEEE Trans. Inf. Theory*, 40:340–349, 1994.

[15] J. R. Treichler and B. G. Agee. A new approach to multipath correction of constant modulus signals. *IEEE Trans. Acousl., Speech, Signal Processing*, 31:459–471, 1983.

[16] A.-J. van der Veen. Algebraic methods for deterministic blind beamforming. *Proceedings of the IEEE*, 86:1987–2008, 1998.

[17] A.-J. van der Veen and A. Paulraj. An analytical constant modulus algorithm. *IEEE Trans. Signal Process.*, 44:1136–1155, 1996.

[18] Y. Wang, Y. C. Pati, Y. M. Cho, A. Paulraj, and T. Kailath. A matrix factorization approach to signal copy of constant modulus signals arriving at an antenna array. In *Proc. 28th Conf. Inform. Sci. Syst.*, Princeton, USA, Mar. 1994.

7

Anti-Collision of Multiple-Tag RFID Systems

7.1 Introduction

In practical systems, we often meet the situation that there are many tags in the interrogation zone of a reader. If multiple tags simultaneously transmit their ID signals to the reader, a collision will occur. In Chapter 6, we have discussed a collision-resolution approach based on blind signal processing technique by using multiple receive antennas at the reader. In this approach, we examine the collision problem from the view of PHY layer, where few communications are needed between the reader and tags. Another approach to the anti-collision problem from PHY layer is to use CDMA (code-division multiple access) [31, 33], where wide band RFID tags must be used. In this chapter, we examine the collision problem from the view of MAC layer, where the collisions are resolved based on several rounds of communications between the reader and tags.

In principle, many advanced multiple channel accessing algorithms in wireless networks can be applied to RFID collision-resolution problem. However, passive RFID systems are highly asymmetric, i.e., the reader is resource-rich, while tags have very limited storage and computing capabilities and are unable to hear the signal transmitted by other tags or to detect collisions, channel access must be arbitrated by the reader [27, 35, 40]. Due to this fact, only basic anti-collision protocols, which can be broadly categorized into Aloha-based algorithms and tree-splitting based algorithms, have been recommended in RFID standards and implemented in practical RFID systems.

In slotted Aloha protocol, the time is discretized into time slots, and all the transmissions will start at the beginning of a time slot. The start of each time slot is signalled by the reader with a short message [27]. The whole interrogation period (maximal backoff time) is divided into several time slots, and each tag will independently, randomly and uniformly select a time slot or backoff time after receiving an interrogation request from the reader. The actual duration of a time slot is chosen according to the amount of data to be transmitted by the tags. The tag will wait till the time reaches the backoff time that it selects and then transmit its tag ID to the reader at that time. The backoff time must be a multiple of the pre-specified slot time. If an ID is received by the reader without collision, it can be identified properly and acknowledged by

Digital Signal Processing for RFID, First Edition. Feng Zheng and Thomas Kaiser.
© 2016 John Wiley & Sons, Ltd. Published 2016 by John Wiley & Sons, Ltd.

the reader. A tag with an acknowledged ID will go to silence. On the other hand, if a collision happens, the tag will be unacknowledged and all the unacknowledged tags will independently select random backoff time again in next round to send their IDs. This process is repeated until all the tags are finally identified and acknowledged by the reader [5, 30, 47].

In tree-splitting (TS) algorithms, the time is also discretized into time slots, and all tags only start to transmit at the beginning of a time slot. For the binary TS algorithm, all the tags will transmit after receiving a query signal from the reader. If a collision happens, then all the tags are divided into two subsets in terms of a randomly assigned integer: 0 or 1. The tags in the first subset, having a random counter 0, will transmit first in next inventory round, while the tags in the second subset, having a random counter 1, will wait until the collisions among the tags in the first subset are completely resolved. If collisions happen among the tags in the first subset, then further splitting inside this subset is made in the same way. This procedure is conducted recursively until all the tags in the first subset are inventoried. Then the reader moves the inventory to the second subset and applies the above procedure to this subset.

Clearly, the above binary TS algorithm can be easily extended to general M-ary tree-splitting algorithm, where the number of splitting branches becomes M instead of 2 as in the binary TS algorithm.

A fundamental difference in the concerns of anti-collision algorithms between wireless networks and RFID systems is that much attention is paid on the stability of the algorithms for wireless networks, while much interest is focused on the transmission or identification efficiency of the algorithms for RFID systems. The reason is that, in wireless networks, data packets often arrive at the common channel continuously but in a bursty manner [16], while in RFID systems, the number of tags to be inventoried is often fixed in typical applications such as the case of a shopping trolley passing through a checking gate. Therefore, if anti-collision algorithms in wireless networks are not well designed, the number of packets to be delivered can be accumulated and goes to divergence quickly, while in RFID systems, all the tags will be finally identified by using an adequate anti-collision algorithm. The only problem is how long it will take to identify all the tags.

In this chapter, we will discuss some basic Aloha-based and TS-based anti-collision algorithms and present some analytical tools to analyse the performance of the algorithms. Our attention is focused on how to analyse and design an anti-collision algorithm to obtain good performance. Therefore, we will not try to walk through all existing anti-collision algorithms for RFID systems in the literature. Instead, we believe that it is more helpful to be familiar with the analytical methods and design principles than the results themselves. Several extensive surveys about anti-collision algorithms for RFID systems can be found in [23, 35, 40].

Another kind of collision in RFID systems is reader-to-reader collision, which occurs when neighbouring readers interrogate a tag simultaneously [12, 15, 18, 37, 45]. Reader-to-reader collision can be dealt with using traditional anti-collision protocols in wireless networks. Therefore, we will not touch this topic in this book.

This chapter is organized as follows. Tree-splitting based anti-collision algorithms are discussed in Section 7.2, Aloha-based anti-collision algorithms are addressed in Section 7.3. Section 7.4 concludes this chapter.

7.2   Tree-Splitting Algorithms

The classical TS algorithm was first reported by Capetanakis [9] and Tsybakov and Mikhailov [43], independently, for general multiple-access contention channels. After that work many studies were devoted to the performance analysis of this algorithm and its variants, see [19, 21, 32] and references therein.

The TS algorithm can be naturally applied to the problem of RFID multiple-tag collision resolution. In this section, we first review the algorithm and then analyse its performance. The performance metrics include transmission efficiency and mean identification delay, whose exact definitions will be given later.

An M-ary TS algorithm works for RFID systems in the following way. The reader is equipped with a trinary feedback device, which tells the tags what has happened in the tags' transmission in the last time slot: *idle, successful* and *colliding*. An idle transmission means that no tags transmit in the last time slot, a successful transmission means that only one tag transmits in the last time slot [1], and a collision means that two or more tags transmit in the last time slot. Suppose that there are N tags. At the start of a new tree algorithm, all the N tags will transmit their IDs to the readers [2]. If an collision happens, then each of the N tags picks a random integer (say m) between (and including) 0 and $M - 1$ with equal probabilities and transmits its ID information at the mth contention time slot. If the transmission at the mth time slot is colliding, which will be informed by the reader's trinary feedback device, then all the tags contained in the mth time slot repeat the same procedure as the tags contained in the root node, i.e., pick independently random integers between (and including) 0 and $M - 1$ and transmit again at the corresponding time slots. This forms a new branch for the tree. This process is repeated and the expansion of the tree branches stops at the node which contains either zero or one tag, resulting an idle or successful transmission. When the transmission at the aforementioned branch is finished, the algorithm moves to next branch until all branches are completed. Upon completion of all the branches of the tree, all the N tags will successfully transmit their IDs to the reader. Thereafter, a new tree algorithm may start again. In the literature, a tree is also called a frame. Figure 7.1 shows an example of the TS algorithm for the case of $N = 15$ and $M = 3$.

In the TS algorithm, we say that the tags are located in the same *layer* if they have experienced the same number of contentions to transmit. The meaning of the concept "layer" can be clearly seen from Figure 7.1.

The TS algorithm can be realized in RFID systems in the following way.

Every tag has a counter, which is initialized to be 0 at the beginning of a frame. The tag transmits its ID when the counter value is 0. Therefore, all the tags within the reader's interrogation zone form a single set and transmit simultaneously at the start of a frame. The reader transmits a feedback to inform tags of the status of the last transmission. All the tags change their counters according to the reader's feedback information. If a collision happens, each tag involved in the last transmission randomly picks an integer between (and including) 0 and $M - 1$ and adds this number to its counter, and all other active tags that did not transmit in the last time slot and are waiting to be inventoried increase their counters by $M - 1$. If a successful transmission happens, the tag involved in the last transmission is put into silence, and

---

[1] When discussing MAC layer problems, it is generally assumed that the propagation errors appeared in the PHY layer can be corrected by a proper coding technique.

[2] In another version of TS algorithm, each tag picks a random integer between (and including) 0 and $M - 1$ at the start of a new tree algorithm before the first contention. In this case, the expected collision number should be modified accordingly, but the overall performance of the algorithm remains the same when N is large enough.

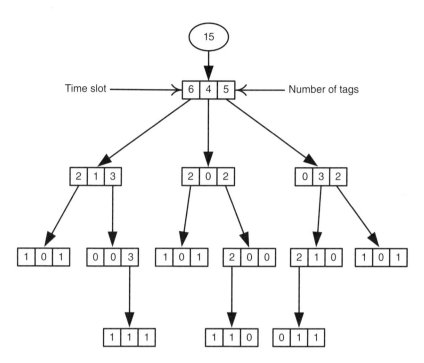

Figure 7.1   An illustration of tree splitting algorithm for the case of $N = 15$ and $M = 3$.

all other tags decrease their counters by 1. If an idle transmission happens, all the other active tags decrease their counters by 1. This procedure moves on: once a tag receives the inventory request signal from the reader and its counter is 0, it transmits. This process is repeated until all the tags in a frame are successfully inventoried.

In order for the reader to determine when to terminate a frame, the reader is also equipped with a counter. At the start of a tree, this counter is initialized to be 0. When a collision happens, the value of this counter is increased by $M - 1$, while when a successful or idle transmission happens, the value of this counter is decreased by 1. It is clear that the value of this counter equals the number of tag sets which wait to be inventoried. Therefore, when the value of this counter is less than 0, all the tags in the tree are successfully inventoried and the frame is terminated.

The above realization method, for the case of binary TS algorithm, was proposed in [35].

The TS algorithm can also be realized in another way. The basic idea of this way is that, once the tags in the first time slot finish their transmit contention, the tags in the next time slot in the same layer will transmit; when the transmit contention is finished in all the time slots in one layer, it moves to the next layer. Technically, this approach works in the following way.

Every tag has two counters. The first counter, denoted by C_1, stores the information that whether or not the tag has transmitted at the current layer. We use the convention

$$C_1 = \begin{cases} 0 & \text{when the tag has transmitted at the current layer,} \\ 1 & \text{otherwise.} \end{cases}$$

The second counter, denoted by C_2, stores the information about which time slot the tag has chosen in the current layer when a collision happens. The reader has two counters too, denoted by C_3 and C_4, respectively, but these two counters are actually two vectors. The counter C_3 stores the time slot sequence chosen by those tags whose transmissions cause collisions in the last layer, and the counter C_4 stores the time slot sequence chosen by those tags whose transmissions cause collisions in the current layer.

Define a vector $\underline{M} := [0 \ 1 \ \cdots \ M - 1]$, and let length stand for the length of a vector.

Algorithm 7.1 TS fair realization

Step 0 Initialization: At the start of a tree, the reader sends out inventory request signal and all the tags transmit their IDs when hearing this inventory request signal. If the transmission is successful or idle, the frame is terminated; if a collision happens, the four counters are updated in the following way:

$$C_1 = 1, \ \text{for all tag,}$$

$$C_2 = m \in \underline{M}: \quad \text{each tag picks its own time slot number,}$$

$$C_3 = \underline{M},$$

$$C_4 = [\ ],$$

where m is the integer that the tag picks.

Step 1 For $k = 1 : length(C_3)$ do: the reader broadcasts signal $C_3(k)$, if $C_2 = C_3(k)$ and $C_1 = 1$, the tag or tags transmit. If the transmission is successful or idle, the counters are updated in the following way:

$$C_1 := \begin{cases} \text{void} & \text{for the tag involved in the transmission and the tag is put into} \\ & \text{silence,} \\ C_1 & \text{otherwise,} \end{cases}$$

$$C_2 := \begin{cases} C_2 - 1 & \text{if } C_1 = 1, \text{ that is, for the tag that has not yet transmitted at} \\ & \text{current layer,} \\ C_2 & \text{otherwise,} \end{cases} \tag{7.1}$$

$$C_3 := C_3 - 1, \tag{7.2}$$

$$C_4 := C_4.$$

If a collision happens, the counters are updated in the following way:

$$C_1 := \begin{cases} 0 & \text{if the tag is involved in the colliding transmission,} \\ C_1 & \text{otherwise,} \end{cases}$$

$$C_2 := \begin{cases} M * C_2 + m & \text{if the tag is involved in the colliding transmission,} \\ C_2 & \text{otherwise,} \end{cases}$$

$$C_3 := C_3,$$

$$C_4 := [C_4 \ \underline{M} * C_3(k)].$$

Step 2 If length(C_4) = 0, stop and the frame is terminated; otherwise let $C_3 := C_4$ and reset $C_1 = 1$ (for all tags), return to Step 1.

Some remarks for Algorithm 7.1 are due. (i) It is seen that the counter C_2 of those tags that have not yet transmitted in the current layer is always updated with the reader's counter C_3 synchronously. (ii) In Step 2, the reader makes the inventory for all tags in terms of their slot positions in the current layer. (iii) The purpose of using equations (7.1) and (7.2) is to avoid the expansion in slot number caused by idle and successful transmissions, which could be very large when the number of tags is large.

The realization procedure as described in Algorithm 7.1 has two advantages. The first is that it can facilitate the performance analysis. In the performance analysis to be followed, we will examine the TS algorithm from this view of point. The second is that it is fairer than to use the first realization procedure to all tags. In multiple tag identification problem, fairness is not an issue. However, for multi-user networks, fairness is indeed a big issue. It is seen that it is not fair for a user who picks a greater slot number if the first TS realization procedure is adopted. For this reason, we call Algorithm 7.1 the TS fair realization algorithm.

Next, let us analyse the performance of the TS algorithm. Two most important performance metrics will be investigated.

The first is the identification delay of a tag. The identification delay of a tag characterizes the number of transmit contentions that the tag has used to finally successfully deliver its ID to the reader. Denote the number of the transmit contentions as d_{tr}. It is clear that d_{tr} is a random variable and depends on tag in a specific realization of a frame. However, the expectation of d_{tr} depends only on the number of tags in a frame and branch number in each tree splitting, i.e. M. Therefore, we define the expectation of d_{tr}, denoted by $\bar{d}_{tr}$, as the mean identification delay of a tag.

The second is the total transmission time of all the tags in a tree. To get this information, we need to know the number of collisions and idle transmissions in a tree. Related to this performance metric is the transmission efficiency, which is defined as the ratio between successful transmission time, i.e. the total time used in successful transmissions, and the total transmission time, i.e. the total time spent on idle, colliding and successful transmissions.

In data networks, saturation analysis is of more interest, i.e. to analyze the percentage of time slots used for successful transmission when the incoming data packets are of a constant speed or distributed according to some known distribution [7], which means that there are always data packets waiting for transmission. In the literature of data networks, the aforementioned percentage of time slots used for successful transmissions in the total transmission time is often called throughput. When the number of tags N approaches infinity, the corresponding transmission efficiency reduces to throughput.

7.2.1 Mean Identification Delay

Consider the case of N tags for an M-ary TS algorithm. Let us examine the TS algorithm from the viewpoint as described in Algorithm 7.1, i.e., the reader inventories the tags by first walking through all the time slots in a layer and then moving to next layer. The following simplifying assumption will be used.

Assumption 7.1 All the N tags, no matter whether or not they have successfully transmitted their IDs, move to the next layer.

Assumption 7.1 is implicitly used in the analysis of TS algorithms in major references, e.g. [17, 19, 21].

Based on Assumption 7.1, at the kth layer, there are in total M^{k-1} branches or time slots for tags to transmit. Let us organize all the branches at the kth layer in an ordered way. The branches from the left to the right as shown in Figure 7.1 are numbered as 1, 2, ..., M^{k-1}, respectively. Similarly, we organize all the tags in an ordered way and let T_n be the slot number that the nth tag picks at the concerned layer. As argued in [9, 19, 21], all the N tags at the kth layer are independently and identically distributed with equal probabilities over the M^{k-1} time slots due to the tree splitting process.

Without loss of generality, we can take the first tag as an example under consideration. Let us now calculate the probability distribution of d_{tr}. First we calculate the probability of $\{\text{Tr}(T_1, k) = \text{succ}\}$, where $\{\text{Tr}(T_1, k) = \text{succ}\}$ stands for the event that the transmission of the first tag (represented by T_1) at the kth layer is successful. Similarly, we use $\{\text{Tr}(T_1, k) = \text{unsucc}\}$ to denote the event that the transmission of the first tag at the kth layer is unsuccessful. A simple calculation yields

$$
\begin{aligned}
\bar{P}_{\text{suc},k} &:= P\{\text{Tr}(T_1, k) = \text{succ}\} \\
&= P\{\{T_1 = 1\} \cap \{T_2 \neq 1\} \cap \{T_3 \neq 1\} \cap \cdots \cap \{T_N \neq 1\}\} \\
&\quad + P\{\{T_1 = 2\} \cap \{T_2 \neq 2\} \cap \{T_3 \neq 2\} \cap \cdots \cap \{T_N \neq 2\}\} \\
&\quad + \cdots \\
&\quad + P\{\{T_1 = M^{k-1}\} \cap \{T_2 \neq M^{k-1}\} \cap \{T_3 \neq M^{k-1}\} \cap \cdots \cap \{T_N \neq M^{k-1}\}\} \\
&= P\{T_1 = 1\} P\{T_2 \neq 1\} P\{T_3 \neq 1\} \cdots P\{T_N \neq 1\} \\
&\quad + P\{T_1 = 2\} P\{T_2 \neq 2\} P\{T_3 \neq 2\} \cdots P\{T_N \neq 2\} \\
&\quad + \cdots \\
&\quad + P\{T_1 = M^{k-1}\} P\{T_2 \neq M^{k-1}\} P\{T_3 \neq M^{k-1}\} \cdots P\{T_N \neq M^{k-1}\} \\
&= \frac{1}{M^{k-1}} \left(\frac{M^{k-1} - 1}{M^{k-1}} \right)^{N-1} + \frac{1}{M^{k-1}} \left(\frac{M^{k-1} - 1}{M^{k-1}} \right)^{N-1} + \cdots \\
&\quad + \frac{1}{M^{k-1}} \left(\frac{M^{k-1} - 1}{M^{k-1}} \right)^{N-1} \\
&= \left(1 - \frac{1}{M^{k-1}} \right)^{N-1}.
\end{aligned}
\tag{7.3}
$$

Therefore, we have

$$
\begin{aligned}
P_{d_{\text{tr}}}(k) &:= P\{d_{\text{tr}} = k\} \\
&= P\{\{\text{Tr}(T_1, 1) = \text{unsucc}\} \cap \cdots \cap \{\text{Tr}(T_1, k-1) = \text{unsucc}\} \cap \{\text{Tr}(T_1, k) = \text{succ}\}\} \\
&= [1 - P\{\text{Tr}(T_1, 1) = \text{succ}\}] \cdots [1 - P\{\text{Tr}(T_1, k-1) = \text{succ}\}] P\{\text{Tr}(T_1, k) = \text{succ}\} \\
&= [1 - \bar{P}_{\text{suc},1}][1 - \bar{P}_{\text{suc},2}] \cdots [1 - \bar{P}_{\text{suc},k-1}] \bar{P}_{\text{suc},k} \\
&= \left[1 - \left(1 - \frac{1}{M} \right)^{N-1} \right] \cdots \left[1 - \left(1 - \frac{1}{M^{k-2}} \right)^{N-1} \right] \left(1 - \frac{1}{M^{k-1}} \right)^{N-1}, \quad k \geq 2.
\end{aligned}
\tag{7.4}
$$

Based on equation (7.4), the mean identification delay can be calculated in terms of the following formula:

$$\bar{d}_{tr} = \begin{cases} 1 & \text{when } N = 1, \\ 1 + \sum\limits_{k=2}^{\infty} kP_{d_{tr}}(k) & \text{when } N \geq 2. \end{cases} \tag{7.5}$$

In [19], the approximate value of $\bar{d}_{tr}$ is given by

$$\bar{d}_{tr} \approx \log_M(N-1) + \frac{1}{2} + \frac{0.5772}{\log M} + \frac{1}{2N \log M} + 1. \tag{7.6}$$

Remark 7.1 In real operation of TS algorithms, the contending tag number decreases with the increasing of contending layer since the tags with successful transmissions will quit the tree. Due to this fact, Assumption 7.1 is somewhat conservative when used in estimating the mean identification delay. Therefore, equation (7.5) actually provides an upper bound for the mean identification delay.

It is seen from equation (7.6) that, the larger the parameter M, the lower the mean identification delay. However, when M is large, the number of idle transmissions becomes great, and hence many time slots are wasted in idle transmissions. Therefore, it is not wise to use a TS algorithm with a very large number of M if the transmission efficiency is concerned.

7.2.2 Collision Analysis and Transmission Efficiency: Approach I

In this and next subsections, we analyse the numbers of colliding, idle and successful transmissions in the TS algorithm and then calculate its transmission efficiency. To this end, we organize the tags and time slots in an ordered way, as described in the preceding subsection.

In this subsection, Assumption 7.1 is adopted. Thus the analysis can be greatly simplified, and the obtained result is an approximation to the true transmission efficiency. In the next subsection, no additional assumptions are used and the exact result for the transmission efficiency will be obtained.

Let $T_{x,j}$ express a transmission at time slot j. Let $P_{idl,k}$, $P_{suc,k}$ and $P_{col,k}$ be the probabilities of the events that the transmission at some specific time slot (say the jth time slot) at the kth layer is idle, successful and colliding, respectively. Then we have

$$P_{idl,k} := P\{T_{x,j} = \text{idle}\}$$

$$= P\{\{T_1 \neq j\} \cap \{T_2 \neq j\} \cap \cdots \cap \{T_N \neq j\}\}$$

$$= P\{T_1 \neq j\}P\{T_2 \neq j\} \cdots P\{T_N \neq j\}$$

$$= \frac{M^{k-1} - 1}{M^{k-1}} \times \frac{M^{k-1} - 1}{M^{k-1}} \times \cdots \times \frac{M^{k-1} - 1}{M^{k-1}}$$

$$= \left(1 - \frac{1}{M^{k-1}}\right)^N,$$

$$P_{\text{suc},k} := P\{\mathrm{T}_{\mathrm{x},j} = \text{successful}\}$$

$$= P\{\{T_1 = j\} \cap \{T_2 \neq j\} \cap \cdots \cap \{T_N \neq j\}\}$$

$$\quad + P\{\{T_2 = j\} \cap \{T_1 \neq j\} \cap \{T_3 \neq j\} \cdots \cap \{T_N \neq j\}\}$$

$$\quad + \cdots$$

$$\quad + P\{\{T_N = j\} \cap \{T_1 \neq j\} \cap \{T_2 \neq j\} \cdots \cap \{T_{N-1} \neq j\}\}$$

$$= P\{T_1 = j\}P\{T_2 \neq j\} \cdots P\{T_N \neq j\}$$

$$\quad + P\{T_2 = j\}P\{T_1 \neq j\}P\{T_3 \neq j\} \cdots P\{T_N \neq j\}$$

$$\quad + \cdots$$

$$\quad + P\{T_N = j\}P\{T_1 \neq j\}P\{T_2 \neq j\} \cdots P\{T_{N-1} \neq j\}$$

$$= \frac{1}{M^{k-1}} \left(\frac{M^{k-1} - 1}{M^{k-1}} \right)^{N-1} + \frac{1}{M^{k-1}} \left(\frac{M^{k-1} - 1}{M^{k-1}} \right)^{N-1} + \cdots$$

$$\quad + \frac{1}{M^{k-1}} \left(\frac{M^{k-1} - 1}{M^{k-1}} \right)^{N-1}$$

$$= N \times \frac{1}{M^{k-1}} \times \left(1 - \frac{1}{M^{k-1}} \right)^{N-1},$$

$$P_{\text{col},k} := P\{\mathrm{T}_{\mathrm{x},j} = \text{collision}\}$$

$$= 1 - P_{\text{idl},k} - P_{\text{suc},k}$$

$$= 1 - \left(1 - \frac{1}{M^{k-1}} \right)^N - \frac{N}{M^{k-1}} \left(1 - \frac{1}{M^{k-1}} \right)^{N-1}.$$

Therefore, the average number of collisions happened at the kth layer is given by

$$N_{\text{col},k} = \sum_{j=1}^{M^{k-1}} P\{\mathrm{T}_{\mathrm{x},j} = \text{collision}\}$$

$$= M^{k-1} \left[1 - \left(1 - \frac{1}{M^{k-1}} \right)^N - \frac{N}{M^{k-1}} \left(1 - \frac{1}{M^{k-1}} \right)^{N-1} \right]. \tag{7.7}$$

The average number of total collisions in all the layers is given by

$$N_{\text{col}} = \sum_{k=1}^{\infty} N_{\text{col},k}$$

$$= \sum_{k=1}^{\infty} M^{k-1} \left[1 - \left(1 - \frac{1}{M^{k-1}} \right)^N - \frac{N}{M^{k-1}} \left(1 - \frac{1}{M^{k-1}} \right)^{N-1} \right]. \tag{7.8}$$

Since each collision generates M branches on the next layer, the average total leaf number, i.e. total time slots used for tag transmissions, in the whole splitting tree is given by

$$N_{\text{leaf}} = MN_{\text{col}}.$$

Obviously, the slot number of successful transmissions is

$$N_{\text{suc}} = N. \tag{7.9}$$

Therefore, the average total number of idle time slots in the whole splitting tree is given by

$$N_{\text{idle}} = N_{\text{leaf}} - N_{\text{col}} - N_{\text{suc}} = (M-1)N_{\text{col}} - N. \tag{7.10}$$

Let T_s, T_c and T_i, respectively, be the time needed for *one* successful, colliding or idle transmission. The following observation generally holds true[3]

$$T_c = T_s, \quad T_i < T_s, \quad T_i < T_c.$$

Here let us normalize the time for colliding and successful transmissions to be unity, that is, $T_s = T_c = 1$. Then T_i will be a fraction of this unit time. Denote this fraction as F_{idle}, i.e., $F_{\text{idle}} = T_i/T_s = T_i/T_c$.

Based on these results, the transmission efficiency of an M-ary TS algorithm is given by

$$E_{\text{TS}}(M, N) = \frac{N_{\text{suc}}T_s}{N_{\text{suc}}T_s + N_{\text{col}}T_c + N_{\text{idle}}T_i} \tag{7.11}$$

$$= \frac{1}{1 + \frac{N_{\text{col}}}{N} + \frac{N_{\text{idle}}}{N}F_{\text{idle}}} = \frac{1}{1 - F_{\text{idle}} + \frac{[1+(M-1)F_{\text{idle}}]N_{\text{col}}}{N}}. \tag{7.12}$$

Substituting equation (7.8) into (7.12) we can readily get the transmission efficiency.

Remark 7.2 Because of the same reason as explained in Remark 7.1, equation (7.8) actually gives an upper bound for the mean identification delay. Therefore, equation (7.12) gives a lower bound for the transmission efficiency.

Due to Remark 7.2, it is meaningful to investigate the exact formulas for the collision number, idle number and total transmission time in a TS algorithm. This is the task of the next subsection.

7.2.3 Collision Analysis and Transmission Efficiency: Approach II

Let $D(n; M)$ be the expectation of the total transmission time of n tags in an M-ary TS algorithm when all the n tags are finally successfully identified. Clearly we have $D(1; M) = T_s$ for any $M \geq 2$. To ease exposition, we use S_i, $i = 1, 2, \cdots, M$, to number the M different time slots.

We first consider several specific cases and then extend the results to general cases.

7.2.3.1 The case of $D(2; 2)$

In this case, each of the two tags can choose any one of S_1 and S_2, producing four possibilities and each with the probability of $\frac{1}{4}$, as shown in Figure 7.2. For patterns (a) and (b), each

---

[3] If Aloha-based anti-collision algorithms are used, a collision can be detected at the stage when tags backscatter their tag handles (i.e. the randomly selected integers in their slot counters after receiving a Query command from the reader), rather than after the relevant tags finish the transmission of their EPC codes. Then T_c can be also much less than T_s. In this case, equation (7.12) should be revised accordingly. In this chapter, we maintain the assumption that $T_c = T_s$ for concision. The general case, considering the practical PHY link parameters of RFID systems, can be addressed similarly and the corresponding results can be found in reference [50].

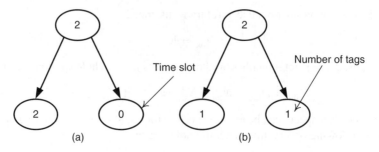

Figure 7.2 An illustration of calculating the total transmission time for TS algorithm for the case of $D(2;2)$.

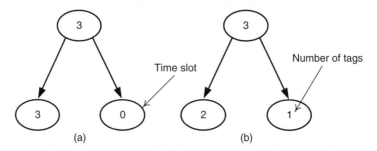

Figure 7.3 An illustration of calculating the total transmission time for TS algorithm for the case of $D(3;2)$.

pattern has two possibilities, corresponding to an exchange of the time slot position for the same pattern.

Therefore, we have

$$D(2;2) = \frac{1}{4}[D(2;2) + T_i] \times 2 + \frac{1}{4}[T_s + T_s] \times 2 + T_c. \tag{7.13}$$

The last term T_c in the right-hand side (RHS) of equation (7.13) stands for the colliding transmission when the two tags compete to transmit at the very first layer. From equation (7.13) it is easy to solve $D(2;2)$ as follows:

$$D(2;2) = 2T_s + 2T_c + T_i. \tag{7.14}$$

7.2.3.2 The case of $D(3;2)$

In this case, each of the three tags can choose any one of S_1 and S_2, producing eight possibilities and each with the probability of $\frac{1}{8}$, as shown in Figure 7.3.

In Figure 7.3, pattern (a) has two possibilities, corresponding to positioning the three tags in two different time slots, and pattern (b) has $\binom{3}{2} \times 2$ possibilities, where the coefficient $\binom{3}{2}$ means that there are $\binom{3}{2}$ ways for the three tags to choose time slot S_1, and the coefficient 2 is due to the permutation of the two time slots holding the same pattern of tags. Therefore, we have

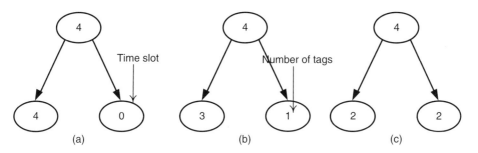

Figure 7.4   An illustration of calculating the total transmission time for TS algorithm for the case of $D(4; 2)$.

$$D(3; 2) = \frac{1}{8}[D(3; 2) + T_i] \times 2 + \frac{1}{8}[D(2; 2) + T_s] \times \binom{3}{2} \times 2 + T_c. \qquad (7.15)$$

From equation (7.15) it is easy to solve $D(3; 2)$ as follows:

$$D(3; 2) = D(2; 2) + T_s + \frac{4}{3}T_c + \frac{1}{3}T_i = 3T_s + \frac{10}{3}T_c + \frac{4}{3}T_i. \qquad (7.16)$$

7.2.3.3   The case of $D(4; 2)$

In this case, each of the four tags can choose any one of S_1 and S_2, producing 16 possibilities and each with the probability of $\frac{1}{16}$, as shown in Figure 7.4.

In Figure 7.4, pattern (a) has two possibilities, pattern (b) has $\binom{4}{1} \times 2$ possibilities and pattern (c) has $\binom{4}{2}$ possibilities. Therefore, we have

$$D(4; 2) = \frac{1}{16}[D(4; 2) + T_i] \times 2 + \frac{1}{16}[D(3; 2) + T_s] \times \binom{4}{1} \times 2$$

$$+ \frac{1}{16}[D(2; 2) + D(2; 2)] \times \binom{4}{2} + T_c$$

$$= \frac{1}{8}[D(4; 2) + T_i] + \frac{4}{8}[D(3; 2) + T_s] + \frac{6}{8}D(2; 2) + T_c. \qquad (7.17)$$

From equation (7.17) it is easy to solve $D(4; 2)$ as follows:

$$D(4; 2) = \frac{1}{7}[4D(3; 2) + 6D(2; 2) + 4T_s + 8T_c + T_i]$$

$$= 4T_s + \frac{100}{21}T_c + \frac{37}{21}T_i$$

$$\approx 4T_s + 4.7619T_c + 1.7619T_i. \qquad (7.18)$$

7.2.3.4   The case of $D(5; 2)$

In this case, each of the five tags can choose any one of S_1 and S_2, producing 32 possibilities and each with the probability of $\frac{1}{32}$, as shown in Figure 7.5.

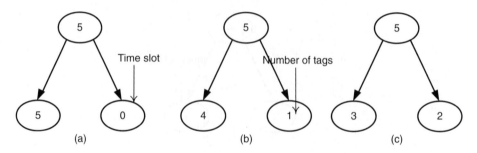

Figure 7.5 An illustration of calculating the total transmission time for TS algorithm for the case of $D(5; 2)$.

In Figure 7.5, pattern (a) has two possibilities, pattern (b) has $\binom{5}{1} \times 2$ possibilities and pattern (c) has $\binom{5}{2} \times 2$ possibilities. Therefore, we have

$$D(5; 2) = \frac{1}{32}[D(5; 2) + T_i] \times 2 + \frac{1}{32}[D(4; 2) + T_s] \times \binom{5}{1} \times 2$$

$$+ \frac{1}{32}[D(3; 2) + D(2; 2)] \times \binom{5}{2} \times 2 + T_c$$

$$= \frac{1}{16}[D(5; 2) + T_i] + \frac{5}{16}[D(4; 2) + T_s]$$

$$+ \frac{10}{16}[D(3; 2) + D(2; 2)] + T_c. \qquad (7.19)$$

From equation (7.19) it is easy to solve $D(5; 2)$ as follows:

$$D(5; 2) = \frac{1}{15}[5D(4; 2) + 10D(3; 2) + 10D(2; 2) + 5T_s + 16T_c + T_i]$$

$$= 5T_s + \frac{652}{105}T_c + \frac{232}{105}T_i$$

$$\approx 5T_s + 6.2095T_c + 2.2095T_i. \qquad (7.20)$$

7.2.3.5 The case of $D(N; 2)$

In this case, each of the N tags can choose any one of S_1 and S_2, producing 2^N possibilities and each with the probability of $\frac{1}{2^N}$. Define

$$N_2 = \begin{cases} \frac{N}{2} & \text{when } N \text{ is even,} \\ \frac{N-1}{2} & \text{when } N \text{ is odd.} \end{cases} \qquad (7.21)$$

Then we have

$$D(N; 2) = \frac{1}{2^N}[D(N; 2) + T_i] \times 2 + \frac{1}{2^N}[D(N-1; 2) + T_s] \times \binom{N}{1} \times 2$$

$$+\frac{1}{2^N}\sum_{k=2}^{N_2-1}\left\{[D(N-k;2)+D(k;2)]\times\binom{N}{k}\times2\right\}$$

$$+\frac{1}{2^N}[D(N_2;2)+D(N_2;2)]\times\binom{N}{N_2}+T_c, \tag{7.22}$$

when $N(\geq6)$ is even; and

$$D(N;2)=\frac{1}{2^N}[D(N;2)+T_i]\times2+\frac{1}{2^N}[D(N-1;2)+T_s]\times\binom{N}{1}\times2$$

$$+\frac{1}{2^N}\sum_{k=2}^{N_2}\left\{[D(N-k;2)+D(k;2)]\times\binom{N}{k}\times2\right\}+T_c, \tag{7.23}$$

when $N(\geq5)$ is odd.

From equations (7.22) and (7.23) we can solve $D(N;2)$ recursively as follows:

$$D(N;2)=\frac{1}{2^{N-1}-1}\left\{ND(N-1;2)+\sum_{k=2}^{N_2-1}\binom{N}{k}[D(N-k;2)+D(k;2)]\right.$$

$$\left.+\binom{N}{N_2}D(N_2;2)+NT_s+2^{N-1}T_c+T_i\right\} \tag{7.24}$$

when N is even; and

$$D(N;2)=\frac{1}{2^{N-1}-1}\left\{ND(N-1;2)+\sum_{k=2}^{N_2}\binom{N}{k}[D(N-k;2)+D(k;2)]\right.$$

$$+NT_s+2^{N-1}T_c+T_i\} \tag{7.25}$$

when N is odd.

7.2.3.6 The case of $D(2;3)$

In this case, each of the two tags can choose any one of S_1, S_2 and S_3, producing nine possibilities and each with the probability of $\frac{1}{9}$, as shown in Figure 7.6.

In Figure 7.6, pattern (a) has three possibilities, corresponding to positioning the two tags in three different time slots, and pattern (b) has $\binom{2}{1}\times\frac{3!}{2}$ possibilities, where the coefficient $\binom{2}{1}$ means that the number of possibilities for the two tags to choose time slot S_1 is $\binom{2}{1}$, the coefficient 3! is due to the permutation of the same pattern in three different time slots and the factor $\frac{1}{2}$ in $\frac{3!}{2}$ is due to the repetition of some patterns in the permutation. Therefore, we have

$$D(2;3)=\frac{1}{9}[D(2;3)+2T_i]\times3+\frac{1}{9}[2T_s+T_i]\times6+T_c. \tag{7.26}$$

From equation (7.26) it is easy to solve $D(2;3)$ as follows:

$$D(2;3)=2T_s+\frac{3}{2}T_c+2T_i. \tag{7.27}$$

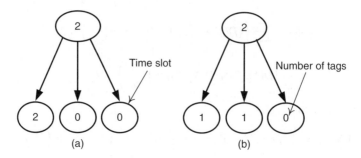

Figure 7.6 An illustration of calculating the total transmission time for TS algorithm for the case of $D(2;3)$.

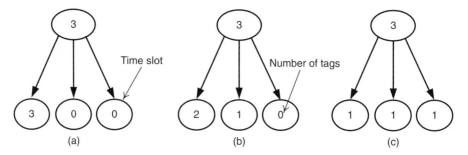

Figure 7.7 An illustration of calculating the total transmission time for TS algorithm for the case of $D(3;3)$.

7.2.3.7 The case of $D(3;3)$

In this case, each of the three tags can choose any one of S_1, S_2 and S_3, producing 27 possibilities and each with the probability of $\frac{1}{27}$, as shown in Figure 7.7.

In Figure 7.7, pattern (a) has three possibilities, pattern (b) has $\binom{3}{2} \times 3!$ possibilities, where the coefficient $\binom{3}{2}$ means that there are $\binom{3}{2}$ ways for the three tags to choose a specific time slot that holds two tags and the coefficient 3! is due to the fact that there are 3! ways (via permutation) for the specific time slots to have 2 tags, 1 tag and 0 tag, respectively, and similarly pattern (c) has 3! (via permutation) possibilities. Therefore, we have

$$D(3;3) = \frac{1}{27}[D(3;3) + 2T_i] \times 3 + \frac{1}{27}[D(2;3) + T_s + T_i] \times \binom{3}{1} \times 3!$$

$$+ \frac{1}{27}[T_s + T_s + T_s] \times 3! + T_c. \tag{7.28}$$

From equation (7.28) it is easy to solve $D(3;3)$ as follows:

$$D(3;3) = \frac{3}{4}D(2;3) + \frac{3}{2}T_s + \frac{9}{8}T_c + T_i = 3T_s + \frac{9}{4}T_c + \frac{5}{2}T_i. \tag{7.29}$$

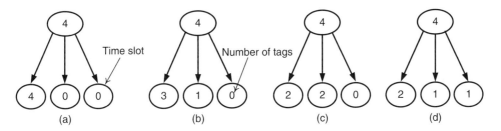

Figure 7.8 An illustration of calculating the total transmission time for TS algorithm for the case of $D(4; 3)$.

7.2.3.8 The case of $D(4; 3)$

In this case, each of the four tags can choose any one of S_1, S_2 and S_3, producing 81 possibilities and each with the probability of $\frac{1}{81}$, as shown in Figure 7.8.

In Figure 7.8, patterns (a), (b), (c) and (d) have 3, $\binom{4}{1} \times 3!$, $\binom{4}{2} \times \frac{3!}{2}$ and $\binom{4}{2} \times \binom{2}{1} \times \frac{3!}{2}$ possibilities, respectively. Therefore, we have

$$D(4; 3) = \frac{1}{81}[D(4; 3) + 2T_i] \times 3 + \frac{1}{81}[D(3; 3) + T_s + T_i] \times \binom{4}{1} \times 3!$$

$$+ \frac{1}{81}[D(2; 3) + D(2; 3) + T_i] \times \binom{4}{2} \times \frac{3!}{2}$$

$$+ \frac{1}{81}[D(2; 3) + T_s + T_s] \times \binom{4}{2} \times \binom{2}{1} \times \frac{3!}{2} + T_c, \qquad (7.30)$$

where in the third and fourth terms of the RHS of equation (7.30), the coefficient $\frac{3!}{2}$ is due to the fact that there are 3! ways (via permutation) for the specific time slots to have 2 tags, 2 tags and 0 tag respectively (or 2 tags, 1 tag and 1 tag, respectively), but half of the patterns are repeated. From equation (7.30) we have

$$D(4; 3) = \frac{1}{27}[D(4; 3) + 2T_i] + \frac{8}{27}[D(3; 3) + T_s + T_i]$$

$$+ \frac{6}{27}[2D(2; 3) + T_i] + \frac{12}{27}[D(2; 3) + 2T_s] + T_c,$$

which further gives

$$D(4; 3) = \frac{1}{26}[8D(3; 3) + 24D(2; 3) + 32T_s + 27T_c + 16T_i]$$

$$= 4T_s + \frac{81}{26}T_c + \frac{42}{13}T_i$$

$$\approx 4T_s + 3.1154T_c + 3.2308T_i. \qquad (7.31)$$

7.2.3.9 The case of $D(5; 3)$

In this case, each of the five tags can choose any one of S_1, S_2 and S_3, producing $3^5 = 243$ possibilities and each with the probability of $\frac{1}{243}$, as shown in Figure 7.9.

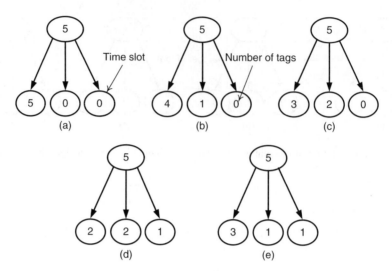

Figure 7.9 An illustration of calculating the total transmission time for TS algorithm for the case of $D(5;3)$.

In Figure 7.9, patterns (a), (b), (c), (d) and (e) have 3, $\binom{5}{4} \times 3!$, $\binom{5}{3} \times 3!$, $\binom{5}{2} \times \binom{3}{2} \times \frac{3!}{2}$ and $\binom{5}{3} \times \binom{2}{1} \times \frac{3!}{2}$ possibilities, respectively. Therefore, we have

$$D(5;3) = \frac{1}{3^5}[D(5;3) + 2T_i] \times 3 + \frac{1}{3^5}[D(4;3) + T_s + T_i] \times \binom{5}{4} \times 3!$$

$$+ \frac{1}{3^5}[D(3;3) + D(2;3) + T_i] \times \binom{5}{3} \times 3!$$

$$+ \frac{1}{3^5}[D(2;3) + D(2;3) + T_s] \times \binom{5}{2} \times \binom{3}{2} \times \frac{3!}{2}$$

$$+ \frac{1}{3^5}[D(3;3) + T_s + T_s] \times \binom{5}{3} \times \binom{2}{1} \times \frac{3!}{2} + T_c$$

$$= \frac{1}{3^4}[D(5;3) + 2T_i] + \frac{10}{3^4}[D(4;3) + T_s + T_i] + \frac{20}{3^4}[D(3;3) + D(2;3) + T_i]$$

$$+ \frac{30}{3^4}[2D(2;3) + T_s] + \frac{20}{3^4}[D(3;3) + 2T_s] + T_c. \tag{7.32}$$

From equation (7.32) it is easy to solve $D(5;3)$ as follows:

$$D(5;3) = \frac{1}{80}[10D(4;3) + 40D(3;3) + 80D(2;3) + 80T_s + 81T_c + 32T_i]$$

$$= 5T_s + \frac{1047}{260}T_c + \frac{527}{130}T_i$$

$$\approx 5T_s + 4.0269T_c + 4.0538T_i. \tag{7.33}$$

7.2.3.10 The case of $D(N; 3)$

In this case, each of the N tags can choose any one of S_1, S_2 and S_3, producing 3^N possibilities and each with the probability of $\frac{1}{3^N}$. Then we have

$$D(N; 3) = \frac{1}{3^N} \sum_{k_1=0}^{N} \sum_{k_2=0}^{N-k_1} \binom{N}{k_1, k_2, k_3} [D(k_1; 3) + D(k_2; 3) + D(k_3; 3)] + T_c$$

$$= \frac{1}{3^N} \sum_{k_1=0}^{N} \sum_{k_2=0}^{N-k_1} \binom{N}{k_1} \binom{N-k_1}{k_2} [D(k_1; 3) + D(k_2; 3)$$

$$+ D(N - k_1 - k_2; 3)] + T_c, \tag{7.34}$$

where $\binom{N}{k_1, k_2, k_3}$ denotes multinomial coefficient with $k_3 = N - k_1 - k_2$. Expanding equation (7.34) yields

$D(N; 3)$

$$= \frac{1}{3^N} \{ D(N; 3) + 2D(0; 3) + \sum_{k_1=0}^{N-1} \sum_{k_2=0}^{N-k_1} \binom{N}{k_1} \binom{N-k_1}{k_2} [D(k_1; 3)$$

$$+ D(k_2; 3) + D(N - k_1 - k_2; 3)] \} + T_c$$

$$= \frac{1}{3^N} \left\{ D(N; 3) + 2D(0; 3) + \sum_{k_1=0}^{N-1} \left[\binom{N}{k_1} [D(k_1; 3) + D(0; 3) + D(N - k_1; 3)] \right. \right.$$

$$\left. \left. + \sum_{k_2=1}^{N-k_1} \binom{N}{k_1} \binom{N-k_1}{k_2} [D(k_1; 3) + D(k_2; 3) + D(N - k_1 - k_2; 3)] \right] \right\} + T_c$$

$$= \frac{1}{3^N} \{ D(N; 3) + 2D(0; 3) + 2D(0; 3) + D(N; 3)$$

$$+ \sum_{k_2=1}^{N} \binom{N}{0} \binom{N}{k_2} [D(0; 3) + D(k_2; 3) + D(N - k_2; 3)]$$

$$+ \sum_{k_1=1}^{N-1} \left[\binom{N}{k_1} [D(k_1; 3) + D(0; 3) + D(N - k_1; 3)] \right.$$

$$\left. + \sum_{k_2=1}^{N-k_1} \binom{N}{k_1} \binom{N-k_1}{k_2} [D(k_1; 3) + D(k_2; 3) + D(N - k_1 - k_2; 3)] \right] \right\} + T_c$$

$$= \frac{1}{3^N} \{ D(N; 3) + 2D(0; 3) + 2D(0; 3) + D(N; 3) + 2D(0; 3) + D(N; 3)$$

$$+ \sum_{k_2=1}^{N-1} \binom{N}{k_2} [D(0; 3) + D(k_2; 3) + D(N - k_2; 3)]$$

$$+ \sum_{k_1=1}^{N-1} \left[\binom{N}{k_1} [D(k_1;3) + D(0;3) + D(N-k_1;3)] \right.$$

$$+ \sum_{k_2=1}^{N-k_1} \binom{N}{k_1} \binom{N-k_1}{k_2} [D(k_1;3)$$

$$\left. + D(k_2;3) + D(N-k_1-k_2;3)]]\} + T_c. \tag{7.35}$$

Notice that in the derivation of equation (7.35) we have used the fact that $D(0;3) = T_i$. Rearranging equation (7.35), we have

$$D(N;3) = \frac{1}{3^N} \left\{ 3D(N;3) + 6T_i + 2\sum_{k=1}^{N-1} \binom{N}{k} [D(k;3) + D(N-k;3) + T_i] \right.$$

$$+ \sum_{k_1=1}^{N-1}\sum_{k_2=1}^{N-k_1} \binom{N}{k_1} \binom{N-k_1}{k_2} [D(k_1;3) + D(k_2;3)$$

$$\left. + D(N-k_1-k_2;3)]\} + T_c. \tag{7.36}$$

From equation (7.36) we can solve $D(N;3)$ recursively as follows:

$$D(N;3) = \frac{1}{3^N - 3} \left\{ 2\sum_{k=1}^{N-1} \binom{N}{k} [D(k;3) + D(N-k;3) + T_i] \right.$$

$$+ \sum_{k_1=1}^{N-1}\sum_{k_2=1}^{N-k_1} \binom{N}{k_1} \binom{N-k_1}{k_2} [D(k_1;3) + D(k_2;3)$$

$$\left. + D(N-k_1-k_2;3)] + 3^N T_c + 6T_i \right\}$$

$$= \frac{1}{3^N - 3} \left\{ 2\sum_{k=1}^{N-1} \binom{N}{k} [D(k;3) + D(N-k;3)] \right.$$

$$+ \sum_{k_1=1}^{N-1}\sum_{k_2=1}^{N-k_1} \binom{N}{k_1} \binom{N-k_1}{k_2} [D(k_1;3) + D(k_2;3)$$

$$\left. + D(N-k_1-k_2;3)] + 3^N T_c + (2^{N+1} + 2)T_i \right\}, \tag{7.37}$$

where we have used the fact that

$$\sum_{k=0}^{N} \binom{N}{k} = 2 + \sum_{k=1}^{N-1} \binom{N}{k} = 2^N.$$

A by-product of equations (7.14), (7.16), (7.18), (7.20), (7.24), (7.25), (7.27), (7.29), (7.31), (7.33) and (7.37) is that we can find the numbers of successful transmissions, colliding transmissions and idle transmissions, respectively, which are equal to the coefficients of T_s, T_c and

T_i in the corresponding equations. Using the symbolic operation in Matlab, we can calculate the corresponding numbers from equations (7.24), (7.25) and (7.37) for general N. Substituting these numbers into equation (7.11), we can obtain the transmission efficiency.

The above approach can be extended to a general M-ary TS algorithm by using multinomial coefficient.

7.2.4   Numerical Results

In this subsection, some numerical and simulation results are presented to show the effectiveness of the analytical results in the preceding subsections and the performance of the TS algorithms.

Figure 7.10 shows the mean identification delay $\bar{d}_{tr}$ for the cases of binary and trinary TS algorithms versus the number of tags, where (a) is for binary TS and (b) for trinary TS. In Figure 7.10, the curves marked with 'Approach I' are calculated based on equation (7.5), and the curves marked with 'Janssen–Jong approach' are calculated based on equation (7.6). As can be seen from Figure 7.10 the Janssen–Jong approach gives a very good approximation for the mean identification delay, while the mean identification delay calculated according to Approach I is higher than the simulated identification delay. The reason for this phenomenon is discussed in Remark 7.1.

It is also illustrated in Figure 7.10 that: (i) the mean identification delay decreases with M (actually, inversely proportional with $\log M$) and (ii) the mean identification delay increases with N almost logarithmically. The quantitative relationship between $\bar{d}_{tr}$ and M or N is more clearly shown in the approximate equation (7.6).

Figure 7.11 shows the transmission efficiency for the cases of binary and trinary TS algorithms versus the number of tags, where (a) is for binary TS and (b) for trinary TS. In Figure 7.11, the curves marked with 'Approach I' are calculated based on the results in Subsection 7.2.2, and the curves marked with 'Approach II' are calculated based on the results in Subsection 7.2.2. As can be seen from Figure 7.11, Approach II gives a very good match to the simulation results for the transmission efficiency, and Approach I also provides a good approximation for the transmission efficiency when N is not too small, for example, when $N \geq 3$ for the cases studied. The reason for this phenomenon is that, when N is too small, the number of idle slots given by equation (7.10) is incorrect. For example, when $N = 1$, clearly $N_{col} = 0$, and thus we have $N_{idle} = -1$ according to (7.10); when $N = 2$ and $M = 2$, we have $N_{col} = 2$ and further $N_{idle} = 0$ according to (7.10), which clearly does not agree with the practical result. Therefore, Approach I applies to the case of large N. Fortunately, N needs not to be too large for Approach I to be valid, as shown in Figure 7.11.

As illustrated in Figure 7.11, for a fixed M, the transmission efficiency approaches to a 'stable' value with the increase of N. For the case of $M = 2$, this 'stable' value is about $E_{TS} \approx 0.35$. For the case of $M = 3$, this 'stable' value is about $E_{TS} \approx 0.37$. However, this 'stable' value is not strictly stable. E_{TS} actually oscillates around this stable value when N is large, even though the magnitude of the oscillation is very small. This oscillation phenomenon is essentially caused by the very nature of tree splitting. Imagine the situation that a deterministic TS algorithm is used, i.e., all the tags are arranged to transmit their IDs sequentially in a frame. Then it is easy to see that the number of idle slots will fluctuate, with N, from zero to $M - 1$ if the deterministic TS algorithm is optimally designed. Therefore, the transmission efficiency will also fluctuate with N around some value.

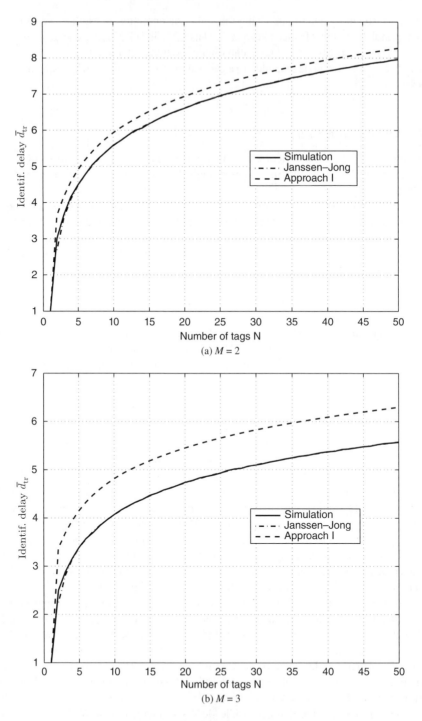

Figure 7.10   The mean identification delay $\bar{d}_{tr}$ for the cases of binary and trinary TS algorithms versus the number of tags. $F_{idle} = 1$.

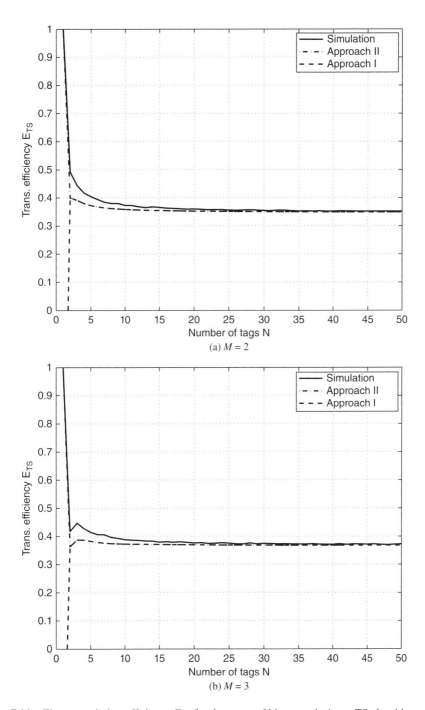

Figure 7.11   The transmission efficiency E_{TS} for the cases of binary and trinary TS algorithms versus the number of tags. $F_{idle} = 1$.

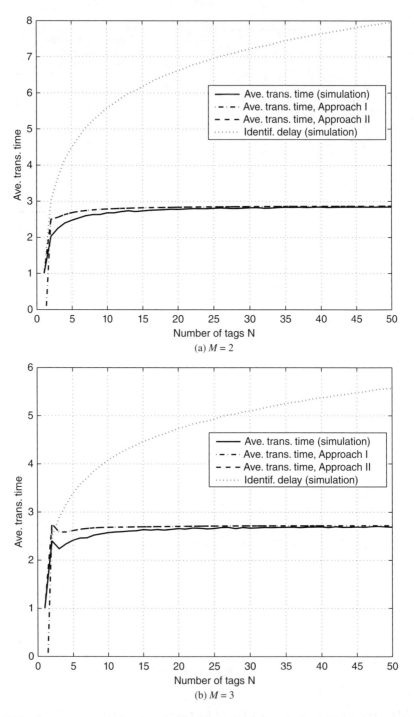

Figure 7.12 Average transmission time for the cases of binary and trinary TS algorithms versus the number of tags. $F_{idle} = 1$.

Figure 7.12 shows the average transmission time (ATT) for the cases of binary and tri-nary TS algorithms versus the number of tags, where (a) is for binary TS and (b) for tri-nary TS. The ATT is defined as the total transmission time, i.e. the sum of time spent in the idle, colliding, and successful transmissions, in a frame divided by the number of tags. In Figure 7.12, the meaning of the legends of 'Approach I' and 'Approach II' is the same as that in Figure 7.11.

As can be seen from Figure 7.12, ATT approaches to a 'stable' value with the increase of N. For the case of $M = 2$, this 'stable' value is about ATT ≈ 2.84. For the case of $M = 3$, this 'stable' value is about ATT ≈ 2.69. However, this 'stable' value is not strictly stable either. The ATT actually oscillates around this stable value when N is large due to the same reason as that for E_{TS}.

In Figure 7.12, the mean identification delay (per tag) is also plotted as a reference. It is seen that ATT is far less than, instead of equal to, the mean identification delay. This is due to the fact that the number of successful transmissions is a highly nonlinear function of the layer (a high layer corresponds to a longer mean identification delay). When N is large, in the first several layers, very few tags can successfully transmit due to collisions and in the higher layers, many tags can successfully transmit because of many time slots available to allocate tags separately.

In the aforementioned numerical results, the transmission time for an idle slot is chosen to be the same as that for a colliding/successful slot, i.e., $F_{idle} = 1$. In practical RFID systems, the former might be a little fraction of the latter, i.e., F_{idle} could be much less than one. In this case, we can choose a larger M to avoid too many colliding transmissions in a frame, so that the transmission efficiency can be improved. Figures 7.13–7.16(a) show how the transmission effi-ciency changes with M and F_{idle} for some fixed N, and Figures 7.13–7.16(b) show the optimal M, denoted by M_{opt}, that maximizes E_{TS} for a corresponding F_{idle}. From Figures 7.13–7.16(a) we can see that the transmission efficiency can be doubled if an optimal M is chosen, compared to the worst M.

Figures 7.13–7.16(b) show that M_{opt} is quite robust against the change in the number of tags N when N is larger than some value (say when $N > 10$). For example, comparing Figure 7.15(b) and Figure 7.16(b) for the cases of $N = 100$ and $N = 1000$, respectively, it is seen that almost the same $M_{opt} \sim F_{idle}$ relationship can be obtained. This observation is also confirmed by the result as shown in Figure 7.11, i.e., the transmission efficiency approaches quickly to a stable value with the increase of N. Therefore, we illustrate the relationship $M_{opt} \sim F_{idle}$ in Table 7.1 only for the case of $N = 100$.

Comparing Figure 7.14(b) and Figure 7.15(b) it can be seen that, if we apply the result of Table 7.1 even to the case of $N = 10$, the performance in the transmission efficiency of the TS algorithm can be kept roughly optimal. From Figure 7.13, where $N = 3$, we see that the optimal M behaves fairly differently to the case of large N, but when N is small, transmission efficiency is not an issue.

Figures 7.13–7.16 show an interesting phenomenon, i.e., the binary TS algorithm works inefficiently for a wide range of N and F_{idle}, while the trinary TS algorithm works efficiently when F_{idle} is larger than 0.5 and N is larger than 10. Therefore, if the optimal M policy is difficult to implement in practical RFID systems, it is recommended to use the trinary TS algorithm.

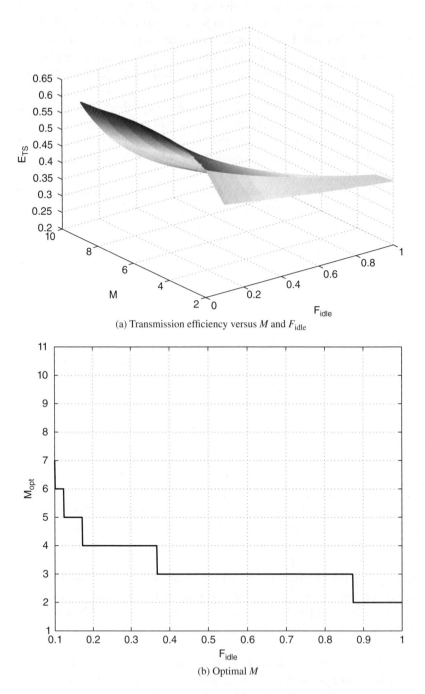

(a) Transmission efficiency versus M and F_{idle}

(b) Optimal M

Figure 7.13   Transmission efficiency E_{TS} versus M and F_{idle}, and optimal M, where N is fixed at $N = 3$.

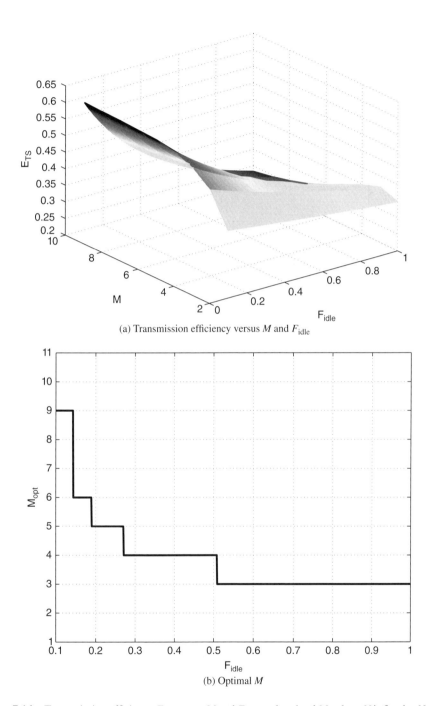

(a) Transmission efficiency versus M and F_{idle}

(b) Optimal M

Figure 7.14 Transmission efficiency E_{TS} versus M and F_{idle}, and optimal M, where N is fixed at $N = 10$.

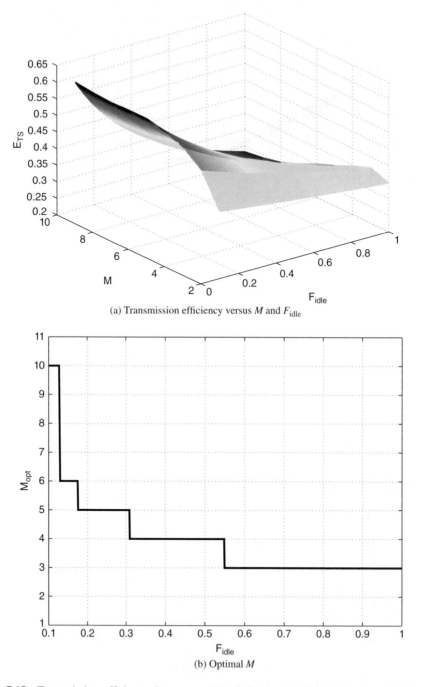

(a) Transmission efficiency versus M and F_{idle}

(b) Optimal M

Figure 7.15 Transmission efficiency E_{TS} versus M and F_{idle}, and optimal M, where N is fixed at $N = 100$.

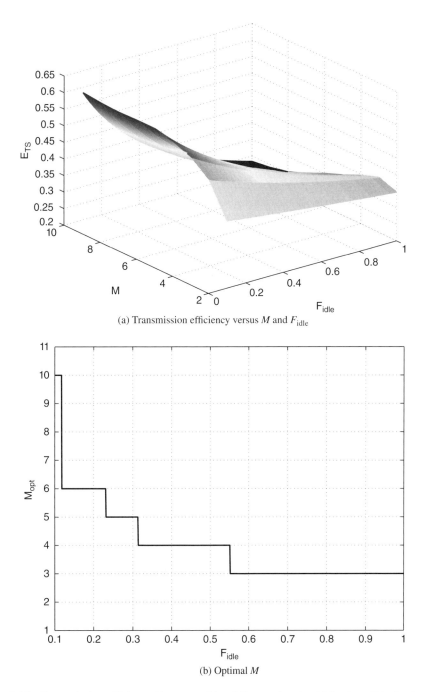

(a) Transmission efficiency versus M and F_{idle}

(b) Optimal M

Figure 7.16 Transmission efficiency E_{TS} versus M and F_{idle}, and optimal M, where N is fixed to be at $N = 1000$.

Table 7.1 The relationship between M_{opt} and F_{idle} for a fixed $N = 100$.

$F_{idle} \in$	M_{opt}
[0.100 0.127]	10
(0.127 0.129)	9,8,7 (linearly decreasing)
[0.129 0.175]	6
[0.176 0.309]	5
[0.310 0.548]	4
[0.549 1.000]	3

7.2.5 Variants of TS Algorithms

One of variants of TS algorithms, which is very easy to be implemented, is the so-called query tree (QT) protocol [28, 34, 47, 49]. In QT, a reader first broadcasts a bit string of a specified length, denoted by s. If the prefix of the ID of a tag matches with s, the tag will send back the remaining bits of its ID. Other tags stay silent. If only one tag responds at a time slot, the tag is identified successfully. If multiple tags respond simultaneously, the transmissions collide and in this case, the reader appends bit '0' or '1' to s, forming a new bit string s0 or s1. The above process is repeated with this new bit string s0 or s1. In this way, the colliding tags are divided into two subgroups, and the reader keeps track of the query strings needed to broadcast with the help of a stack. Finally all the tags can be identified.

An advantage of QT is that tags do not have to remember their inquiring history. Therefore, tags are not required to be equipped with additional writable on-chip memory. A disadvantage of QT is that the inventory delay is affected by the distribution and length of tag IDs. One can easily see this point by examining an extreme case: two tags with only the last digits in their IDs being different. A similar situation often happens when the same kind of products, i.e. the products produced by the same production line, are to be inventoried. In this case, the IDs of the products differ often only in the last several digits.

Another variant of TS algorithms is bit-by-bit binary tree (BBT) protocol [10, 47]. It differs from QT in that, once a tag receives the inventory request from the reader which broadcasts a bit string, the tag with a matching prefix in its ID will send back the next bit in its ID, instead of all the remaining bits of its ID as in QT, to the reader. All the other process keeps the same as that in QT.

One problem with the TS algorithm is how to deal with the new tag arrivals that come in while a collision is being resolved.

7.3 Aloha-Based Algorithm

The pure Aloha protocol was developed in the late 1960s and early 1970s at the University of Hawaii [5, 6]. The basic idea of an Aloha network is simple: If a node or client has data to transmit, let it transmit; if the data is received correctly at the hub, a short acknowledgement packet will be sent to the client; if a collision happens (without receiving an acknowledgement by the client after a short waiting time), it will wait for a random amount of time and retransmit the data again. Note that in pure Aloha, the starting time of data transmission can be at any moment of time, i.e., both the waiting time and transmission time are continuous variables.

In 1975, Roberts [38] published a method, called slotted Aloha, where the time is divided into discrete time intervals, called time slots. Typically, the length of a time slot is set equal to the time needed to transmit a data packet. A client can only transmit at the beginning of a time slot and if a collision happens, the client will wait for an integer but random multiples of the interval of a time slot to retransmit. Slotted Aloha doubles the throughput of a pure Aloha network [38, 41]. The maximal throughput of a slotted Aloha network is $1/e \approx 0.368$.

Soon after Aloha was invented, its property study attracted a great attention of both practice engineers and academia in data networks and it was soon found that both Aloha and slotted Aloha are fundamentally unstable [16, 24, 42]. Many new variants of random-access protocols, such as carrier sense multiple access with collision avoidance (CSMA/CA), popularly used in IEEE 802.11 LANs [1, 2, 3, 4], and multipacket-reception-based Aloha [36] etc. have been developed based on the principle of slotted Aloha.

In this chapter, we consider a framed Aloha scheme. In a framed Aloha, one frame consists of several, say L, time slots, and a tag randomly chooses a time slot among these L time slots to transmit. Once a collision happens, it will again randomly choose a time slot and wait in the next frame (which is also called next layer in the sequel analysis) to transmit. The major difference between framed Aloha and standard slotted Aloha is that, in the former scheme, a tag transmits its tag ID once per frame. There are several reasons for RFID to adopt framed Aloha [46]. For example, it is convenient for a reader to send acknowledgmentes periodically (once per frame) rather than after every slot. Secondly, the use of a frame structure imposes a constraint on retransmission probability, which may be useful in managing the queue length of incoming tags.

In a framed Aloha algorithm, we say that the tags are located in the same *layer* if they have experienced the same number of contentions to transmit.

A by-product of the framed Aloha is that it is easy for the reader to calculate the number of colliding slots, which will enable many adaptive random-access protocols. As will be discussed next, we propose two simple adaptive Aloha algorithms, which can greatly reduce the tag's mean identification delay or inventory time.

In a framed Aloha scheme, the frame length may be either fixed or variable depending on the particular system implementation. The former is called static framed Aloha, while the latter is called dynamic framed Aloha. To easy analysis, we first consider static framed Aloha, and then we point out several other variants of the dynamic framed Aloha and propose two adaptive Aloha schemes. In the proposed algorithms, the collision rate (to be defined later) in a frame is used to update the frame size in real time.

In Subsections 7.3.1–7.3.3, it is assumed that the frame size, i.e. the number of time slots in one frame, is L.

7.3.1 Mean Identification Delay

We use Assumption 7.1 to derive the mean identification delay.

At the kth layer, there are L time slots in total, numbered $1, 2, \cdots, L$, for tags to transmit. As in Section 7.2.1, we organize all the tags in an ordered way and let T_n be the slot number that the nth tag picks at the concerned layer.

Similar to the TS algorithm, we use d_{tr} to denote the identification delay of a tag, that is, after d_{tr} transmit contention, the tag's ID is finally successfully delivered to the reader.

Without loss of generality, we can take the first tag as an example for consideration. Let us now calculate the probability distribution of d_{tr}. First we calculate the probability of $\{\text{Tr}(T_1, k) = \text{succ}\}$, where $\{\text{Tr}(T_1, k) = \text{succ}\}$ stands for the event that the transmission of the first tag (represented by T_1) at the kth layer is successful. Similarly, we use $\{\text{Tr}(T_1, k) = \text{unsucc}\}$ to denote the event that the transmission of the first tag at the kth layer is unsuccessful. Following the same procedure as that in equations (7.3) and (7.4), respectively, we have

$$\bar{P}_{\text{suc},k} := P\{\text{Tr}(T_1, k) = \text{succ}\}$$
$$= P\{\{T_1 = 1\} \cap \{T_2 \neq 1\} \cap \{T_3 \neq 1\} \cap \cdots \cap \{T_N \neq 1\}\}$$
$$+ P\{\{T_1 = 2\} \cap \{T_2 \neq 2\} \cap \{T_3 \neq 2\} \cap \cdots \cap \{T_N \neq 2\}\}$$
$$+ \cdots$$
$$+ P\{\{T_1 = L\} \cap \{T_2 \neq L\} \cap \{T_3 \neq L\} \cap \cdots \cap \{T_N \neq L\}\}$$
$$= \left(1 - \frac{1}{L}\right)^{N-1}$$

and

$$P_{d_{tr}}(k) := P\{d_{tr} = k\}$$
$$= P\{\{\text{Tr}(T_1, 1) = \text{unsucc}\} \cap \cdots \cap \{\text{Tr}(T_1, k-1) = \text{unsucc}\} \cap \{\text{Tr}(T_1, k) = \text{succ}\}\}$$
$$= [1 - P\{\text{Tr}(T_1, 1) = \text{succ}\}] \cdots [1 - P\{\text{Tr}(T_1, k-1) = \text{succ}\}]P\{\text{Tr}(T_1, k) = \text{succ}\}$$
$$= [1 - \bar{P}_{\text{suc},1}][1 - \bar{P}_{\text{suc},2}] \cdots [1 - \bar{P}_{\text{suc},k-1}]\bar{P}_{\text{suc},k}$$
$$= \left[1 - \left(1 - \frac{1}{L}\right)^{N-1}\right]^{k-1} \left(1 - \frac{1}{L}\right)^{N-1}, \quad k \geq 1. \tag{7.38}$$

Based on equation (7.38), the mean identification delay can be obtained as follows:

$$\bar{d}_{tr} = \sum_{k=1}^{\infty} k P_{d_{tr}}(k)$$
$$= \sum_{k=1}^{\infty} k \left[1 - \left(1 - \frac{1}{L}\right)^{N-1}\right]^{k-1} \left(1 - \frac{1}{L}\right)^{N-1}$$
$$= \frac{1}{\left(1 - \frac{1}{L}\right)^{N-1}}, \tag{7.39}$$

where we have used the fact that

$$\sum_{k=1}^{\infty} k(1 - c)^{k-1} = \frac{1}{c^2}$$

for any constant c satisfying $|c| < 1$.

Remark 7.3 In deriving equation (7.39), we have used Assumption 7.1. In practical Aloha, the number of tags will gradually decrease with the number of the layer. Due to this fact, the first term of $P_{d_{tr}}(k)$ in equation (7.38), i.e. $\left[1 - \left(1 - \frac{1}{L}\right)^{N-1}\right]^{k-1}$ – the unsuccessful trans-

mission probability of a tag at the preceding $k-1$ layers, is greater than the corresponding probability in the practical situation, while the second term of $P_{d_{tr}}(k)$ in equation (7.38), i.e. $\left(1 - \frac{1}{L}\right)^{N-1}$ – the successful transmission probability of a tag at the kth layer, is less than the corresponding probability in the practical situation. Therefore, equation (7.39) only provides an approximation to the mean identification delay. When N is significantly greater than L, the first term in (7.38) becomes dominant, and then equation (7.39) gives an upper bound for $\bar{d}_{tr}$. On the other hand, when L is significantly greater than N, the second term in (7.38) becomes dominant and then equation (7.39) provides a lower bound for $\bar{d}_{tr}$.

7.3.2   Collision Analysis and Transmission Efficiency

In a frame with N tags and L time slots, let ξ denote the number of tags being allocated in any a given time slot. Clearly, ξ is a random variable and its distribution is given by

$$P(\xi = n | N, L) = \binom{N}{n} \left(\frac{1}{L}\right)^n \left(1 - \frac{1}{L}\right)^{N-n}, \quad 0 \le n \le N. \tag{7.40}$$

The number of tags in a particular slot is called the occupancy number of that slot [20]. The distribution (7.40) applies to all L time slots. Therefore, the expected value of the number of slots with occupancy number n is given by [20, p. 114]

$$\eta(n | N, L) = L P(\xi = n | N, L) = L \binom{N}{n} \left(\frac{1}{L}\right)^n \left(1 - \frac{1}{L}\right)^{N-n}, \quad 0 \le n \le N. \tag{7.41}$$

Especially, the expected value of the number of idle slots, denoted by $\mu_0(N, L)$, and the expected value of the number of successful transmission slots, denoted by $\mu_1(N, L)$, are given by

$$\mu_0(N, L) = \eta(0 | N, L) = L \binom{N}{0} \left(\frac{1}{L}\right)^0 \left(1 - \frac{1}{L}\right)^N = L \left(1 - \frac{1}{L}\right)^N, \tag{7.42}$$

$$\mu_1(N, L) = \eta(1 | N, L) = L \binom{N}{1} \left(\frac{1}{L}\right)^1 \left(1 - \frac{1}{L}\right)^{N-1} = N \left(1 - \frac{1}{L}\right)^{N-1}. \tag{7.43}$$

In order to further calculate the numbers of idle and successful transmissions, we need to know the probability of *exactly* k successful transmissions in a contention frame with N tags and the frame size being L. We have obtained this probability in Appendix 7.C. Denote this probability as $\alpha(k; N, L)$, i.e., $\alpha(k; N, L) := P(Y = k)$ by using the notation in Appendix 7.C. For convenience of reading, we summarize the results as follows:

$$\alpha(k; N, L) = \binom{L}{k} P_s(k; N, L)[1 - P_S(N - k, L - k)],$$

$$P_s(k; N, L) = \begin{cases} \frac{N!}{(N-k)!} \frac{(L-k)^{N-k}}{L^N} & \text{when } k \le N \text{ and } k \le k_{\max}, \\ 0 & \text{otherwise,} \end{cases}$$

$$P_S(N, L) = N \left(1 - \frac{1}{L}\right)^{N-1} - \binom{L}{2} P_s(2; N, L) + \binom{L}{3} P_s(3; N, L) - \cdots$$
$$+ (-1)^{L-1} P_s(L; N, L),$$

where

$$k_{\max} = \begin{cases} L-1 & \text{when } N > L, \\ L & \text{when } N \leq L. \end{cases}$$

Let $D(n; L)$ be the expectation of the total transmission time of n tags in a framed Aloha scheme with the frame size being L when all the n tags are finally successfully identified. Clearly we have $D(1; L) = T_s$ and $D(0; L) = 0$ for any $L \geq 1$. Then we have

$$D(N; L) = \mu_0(N, L)T_i + [L - \mu_0(N, L) - \mu_1(N, L)]T_c$$

$$+ \alpha(0; N, L)D(N; L) + \sum_{k=1}^{N} \alpha(k; N, L)[kT_s + D(N - k; L)]. \qquad (7.44)$$

From equation (7.44) we can solve

$$D(N; L) = \frac{1}{1 - \alpha(0; N, L)} \{ \mu_0(N, L)T_i + [L - \mu_0(N, L) - \mu_1(N, L)]T_c$$

$$+ \sum_{k=1}^{N} \alpha(k; N, L) \left[kT_s + D(N - k; L) \right] \}. \qquad (7.45)$$

From equation (7.45), we can find the numbers of successful transmissions, colliding transmissions, and idle transmissions, respectively, which are equal to the coefficients of T_s, T_c and T_i in (7.45). These coefficients can be calculated using symbolic operation in Matlab. Substituting these numbers into equation (7.11) we can obtain the transmission efficiency. Note that the transmission efficiency for Aloha is also defined by the right hand side of equation (7.11). Here we denote it as E_{Aloha}.

7.3.3 Numerical Results

In this subsection, some numerical and simulation results are presented to show the performance of the discussed Aloha algorithm, which will motivate us to consider some other improved Aloha algorithms.

Figure 7.17 shows the mean identification delay $\bar{d}_{tr}$ for different L, where in (a) $L = 2, 4, 8, 16$ and in (b) $L = 32, 64, 128, 256$. The analytical result is obtained from equation (7.39). From this figure it is seen that there is some gap between the analytical result and simulation result. The reason is explained in Remark 7.3. However, the analytical result can well predict the asymptotical behavior of the mean identification delay.

Both simulation and analytical results show that the mean identification delay exponentially increases with the number of tags. For example, consider two cases: case (i): $L = 2$ and $N = 25$, then $\bar{d}_{tr} = 1.4084 \times 10^6$ time slots; case (ii): $L = 4$ and $N = 50$, then $\bar{d}_{tr} = 1.0583 \times 10^5$ time slots. Selecting an 'optimistic' parameter setting of RFID systems from those parameter ranges as presented in Section 2.7.1 of Chapter 2[4]: $T_0 = 6.25$ μs, $T_1 = 1.5T_0 = 9.375$ μs, $T_{\text{RTcal}} = T_0 + T_1 = 15.625$ μs, $T_{\text{TRcal}} = 1.1T_{\text{RTcal}} = 17.1875$ μs and DR = 64/3, this choice yields a data rate DR/$T_{\text{TRcal}} = 1.242$ Mb/s. Then a time slot needs about $96/(1.242 \times 10^6) = 77.3 \times 10^{-6}$ s of transmission time. Thus in cases (i) and (ii), each tag needs about 109 s and 8.2 s,

---

[4] The meaning of the symbols T_0, T_1, T_{RTcal}, and T_{TRcal} here is referred to Section 2.7.1 of Chapter 2.

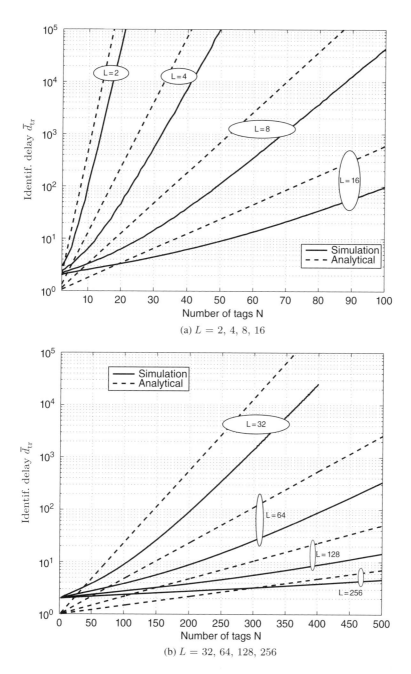

(a) $L = 2, 4, 8, 16$

(b) $L = 32, 64, 128, 256$

Figure 7.17　The mean identification delay $\bar{d}_{\mathrm{tr}}$ of the static Aloha versus the number of tags for different L. $F_{\mathrm{idle}} = 1$.

respectively, of waiting time to be identified. This is too long. If a perfect time-division multiple access could be realized, the mean waiting time would be about $25/2 \times 77.3 \times 10^{-6}$ s $\approx$ 0.97 ms and $50/2 \times 77.3 \times 10^{-6}$ s $\approx$ 1.93 ms. If the number of tags further increases to, say, 100, while the frame size is kept unchanged, the mean identification delay will be prohibitively long.

Figure 7.18 illustrates the transmission efficiency E_{Aloha} versus the number of tags for different frame size L, where in (a) $L = 2, 4, 8, 16$ and in (b) $L = 32, 64, 128, 256$. The analytical results, marked in the solid curves, are obtained via equation (7.45). Due to the fact that the binomial coefficient can be calculated only up to a limited range of N in Matlab, the analytical results are only illustrated in this limited range of N. As can be seen from Figure 7.18 that the theoretical analysis matches with the simulation results very well in aforementioned range of N.

Figure 7.18 shows that, when the frame size L is less than 64, the transmission efficiency is greater than 0.2 only in a very limited range of N. This range depends on L. For example, if $L = 8$, $E_{Aloha} \geq 0.2$ only when $5 \leq N \leq 58$; and if $L = 64$, $E_{Aloha} \geq 0.2$ only when $39 \leq N \leq$ 232. When N becomes further large, the transmission efficiency decreases rapidly to some very low values. This is not satisfactory in practical RFID applications.

From Figure 7.18 we can see that for a fixed L, there exists a corresponding N, denoted by N_{opt}, which makes the transmission efficiency achieves its highest value. It is interesting to find the relationship between N_{opt} and L. Figure 7.19 shows this relationship. It is found that this relationship can be well approximated by a linear function. Through curve fitting, it can be found that

$$N_{opt} \approx 1.9373L - 5.2502. \qquad (7.46)$$

For a comparison, the linear approximation characterized by equation (7.46) is also plotted in Figure 7.19.

Equation (7.46) can be used to adjust the frame size in dynamic framed Aloha protocol, if the number of tags is known by some way. From equation (7.46), one can solve the optimal frame size, denoted by L_{opt}, for a given N as follows:

$$L_{opt} \approx \frac{N + 5.2502}{1.9373}. \qquad (7.47)$$

Notice that equation (7.47) is obtained under the condition that a static Aloha frame is finished, i.e., all the tags are finally identified after several rounds or layers of transmissions by using the same frame-size Aloha. If only one round or layer of contention is considered, then the transmission efficiency is actually reduced to the successful transmission probability in one layer of a frame, which reads $p_1 = \frac{N}{L}\left(1 - \frac{1}{L}\right)^{N-1}$ (see equation (7.60) in Appendix 7.B). By maximizing p_1 with respect to L, we can obtain $L_{opt} = N$. In the literature, it is this kind of relationship between optimal L and N that is often addressed.

7.3.4   Adaptive Frame Size Aloha Algorithms

Both analytical and simulation results show that the mean identification delay of a tag becomes very large even for a moderate number of tags when the frame size is fixed to be small, while the transmission efficiency becomes very low when the number of tags is small (due to idle

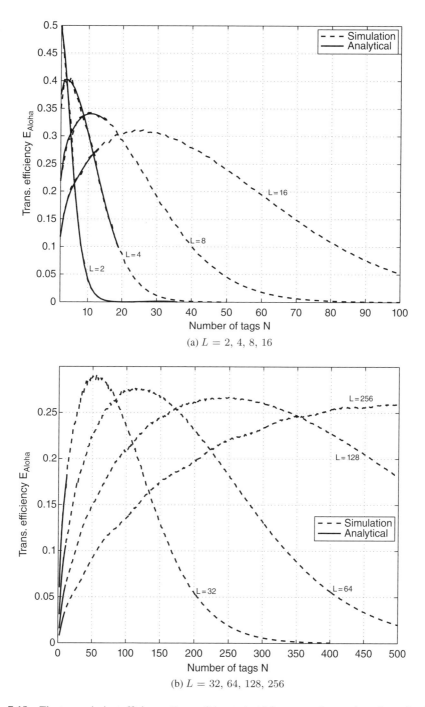

(a) $L = 2, 4, 8, 16$

(b) $L = 32, 64, 128, 256$

Figure 7.18 The transmission efficiency E_{Aloha} of the static Aloha versus the number of tags for different frame size. $F_{\text{idle}} = 1$.

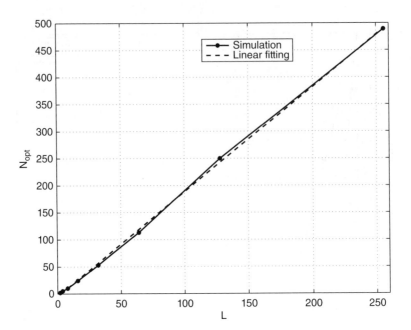

Figure 7.19 The relationship between N_{opt} and L, where $F_{idle} = 1$.

transmissions) or large (due to colliding transmissions) for a fixed frame size. Therefore, it is beneficial if the frame size can be adjusted according to practical situations. The framed Aloha scheme makes it easy to solve this problem.

It is observed from Figure 7.18 and equation (7.47) that, when the number of time slots in a frame matches properly with the number of tags in the interrogation zone, the transmission efficiency approaches to its optimal value. Based on this idea, a dynamic frame slotted Aloha protocol was first proposed in [39] for general data networks and then applied to RFID systems in [11, 25, 26, 29, 44, 48] with different tag-number estimation methods, where the frame size is dynamically adjusted according to the estimated number of tags or stations. Therefore, they yield better performance than the static Aloha protocol. But many rounds of communications are needed to optimize the frame size before the identification process [26]. Note that several methods have been developed to estimate the number of tags. For example, in [22], it is shown that exact number of tags (up to four tags) can be reliably identified by analysing the histograms of the received waveform at the reader during collisions; in [25], a maximum likelihood estimator is used to estimate the number of tags based on partially observed frame contention data; in [14], a Bayesian approach is developed to estimate the number of tags, which is computation-expensive and in [44], a mean-square estimator is proposed to estimate tag number.

One issue for the dynamic frame slotted Aloha protocol is that generally it is very difficult to estimate accurately the number of tags, even though the number of tags can be estimated from the number of colliding transmissions in principle since the three parameters N, L and the *expected* number of colliding transmissions, which can be measured, are related with each other through equations (7.42)–(7.43). However, to measure the *expected* number of colliding transmissions needs many rounds of communications.

Based on the above consideration, we propose two algorithms, called adaptive frame size Aloha 1 (AFSA1) and adaptive frame size Aloha 2 (AFSA2), to solve the aforementioned problem.

Algorithm 7.2   AFSA1

Step 0 Initial setting: given numbers L_0, L_{max} and R_{c0}, where R_{c0} is a pre-defined threshold for the ratio between the number of colliding transmissions and the frame size of the current layer, L_{max} is a pre-defined maximal frame size, and L_0 is the frame size of the initial frame. Typically, L_0 and L_{max} are set to be some powers of two. Set $L = L_0$.

Step 1 Checking frame status: at every frame, the reader calculates the collision rate, denoted by R_c:

$$R_c = \frac{N_c}{L},\tag{7.48}$$

where N_c is the number of colliding transmissions in the *current* layer and L is the frame size of the *current* layer.

Step 2 Adjusting the frame size according to the following law:

$$L := \begin{cases} 2L & \text{if } R_c > R_{c0} \text{ and } L < L_{max}, \\ L & \text{otherwise, i.e., } L \text{ is kept unchanged.} \end{cases}\tag{7.49}$$

Go to the next layer.

Algorithm 7.3   AFSA2

Step 0 Initial setting: given numbers L_0, L_{max} and R_{c0}, where R_{c0}, L_0 and L_{max} are of the same meaning as in AFSA1. Set $L = L_0$.

Step 1 Checking frame status: at every frame, the reader calculates the collision rate R_c according to equation (7.48).

Step 2 Adjusting the frame size according to the following law:

$$L := \begin{cases} 2L & \text{if } R_c > R_{c0} \text{ and } L < L_{max}, \\ L/2 & \text{if } R_c < R_{c0}/2 \text{ and } L > L_0, \\ L & \text{otherwise, i.e., } L \text{ is kept unchanged.} \end{cases}\tag{7.50}$$

Go to the next layer.

The difference between AFSA1 and AFSA2 lies in that the frame size will not be reduced in AFSA1 once it is increased to L_{max}, while in AFSA2 the frame size can be increased or decreased between L_0 and L_{max} depending on the collision rate. Therefore, the number of idle slots is also well controlled in AFSA2 if the parameter R_{c0} is selected properly.

The exponential frame size adjustment method in AFSA1 and AFSA2 is borrowed from the idea of distributed coordination function (DCF) in CSMA/CA algorithm used in IEEE 802.11 [8]. It is also motivated by the fact that the mean identification delay increases exponentially with the number of tags. Therefore, an exponential increase in the frame size might be able to mitigate the effect of mean identification delay increase. Another idea is to increase the frame size by a fixed quantity (say Δ) when $R_c > R_{c0}$. However, this approach cannot adjust the system behavior rapidly.

Another reason for us to choose to double or half the frame size in AFSA1 and AFSA2 in the corresponding cases is that it can be easily implemented in RFID using the Q parameter in RFID protocol [13].

Note that the overhead to implement AFSA1 or AFSA2 in RFID is negligible: the reader needs only to measure the collision rate in the current frame and then to double or half the frame size accordingly. No additional computational burden is needed at the tag.

Intuitively, when the frame size is to be decreased in AFSA2 should be determined according to the idle transmission rate at a layer. Since the collision rate and idle transmission rate can be mutually determined theoretically, we choose to use the collision rate to decide when the frame size should be decreased to ease the implementation.

The transmission efficiency and mean identification delay for AFSA1 and AFSA2 are illustrated in Figures 7.20–7.22, corresponding to different R_{c0}. In Figures 7.20–7.22, we choose $L_0 = 16$, $L_{max} = 256$, and the parameter F_{idle} is fixed to be the unity. Comparing Figures 7.20–7.22(b) and Figure 7.17 we can see that the mean identification delay of a tag is greatly reduced by using AFSA1 or AFSA2 (in several orders of magnitude). For example, when $N = 200$, the mean identification delay $\bar{d}_{tr}$ is around 30 039 time slots by using the static Aloha with fixed $L = 16$ (simulated but not plotted in Figure 7.17(a)), while $\bar{d}_{tr}$ is reduced to $6 \sim 10$ time slots by using AFSA1 or AFSA2 with $R_{c0} = 0.5 \sim 0.8$ (see Figures 7.20–7.21(b)). When $N = 500$, $\bar{d}_{tr}$ is around 335 time slots by using the static Aloha with $L = 64$, while $\bar{d}_{tr}$ is maintained to be $8 \sim 10$ time slots by using AFSA1 and AFSA2 with $R_{c0} = 0.5 \sim 0.8$.

From Figures 7.20–7.21 (a) and Figure 7.18 we see that, in the aforementioned range of N, i.e. $N \leq 500$, AFSA1 and AFSA2 schemes can maintain the transmission efficiency well above 0.2 when R_{c0} is chosen to be $0.5 \sim 0.8$, while the transmission efficiency of the static Aloha decreases far below 0.1 except for a small range of N for a corresponding frame size L. Remarkably, AFSA2 scheme can maintain the transmission efficiency well above 0.28 when R_{c0} is chosen to be $0.5 \sim 0.7$.

It is seen from Figures 7.22 that, when R_{c0} is chosen to be too large, for example, 0.9 or 0.95, both transmission efficiency and mean identification delay exhibit a strong oscillating tendency with respect to N, even though the transmission efficiency is much higher than that of the static Aloha, and the mean identification delay is much lower than that of the static Aloha. The performance of AFSA1 and AFSA2 deteriorates when R_{c0} is chosen to be too large. Therefore, we will focus our attention on the middle range of R_{c0}, for example, when $05 \leq R_{c0} \leq 0.8$.

Next let us investigate the performance of AFSA1 and AFSA2 when the transmission time of an idle slot is less than that of a successful or colliding transmission. Figures 7.23–7.26 illustrate the transmission efficiency of AFSA1 and AFSA2 schemes for $F_{idle} = 0.1, 0.3, 0.5$ and 0.7, respectively. Comparing the results in Figures 7.23–7.26 and Figures 7.20 and 7.21(a) we can obtain the following conclusions. (i) When F_{idle} is small, e.g. when $0.1 \leq F_{idle} \leq 0.3$, AFSA1 and AFSA2 produce almost the same performance. In most cases, AFSA2 yields slightly better performance than AFSA1, but in some other cases, e.g. when $F_{idle} = 0.1$ and $R_{c0} = 0.6, 0.7$ and 0.8 or $F_{idle} = 0.3$ and $R_{c0} = 0.8$, AFSA1 gives a slightly better performance than AFSA2. (ii) When F_{idle} is large, e.g. when $0.5 \leq F_{idle} \leq 1.0$, AFSA2 provides

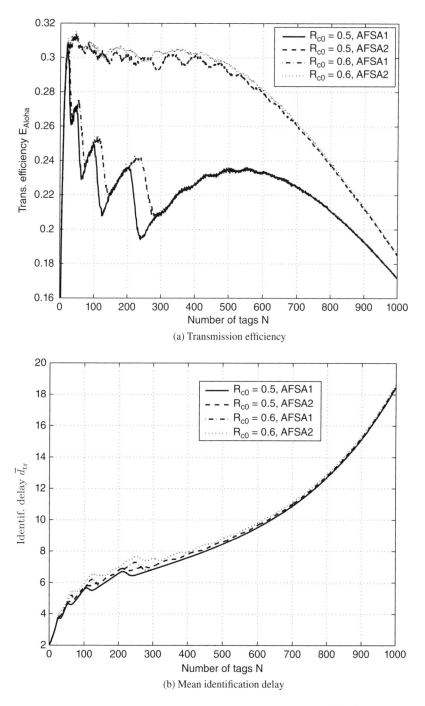

(a) Transmission efficiency

(b) Mean identification delay

Figure 7.20 The transmission efficiency E_{Aloha} and mean identification delay $\bar{d}_{tr}$ of AFSA1 and AFSA2 versus the number of tags. $R_{c0} = 0.5, 0.6$, $F_{idle} = 1$.

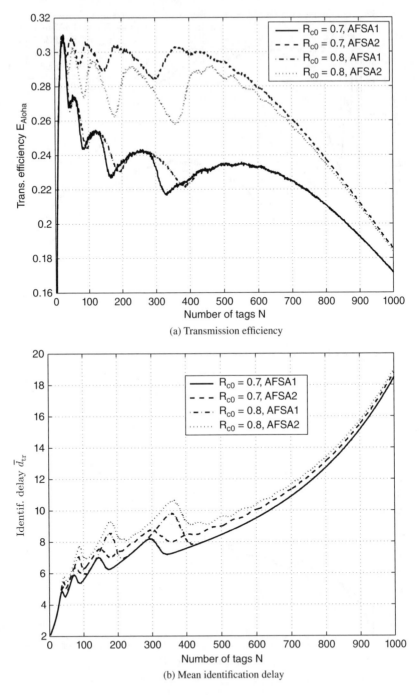

(a) Transmission efficiency

(b) Mean identification delay

Figure 7.21 The transmission efficiency E_{Aloha} and mean identification delay $\bar{d}_{\text{tr}}$ of AFSA1 and AFSA2 versus the number of tags. $R_{c0} = 0.7, 0.8, F_{\text{idle}} = 1$.

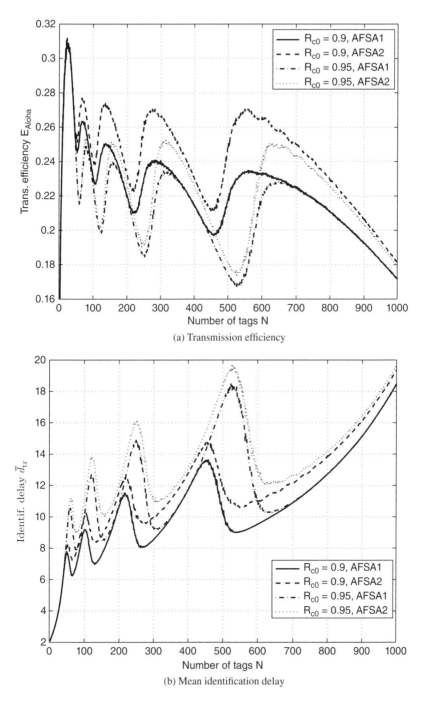

Figure 7.22   The transmission efficiency E_{Aloha} and mean identification delay $\bar{d}_{tr}$ of AFSA1 and AFSA2 versus the number of tags. $R_{c0} = 0.9, 0.95, F_{idle} = 1$.

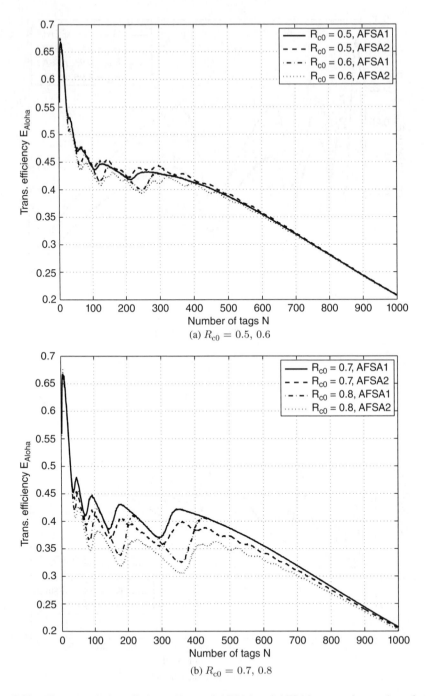

Figure 7.23 The transmission efficiency E_{Aloha} of AFSA1 and AFSA2 versus the number of tags for different R_{c0}, where $F_{\text{idle}} = 0.1$.

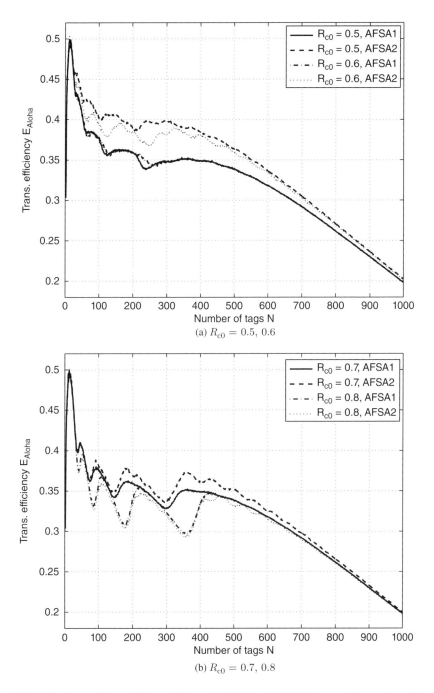

Figure 7.24 The transmission efficiency E_{Aloha} of AFSA1 and AFSA2 versus the number of tags for different R_{c0}, where $F_{idle} = 0.3$.

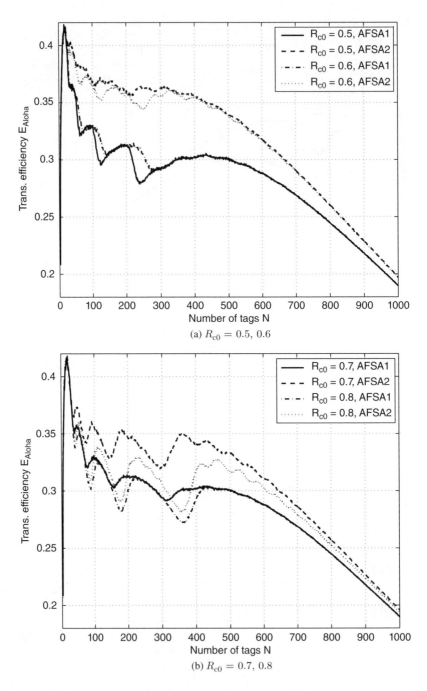

(a) $R_{c0} = 0.5, 0.6$

(b) $R_{c0} = 0.7, 0.8$

Figure 7.25   The transmission efficiency E_{Aloha} of AFSA1 and AFSA2 versus the number of tags for different R_{c0}, where $F_{idle} = 0.5$.

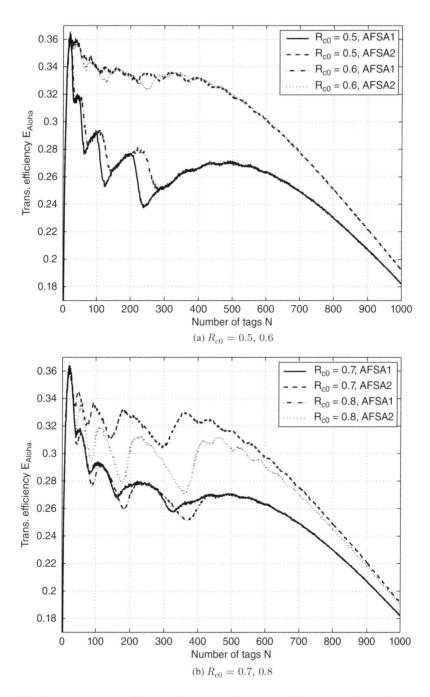

(a) $R_{c0} = 0.5, 0.6$

(b) $R_{c0} = 0.7, 0.8$

Figure 7.26 The transmission efficiency E_{Aloha} of AFSA1 and AFSA2 versus the number of tags for different R_{c0}, where $F_{idle} = 0.7$.

far better performance than AFSA1. Since the computational burden of AFSA2 is negligible compared to that of AFSA1 (and even to the static Aloha), it is recommended to use AFSA2 in general.

An interesting topic for dynamic/adaptive framed Aloha is the in-frame adjustment of the frame size, which means that the the frame size is adjusted by observing only a part of a contention frame if the number of collisions or empty slots exceeds a certain value. This technique is enabled by using the QueryAdjust command in the EPC Global HF Class Gen 2 protocol [13]. The performance of this kind of Aloha was investigated in [14, 25].

Finally, it is worth pointing out that Aloha-based protocols are simple, but they have the tag starvation problem, i.e., a tag may never be successfully identified because of collisions with other tags [47].

7.4   Summary

In this chapter, we have discussed tree-splitting and Aloha based anti-collision algorithms for multi-tag RFID systems. For tree-splitting based anti-collision algorithms, a fair realization approach has been proposed, the analytical result for the mean identification delay has been obtained and an analytical approach for analysing the expected total transmission time has been presented. Using this analytical approach, we can easily calculate the numbers of idle, colliding, and successful transmissions in a tree-splitting algorithm. Based on this result, we can further calculate the transmission efficiency of TS algorithm. It is shown that the simulation results agree with the analytical results very well.

The performance of the transmission efficiency of M-ary TS algorithm is compared according to different M, N (number of tags) and F_{idle} (the ratio between the length of an idle transmission and the length of a successful transmission). Depending on N and F_{idle}, an optimal M exists to achieve maximal transmission efficiency. It is found that the binary TS algorithm works inefficiently for a wide range of N and F_{idle}, while the trinary TS algorithm works efficiently when F_{idle} is larger than 0.5 and N is larger than 10. Therefore, if the optimal M policy is difficult to implement in practical RFID systems, it is recommended to use the trinary TS algorithm.

For Aloha based anti-collision algorithms, a closed-form formula for the mean identification delay has been also obtained and an analytical approach to calculating the expected total transmission time has been presented. Based on this approach, we can calculate the numbers of idle, colliding and successful transmissions and further the transmission efficiency of the static Aloha algorithm. It is shown that the closed-form formula for the mean identification delay can predict well the asymptotical tendency of the mean identification delay and the analytical results for the transmission efficiency agree very accurately with the simulation results.

It is found that the static Aloha yields very poor performance in both mean identification delay and transmission efficiency. Therefore, two adaptive frame size Aloha algorithms, namely AFSA1 and AFSA2, have been proposed. Very light computational burden at the reader is needed: the reader needs only to measure the collision rate in the current frame and then to double or half the frame size accordingly. No additional computational burden is required at the tag.

Simulation results show that AFSA1 and AFSA2 can significantly improve the performance of Aloha in both mean identification delay and transmission efficiency. The mean identification delay of a tag can be reduced to less than 10 time slots by using AFSA1 and AFSA2, compared to several hundreds or even more than tens of thousands of time slots in static Aloha. It is worth noting that AFSA2 scheme can maintain the transmission efficiency well above 0.28 when R_{c0} is chosen to be $0.5 \sim 0.7$ for a wide range of N and F_{idle}. Therefore, it is recommended to use AFSA2 in general.

In AFSA1 and AFSA2, the parameters L_0 and L_{max} are generally determined in correspondence to a system setup, but two other parameters, the collision rate thresholds R_{c0} (a threshold used to double the frame size) and $\frac{1}{2}R_{c0}$ (a threshold used to half the frame size) can be further optimized. Actually one can choose another independent parameter, instead of $\frac{1}{2}R_{c0}$, as a threshold to half the frame size. A study in this aspect is presented in reference [51].

In this chapter, the capturing effect is not considered. Capture means that some tags' IDs can be identified even if the transmissions of several tags are colliding. The capturing effect can affect the system performance dramatically [36, 46]. It is interesting to investigate how AFSA1 and AFSA2 perform considering the capturing effect.

The transmission efficiency of TS algorithms is a little bit higher than that of Aloha based anti-collision algorithms, but in TS algorithms, the tag needs a rewritable counter to remember its historical status in the tree if a 'traditional' realization (not QT or BBT) method is used. In Aloha (even for both dynamic and adaptive Aloha), no additional memory device is needed.

Appendix 7.A Inclusion-Exclusion Principle

The purpose of the following three appendices is to derive the probability of *exactly k* successful transmissions in a contention frame with N tags and the frame size being L. To this end, let us first introduce the inclusion-exclusion principle, which is used to count the number of elements in a union of a finite number of sets. Let $|A|$ denote the cardinality of a finite set A. It is well known that for any two finite sets A_1 and A_2, we have

$$|A_1 \cup A_2| = |A_1| + |A_2| - |A_1 \cap A_2|, \tag{7.51}$$

and for any three finite sets A_1, A_2 and A_3, we have

$$|A_1 \cup A_2 \cup A_3| = |A_1| + |A_2| + |A_3| - (|A_1 \cap A_2| + |A_1 \cap A_3| + |A_2 \cap A_3|)$$
$$+ |A_1 \cap A_2 \cap A_3|. \tag{7.52}$$

Equations (7.51) and (7.52) can be extended to the following general case. Consider n finite sets A_i, $i = 1, 2 \dots n$. Then we have

$$\left| \bigcup_{i=1}^{n} A_i \right| = \sum_{i=1}^{n} |A_i| - \sum_{1 \leq i_1 < i_2 \leq n} |A_{i_1} \cap A_{i_2}| + \sum_{1 \leq i_1 < i_2 < i_3 \leq n} |A_{i_1} \cap A_{i_2} \cap A_{i_3}| - \cdots$$
$$+ (-1)^{n-1} |A_1 \cap A_2 \cap \cdots \cap A_n|. \tag{7.53}$$

In the probability problem, we have similar formulas to $(7.51)-(7.53)$. Let A_i, $i = 1, 2, \cdots, n$, be any events in a probability space, with P denoting its probability measure. Then we have

$$P(A_1 \cup A_2) = P(A_1) + P(A_2) - P(A_1 \cap A_2), \tag{7.54}$$

$$P(A_1 \cup A_2 \cup A_3) = P(A_1) + P(A_2) + P(A_3)$$
$$-[P(A_1 \cap A_2) + P(A_1 \cap A_3) + P(A_2 \cap A_3)]$$
$$+P(A_1 \cap A_2 \cap A_3), \tag{7.55}$$

$$P\left(\bigcup_{i=1}^{n} A_i\right) = \sum_{i=1}^{n} P(A_i) - \sum_{1 \le i_1 < i_2 \le n} P(A_{i_1} \cap A_{i_2})$$
$$+ \sum_{1 \le i_1 < i_2 < i_3 \le n} P(A_{i_1} \cap A_{i_2} \cap A_{i_3}) - \cdots$$
$$+(-1)^{n-1} P(A_1 \cap A_2 \cap \cdots \cap A_n). \tag{7.56}$$

Equations $(7.54)-(7.56)$ are useful when it is difficult to calculate the probability of events like $\bigcup_{i=1}^{n} A_i$, whereas the probability of events like $\bigcap_{i=1}^{n} A_i$ is accessible.

In the following, we will use the inclusion-exclusion principle to calculate the probability of an exact number of successful transmissions in a contention of an Aloha frame. Equations (7.53) and (7.56) are called inclusion-exclusion principle.

Appendix 7.B Probability of Successful Transmissions in Some Particular Time Slots in Aloha

In this appendix, we investigate the following issue. Assume that each of N tags independently and randomly picks one of L time slots, numbered as 1, 2, ..., L, for transmission. The probability of any one time slot to be chosen by any one tag is $\frac{1}{L}$. If a time slot is not chosen by any tags, then this time slot is idle; if a time slot is chosen by more than one tag, then this time slot is in collision; if a time slot is chosen by only one tag, then the transmission at this time slot is successful. We aim to find the probability that N_1 *particular* time slots are idle and N_2 *particular* time slots are successful in a contention frame. Denote this probability as $P_{\text{is}}(N_1, N_2; N, L)$, where $N_1 \ge 0$, $N_2 \ge 0$ and both N_1 and N_2 are integers.

Without loss of generality (because of the symmetry of transmission probability in different time slots) and to facilitate following exposition, we can assume that the time slots numbered 1, ..., N_1 are idle and the time slots numbered $N_1 + 1$, ..., $N_1 + N_2$ are successful. The tags are not organized in an ordered way, but to easy the exposition, we can name them in an ordered way. Let

$$S_{\text{idle}} := \begin{cases} \{1, \cdots, N_1\} & \text{if } N_1 \ge 1, \\ \emptyset & \text{if } N_1 = 0, \end{cases}$$

$$S_{\text{suc}} := \begin{cases} \{N_1 + 1, \cdots, N_1 + N_2\} & \text{if } N_2 \ge 1, \\ \emptyset & \text{if } N_2 = 0, \end{cases}$$

where $\emptyset$ stands for the empty set.

Let T_n denote the slot number that the nth tag picks. Let us argue in the following way. Obviously, none of T_n's can choose the elements in set S_{idle}, only N_2 tags, say $T_{j_1}, \ldots, T_{j_{N_2}}$, can choose the elements in set S_{suc}, and each element in set S_{suc} must be chosen once. Let the j_1th tag first choose. It can have N_2 choices out of the total L choices. Then the j_2th tag makes its choice. It can have $N_2 - 1$ choices out of the total L choices. This process repeats until the j_{N_2}th tag makes its choice. Therefore, we have

$$P_{\text{is}}(N_1, N_2; N, L) = \binom{N}{N_2} \cdot \frac{N_2}{L} \cdot \frac{N_2 - 1}{L} \cdots \cdots \frac{1}{L} \cdot \left(\frac{L - N_1 - N_2}{L} \right)^{N - N_2} \quad (7.57)$$

$$= \frac{N!}{(N - N_2)!} \frac{(L - N_1 - N_2)^{N - N_2}}{L^N}, \quad (7.58)$$

where the last term $\left(\frac{L - N_1 - N_2}{L} \right)^{N - N_2}$ in equation (7.57) is the probability that the remaining $N - N_2$ tags chose the time slots in the set $\{N_1 + N_2 + 1, \cdots, L\}$.

From equation (7.58) we can obtain the following probabilities. The probability that N_1 particular time slots are idle (denoted by $P_i(N_1; N, L)$) is:

$$P_i(N_1; N, L) = \left(1 - \frac{N_1}{L} \right)^N.$$

The probability that the transmissions at N_2 particular time slots are successful (denoted by $P_s(N_2; N, L)$) is:

$$P_s(N_2; N, L) = \begin{cases} \frac{N!}{(N - N_2)!} \frac{(L - N_2)^{N - N_2}}{L^N} & \text{when } N_2 \leq N \text{ and } N_2 \leq N_{2,\text{max}}, \\ 0 & \text{otherwise,} \end{cases} \quad (7.59)$$

where

$$N_{2,\text{max}} = \begin{cases} L - 1 & \text{when } N > L, \\ L & \text{when } N \leq L. \end{cases}$$

Especially, the probability of a successful transmission at any a given time slot is given by

$$p_1 := P_s(1; N, L) = \frac{N}{L} \left(1 - \frac{1}{L} \right)^{N-1}. \quad (7.60)$$

Appendix 7.C Probability of an Exact Number of Successful Transmissions in Aloha

In this appendix, we calculate the probability of *exactly* k successful transmissions in a contention of an Aloha frame, where k is an integer.

Let X_i, $i = 1, 2, \cdots, L$, denote the event of a successful transmission at the ith time slot. Clearly, $P(X_i)$ is given by equation (7.60). Let Y denote the number of successful transmissions in all the time slots in a contention frame. Therefore, $Y = k$ means that *exactly* k of X_i's happen in a contention.

First, let us consider the probability of the event $\{Y = 0\}$. We have

$$
\begin{aligned}
P(Y = 0) &= P\{\bar{X}_1 \cap \bar{X}_2 \cap \cdots \cap \bar{X}_L\} \\
&= P\{\overline{X_1 \cup X_2 \cup \cdots \cup X_L}\} \\
&= 1 - P\{X_1 \cup X_2 \cup \cdots \cup X_L\}.
\end{aligned} \tag{7.61}
$$

Based on the inclusion-exclusion principle and the results in Appendix 7.B, we have

$$
\begin{aligned}
&P\{X_1 \cup X_2 \cup \cdots \cup X_L\} \\
&= \sum_{i=1}^{L} P(X_i) - \sum_{1 \le i_1 < i_2 \le L} P(X_{i_1} \cap X_{i_2}) \\
&\quad + \sum_{1 \le i_1 < i_2 < i_3 \le L} P(X_{i_1} \cap X_{i_2} \cap X_{i_3}) - \cdots + (-1)^{L-1} P(X_1 \cap X_2 \cap \cdots \cap X_L) \\
&= N\left(1 - \frac{1}{L}\right)^{N-1} - \binom{L}{2} P_s(2; N, L) + \binom{L}{3} P_s(3; N, L) - \cdots \\
&\quad + (-1)^{L-1} P_s(L; N, L).
\end{aligned} \tag{7.62}
$$

Since equation (7.62) itself is useful, we denote the probability as $P_S(N, L)$, i.e.,

$$
\begin{aligned}
P_S(N, L) &:= P\{X_1 \cup X_2 \cup \cdots \cup X_L\} \\
&= N\left(1 - \frac{1}{L}\right)^{N-1} - \binom{L}{2} P_s(2; N, L) + \binom{L}{3} P_s(3; N, L) - \cdots \\
&\quad + (-1)^{L-1} P_s(L; N, L).
\end{aligned} \tag{7.63}
$$

It is the probability of at least one successful transmission of N tags in an L contention frame. Substituting equations (7.59) and (7.63) into (7.61), we can get $P(Y = 0)$:

$$
P(Y = 0) = 1 - P_S(N, L).
$$

Next, let us investigate the probability of $\{Y = k\}$. Due to the symmetric nature of successful transmissions at different time slots, we have

$$
\begin{aligned}
P(Y = k) &= \binom{L}{k} P\left\{ X_{l_1} \cap X_{l_2} \cap \cdots \cap X_{l_k} \bigcap_{i \ne l_1, l_2, \cdots, l_k} \bar{X}_i \right\} \\
&= \binom{L}{k} P\left\{ X_1 \cap X_2 \cap \cdots \cap X_k \bigcap_{i=k+1}^{L} \bar{X}_i \right\},
\end{aligned} \tag{7.64}
$$

where $l_1, l_2 \ldots l_k \in \{1, 2 \ldots L\}$ are k different integers. Let $Z_1 = X_1 \cap X_2 \cap \cdots \cap X_k$ and $Z_2 = \bigcap_{i=k+1}^{L} \bar{X}_i$. Then we have

$$
\begin{aligned}
P(Y = k) &= \binom{L}{k} P\{Z_1 \cap Z_2\} \\
&= \binom{L}{k} P\{Z_1\} P\{Z_2 | Z_1\}.
\end{aligned} \tag{7.65}
$$

From Appendix 7.B, we have

$$P\{Z_1\} = P_s(k; N, L).   \tag{7.66}$$

Using the result (7.63), we have

$$
\begin{aligned}
P\{Z_2|Z_1\} &= P\{\bar{X}_{k+1} \cap \bar{X}_{k+2} \cap \cdots \cap \bar{X}_L|Z_1\} \\
&= P\{\overline{X_{k+1} \cup X_{k+2} \cup \cdots \cup X_L} \,|Z_1\} \\
&= 1 - P\{X_{k+1} \cup X_{k+2} \cup \cdots \cup X_L|Z_1\} \\
&= 1 - P_S(N - k, L - k).   \tag{7.67}
\end{aligned}
$$

Substituting equations (7.66) and (7.67) into (7.65), we can get the probability $P(Y = k)$:

$$P(Y = k) = \binom{L}{k} P_s(k; N, L)[1 - P_S(N - k, L - k)].   \tag{7.68}$$

References

[1] *IEEE Standard for Wireless Lan Medium Access Control (MAC) and Physical Layer(PHY) Specification*, 1999 Edition (R2003).

[2] *IEEE Standard for Wireless LAN Medium Access Control (MAC) and Physical Layer (PHY) specifications: High-speed Physical Layer in the 5 GHZ Band*, Sept. 1999.

[3] *IEEE Standard for Wireless LAN Medium Access Control (MAC) and Physical Layer (PHY) specifications: Higher-Speed Physical Layer Extension in the 2.4 GHz Band*, Sept. 1999.

[4] *IEEE Standard for Wireless LAN Medium Access Control (MAC) and Physical Layer (PHY) specifications, Amendment 4: Further Higher Data Rate Extension in the 2.4 GHz Band*, June 2003.

[5] N. Abramson. The ALOHA system - another alternative for computer communications. In *Proc. 1970 Fall Joint Computer Conf., AFIPS Press*, Nov. 1970.

[6] N. Abramson. Development of the ALOHANET. *IEEE Trans. Inform. Theory*, 31:119–123, 1985.

[7] D. Bertsekas and R. G. Gallager. *Data Networks*. Prentice-Hall, Upper Saddle River, NJ, 1992.

[8] G. Bianchi. Performance analysis of the IEEE 802.11 distributed coordination function. *IEEE J. Sel. Areas Commun.*, 18:535–547, 2000.

[9] J. I. Capetanakis. Tree algorithms for packet broadcast channels. *IEEE Trans. Inform. Theory*, 25:505–515, 1979.

[10] H.-S. Choi, J.-R. Cha, and J.-H. Kim. Fast wireless anti-collision algorithm in ubiquitous ID system. In *Proc. 2004 IEEE 60th Vehicular Technology Conf.*, pages 4589–4592, Los Angeles, CA, USA, 26–29 Sept. 2004.

[11] J. S. Choi, H. Lee, D. W. Engels, and R. Elmasri. Robust and dynamic bin slotted anti-collision algorithms in RFID system. In *2008 IEEE Int. Conf. on RFID*, pages 191–198, Las Vegas, USA, 16–17 Apr. 2008.

[12] D. Engels and S. Sarma. The reader collision problem. Technical Report MIT-AUTOID-WH007, Auto-ID Center, Nov. 2001.

[13] EPCglobal. EPC radio-frequency identity protocols - Class-1 Generation-2 UHF RFID protocol for communications at 860 MHz - 960 MHz, version 1.2.0. 2008. Available: www.gs1.org/gsmp/kc/epcglobal/uhfc1g2/uhfc1g2&uscore;1&uscore;2&uscore;0-standard-20080511.pdf.

[14] C. Floerkemeier. Bayesian transmission strategy for framed ALOHA based RFID protocols. In *2007 IEEE Int. Conf. on RFID*, pages 228235, Grapevine, Texas, USA, 26–28 Mar. 2007.

[15] C. Galiotto, K. Cetin, S. Frattasi, N. Marchetti, N. R. Prasad, and R. Prasad. High fairness reader anti-collision protocol in passive RFID systems. In *2011 IEEE Int. Conf. on RFID*, pages 113–120, Orlando, Florida, USA, 12-14 Apr. 2011.

[16] R. G. Gallager. A perspective on multiaccess channels. *IEEE Trans. Inform. Theory*, 31:124–142, 1985.

[17] H. Guo, V. C. M. Leung, and M. Bolic. *M*-ary RFID tags splitting with small idle slots. *IEEE Trans. Automation Science and Engineering*, 9:177–181, 2012.

[18] E. Hamouda, N. Mitton, and D. Simplot-Ryl. Reader anti-collision in dense RFID networks with mobile tags. In *2011 IEEE Int. Conf. on RFID - Technologies and Applications*, pages 327–334, Sitges, Spain, 15–16 Sept. 2011.

[19] A. J. E. M. Janssen and M. J. M. de Jong. Analysis of contention tree algorithms. *IEEE Trans. Inform. Theory*, 46:2163–2172, 2000.

[20] N. L. Johnson and S. Kotz. *Urn Models and Their Applications*. John Wiley & Sons, Inc., New York, 1977.

[21] M. A. Kaplan and E. Gulko. Analytic properties of multiple-access trees. *IEEE Trans. Inform. Theory*, 31:255–263, 1985.

[22] R. S. Khasgiwale, R. U. Adyanthaya, and D. W. Engels. Extracting information from tag collisions. In *2009 IEEE Int. Conf. on RFID*, pages 131138, Orlando, Florida, USA, 27–28 Apr. 2009.

[23] D. K. Klair, K.-W. Chin, and R. Raad. A survey and tutorial of RFID anti-collision protocols. *IEEE Communications & Tutorials*, 12:400–421, 2010.

[24] L. Kleinrock. On queueing problems in random-access communications. *IEEE Trans. Inform. Theory*, 31:166–175, 1985.

[25] B. Knerr, M. Holzer, C. Angerer, and M. Rupp. Slot-wise maximum likelihood estimation of the tag population size in FSA protocols. *IEEE Trans. Commun.*, 58:578–585, 2010.

[26] M. Kodialam and T. Nandagopal. Fast and reliable estimation schemes in RFID systems. In *Proc. 12th Annual Int. Conf. Mobile Computing and Networking*, pages 322–333, Los Angeles, CA, USA, 24–29 Sept. 2006.

[27] T. F. La Porta, G. Maselli, and C. Petrioli. Anticollision protocols for single-reader RFID systems: Temporal analysis and optimization. *IEEE Trans. Mobile Computing*, 10(2):267–279, 2011.

[28] C. Law, K. Lee, and K. Y. Siu. Efficient memoryless protocol for tag identification. In *Proc. 4th Int. Workshop on Discrete Algorithms and Methods for Mobile Computing and Communications*, pages 75–84, Boston, Massachusetts, USA, August 2000.

[29] S.-R. Lee, S.-D. Joo, and C.-W. Lee. An enhanced dynamic framed slotted ALOHA algorithm for RFID tag identification. In *Proc. 2nd Annual Int. Conf. Mobile and Ubiquitous Systems: Networking and Services*, pages 166–172, San Diego, CA, USA, 17–21 Jul. 2005.

[30] L. Liu and S. Lai. ALOHA-based anti-collision algorithms used in RFID system. In *Proc. Int. Conf. Wireless Communications, Networking and Mobile Computing*, Wuhan, China, 22–24 Sept. 2006.

[31] A. Loeffler. Using CDMA as anti-collision method for RFID–research & applications. In *Current Trends and Challenges in RFID* (C. Turcu ed.). InTech, Rijeka, pages 306–328, 2011. Available from www.intechopen.com/books/current-trends-and-challenges-in-rfid/using-cdma-as-anti-collision-methodfor-rfid-research-applications.

[32] P. Mathys and P. Flajolet. *Q*-ary collision resolution algorithms in random-access systems with free or blocked channel access. *IEEE Trans. Inform. Theory*, 31:217–243, 1985.

[33] C. Mutti and C. Floerkemeier. CDMA-based RFID systems in dense scenarios: Concepts and challenges. In *2008 IEEE Int. Conf. on RFID*, pages 215–222, Las Vegas, USA, 16–17 Apr. 2008.

[34] J. Myung, W. Lee, and T. K. Shih. An adaptive memoryless protocol for RFID tag collision arbitration. *IEEE Trans. Multimedia*, 8:1096–1101, 2006.

[35] J. Myung, W. Lee, J. Srivastava, and T. K. Shih. Tag-splitting: Adaptive collision arbitration protocols for RFID tag identification. *IEEE Trans. Parallel and Distributed Systems*, 18:763–775, 2007.

[36] V. Naware, G. Mergen, and L. Tong. Stability and delay of finite-user slotted ALOHA with multipacket reception. *IEEE Trans. Inform. Theory*, 51:2636–2656, 2005.

[37] C.-H. Quan, J.-C. Choi, G.-Y. Choi, and C.-W. Lee. The slotted-LBT: A RFID reader medium access scheme in dense reader environments. In *2008 IEEE Int. Conf. on RFID*, pages 207–214, Las Vegas, USA, 16–17 Apr. 2008.

[38] L G. Roberts. ALOHA packet system with and without slots and capture. *ACM SIGCOMM Computer Communications Review*, 5(2):28–42, 1975.

[39] F. Schoute. Dynamic frame length ALOHA. *IEEE Trans. Commun.*, 31:565–568, 1983.

[40] D.-H. Shih, P.-L. Sun, D. C. Yen, and S.-M. Huang. Taxonomy and survey of RFID anti-collision protocols. *Computer Communications*, 29:2150–2166, 2006.

[41] A. S. Tanenbaum and D. J. Wetherall. *Computer Networks*. Prentice-Hall, Boston, 5th edition, 2011.

[42] B. S. Tsybakov. Survey of USSR contributions to random multiple-access communications. *IEEE Trans. Inform. Theory*, 31:143–165, 1985.

[43] B. S. Tsybakov and V. A. Mikhailov. Free synchronous packet access in a broadcast channel with feedback. *Problemy Peredachi Informatsii*, 14(4):32–59, 1978 (English version: in *Problems of Information Transmission*, vol.14, no.4, pp. 259–280, 1978).

[44] H. Vogt. Efficient object identification with passive RFID tags. In *Proc. 1st Int. Conf. Pervasive Computing*, pages 98–113, Zurich, Switzerland, 26–28 Aug. 2002.

[45] J. Waldrop, D. Engels, and S. Sarma. Colorwave: An anticollision algorithm for the reader collision. In *Proc. IEEE Int. Conf. Comm.*, pages 1206–1210, Anchorage, Alaska, USA, 11–15 May 2003.

[46] J. E. Wieselthier, A. Ephremides, and L. A. Michaels. An exact analysis and performance evaluation of framed ALOHA with capture. *IEEE Trans. Commun.*, 37:125–137, 1989.

[47] M.-K. Yeh, J.-R. Jiang, and S.-T. Huang. Adaptive splitting and pre-signaling for RFID tag anti-collision. *Computer Communications*, 32:1862–1870, 2009.

[48] B. Zhen, M. Kobayashi, and M. Shimizu. Framed ALOHA for multiple RFID objects identification. *IEICE Trans. Commun.*, E88–B:991–999, 2005.

[49] F. Zhou, C. Chen, D. Jin, C. Huang, and H. Min. Evaluating and optimizing power consumption of anti-collision protocols for applications in RFID systems. In *Proc. 2004 Int. Symp. on Low Power Electronics and Design*, pages 357–362, Newport Beach, CA, USA, 9–11 Aug. 2004.

[50] F. Zheng and T. Kaiser. Adaptive Aloha anti-collision algorithms for RFID systems. Submitted for publication.

[51] F. Zheng and T. Kaiser. An optimal adaptive Aloha anti-collision algorithm for RFID systems. Submitted for publication.

8

Localization with RFID

8.1 Introduction

Localization based on radio technology has been a long standing issue, which can be traced back to the invention of radar and is still a hot research field for different applications. Currently, one of most successful and widely used radio-based localization systems is the global positioning system (GPS), which is a space-based satellite navigation system providing location information for anywhere on or near Earth with an unobstructed line of sight (LoS) to four or more GPS satellites. However, GPS cannot be applied to some scenarios such as indoor localization. Instead, the technology of wireless networks or sensors, such as RFID, cellular networks, wireless local area networks (WLANs), Bluetooth and so on can be exploited for indoor-like localization [31]. Indeed, this kind of localization technique has found wide applications in logistics, security tracking, health care, location-sensitive billing and production process control to name a few. [16, 18, 46, 69]. Comprehensive surveys on the localization issue using different wireless technology have constantly appeared in the literature. See, for example [6, 16, 18, 19, 20, 21, 31, 38, 45, 46, 60, 62, 69] for recent reviews. In [20, 31], comparisons in cost, performance, robustness, security, complexity and limitations among different types of indoor positioning systems are documented in full detail and the report [15] provides a comparison for the accuracy of several commercial position-sensing systems available in the market.

The main challenge for indoor localization by using wireless networks or sensors is caused by two factors due to radio wave propagation environments: non line of sight (NLoS) and multipaths. Due to the NLoS, the measured ranging information is always positively biased from the true range between the object to be localized and the sensor, and the bias is difficult to estimate. Due to the multipaths, it is difficult to identify which path comes from the object to be localized when the object is not in LoS.

The approaches for indoor-like localization based on wireless technologies can be broadly categorized into three classes [6, 18, 25]:

1. Geometric or parameterized class. In this approach, the position of an object is solved from the information such as received signal strength (RSS), angle of arrival (AoA)[1], or time

---

[1] In the literature, AoA is often referred to as direction of arrival (DoA).

Digital Signal Processing for RFID, First Edition. Feng Zheng and Thomas Kaiser.
© 2016 John Wiley & Sons, Ltd. Published 2016 by John Wiley & Sons, Ltd.

of arrival (ToA) of the received signal at the sensors. This approach generally involves two steps. The first step is to obtain location-related information such as RSS, AoA and/or ToA. The second step is to use various kinds of algorithms to solve the position of the object from the obtained location-related information.

2. Mapping (or fingerprinting) class. A basic assumption of this approach is that the signal characteristics, such as channel impulse responses (CIRs) or power delay profiles (PDPs), as a function of the measured location, are one-to-one mappings. The CIRs or PDPs can be measured and hence the position of the object can be solved from these mappings.

3. Proximity class. In this approach, dense sensors are often deployed in the area of object to be localized. When the object enters in the sensing range of a sensor, its location is assumed to be the same as that of the sensor. If multiple sensors receive the signals from the object, the object can be assumed to be co-located with the sensor that receives the strongest signal or its position can be solved as the geometric centre of those sensors, weighted somehow by the strength of their received signals.

The mapping approach is especially useful in wideband or ultra wideband (UWB) systems [18, 26]. In the mapping approach, multipaths act a positive role. It is the rich multipaths that make this approach possible. The more scatterers the environment has, or the richer the multipaths, the better localization accuracy the system can achieve. NLoS is not an issue in this approach. For the geometric or parameterized approach, multipaths and NLoS generally play an negative role. Since major RFID systems use narrow band technology, we will not discuss the mapping approach in this chapter.

A typical example of the proximity approach is to localize a mobile in a cellular network through the position of the base station in a cell that the mobile is connected to. Clearly, the localization accuracy is on the order of the size of the cells. For localization methods based on general wireless technologies, this approach might not be attractive because of poor accuracies, but it becomes increasingly interesting when using RFID technology because dense RFID sensors can be deployed.

In the geometric approach, the most popularly used location-related information is ToA and AoA in 'traditional' radio-based localization methods such as radar and GPS. In traditional localization systems, RSS method is not popularly used. Even though RSS is easy to measure, it provides less accurate positioning [38] in 'traditional' localization systems, because accurate channel models for the RSS approach are needed [49, 50, 51, 52, 53]. However, RSS method is attracting more and more interest in RFID-based localization systems [7, 11, 21, 23, 28, 34, 65, 67, 68]. This is because RSS information in the form of the received signal strength indicator (RSSI) is commonly built into the transceiver chips in commercial RFID readers [7].

Several types of indoor localization systems have been extensively studied in the literature and some are available in the market. Infrared, WiFi, ultrasonic and RFID are some examples of these systems. Each of these systems has its own strengths as well as limitations. For example, WiFi devices are much larger in size and have much more strict power requirements than RFID tags, which makes RFID tags an attractive choice for localization [24].

RFID was invented mainly for object inventory, but recently it has been found that RFID can be used to deal with some major challenges in localization for indoor-like scenarios with sufficiently high accuracies and reliability. Therefore, RFID-based localization could possibly become another important application area of RFID in the near future.

In principle, the problem of localization with the help of RFID is similar to the radar ranging problem. However, RFID ranging has its peculiar concerns. Since the distance between the reader and tag is usually short (typically on the order of less than 10 m), the round-trip signal delay is on the order of a few tens of nanoseconds. Because typical RFID systems use narrowband radios, it is difficult to measure the time of arrival or time difference of arrival of the RFID signal. Thus phase information is extremely useful for the RFID localization problem.

In the literature, there are a number of research reports on RFID-based localization.

The approaches in [17, 22, 34, 57, 63, 65, 68] belong to the proximity class. LANDMARC developed in [34] localizes RFID tags through comparing the power profile of the tag to be localized with the power profiles of a number of reference tags with known locations. In this system, nine readers with indications of eight different received power levels (not the power itself) are used and a number of reference tags (i.e. tags with fixed and known positions) are deployed for localization. To localize a tag, its received signal strengths from all the readers are compared to the received signal strengths of all the reference tags. The estimated tag location is given by a weighted average of the k-nearest neighbouring (k-NN) reference tags. The system performance is robust against some environmental factors. However, it is sensitive to orientations of both reference tags and the tag to be localized, especially when the tag is used to track moving objects [24]. Another issue for LANDMARC is that the reference tags cannot be too densely arranged, since otherwise the resultant interference power would be too strong. Some improvement methods for LANDMARC are discussed in [65, 68]. In [68], the number of usable reference positions is increased by adding some virtual reference tags whose RSSI values are calculated through a linear interpolation algorithm by using the RSSI values of the real reference tags. In [65], the effect of the tag's diversity derived from different manufacturer types and used time of built-in batteries on the RSS values of tags is taken into account. In [17], a number of tags are deployed at the ceiling of a room in a rectangular grid format with tags' positions being known and the position of a robot carrying a reader on the floor is to be found. It is shown that this system can achieve an average positioning error of about 0.13 m (and 0.1 m, respectively) with a tag density of about 0.7 tags/m^2 (and 1 tags/m^2, respectively). In [63], a scheme of sparsely distributed tags is proposed to improve efficiency of RFID localization by using the similar approach as in [17].

The approaches in [11, 13, 14, 24, 28, 29, 30, 44, 59, 66] belong to the geometric class. Traditionally, range information can be obtained from the RSS, ToA and time difference of arrival (TDoA). However, it is pointed out in [30, 35] that the RSS approach is inaccurate, particularly in a complicated propagation environment and it is noted in [2, 24, 35] that the ToA and TDoA approaches are improper for using in RFID-based localization as these measurements are difficult and expensive to implement. In [24], phase difference among two or more receive antennas at the reader is used for localization. The experiments in [24] show that the phase difference approach can be used in three-dimensional positioning, motion estimation and tracking with high accuracy (within 0.57°). Exploiting multiple frequencies may further improve the range estimation performance [30]. In [30], the latter approach is extensively studied and several important advantages in the range estimation of passive RFID tags by using multi-frequency-based techniques are illustrated. For example, using well designed multiple frequencies can effectively mitigate the phase wrapping and range ambiguity problem that may be encountered in the phase difference approach. In [66], a very simple antenna array with only two antennas at the reader is illustrated, via analysis and simulation, to be able to measure the AoA of the tag's backscattered signal with a reasonable level of high accuracy. In [59], the experimental results show that a two-element antenna array equipped at the reader can achieve an average error of 16 cm in a controlled indoor environment. In [13], the authors use a circularly polarized three-element patch antenna array at the reader to localize a tag in a 3 m × 3 m test grid with 25 equally spaced measurement points. It is shown that the system can achieve a measured mean error of 9 cm and a measured mean standard deviation of 7 cm. The study [57] shows that using RSSI measurement can yield localization errors within 35 cm in an outdoor (campus) environment. Other case studies are reported by using RSS [7, 11, 28, 44], AoA [2, 32] and ToA based on wideband/UWB RFID [14, 33], respectively.

In [29], a frequency-modulated continuous-wave based RFID localization technique is discussed. The dependence of the ranging resolution on the bandwidth in this approach is investigated. The principle behind the approach in [29] is the same as that of radar ranging systems, i.e. using the ToA information. Note that the charging time is not considered in [29]. If this were considered, one would find that the ranging resolution given in [29] would be too optimistic.

The report [23] studied the path-loss model of RFID for indoor localization. It shows via experiments that the two-parameter path-loss models are suitable for characterizinge the RFID RSS fluctuations and RFID RSS values are slightly more stable than the WLAN RSS values.

To the best of our knowledge, there are no research reports using the mapping (or finger-printing) class in a strict sense in RFID localization.

RFID localization technique has also found a wide range of applications, for example, in healthcare [36, 37], surgical operation [5], robot navigation [22, 39], automatic guided vehicles [1], mining safety [43, 64], sediment analysis [41] and so on.

In the above, we have given a brief review for RFID localization technique. More reviews can be found in [6, 35, 44, 45, 69]. In particular, an excellent review for phase-based localization techniques by using UHF RFID tags is provided in [35]. From the above discussion, we can see that RFID localization technique possesses its own pros and cons and due to this fact, several hot issues should be addressed. In this chapter, we will present some basic localization algorithms based on RFID systems. Starting from these basic algorithms, more advanced techniques can be developed if new capabilities of RFID readers and tags are exploited.

In typical RFID localization problems, it is generally assumed that the object to be localized is attached with a tag and there are several readers or tags whose positions are known. Of course, the identities of all the tags are assumed to be known. Therefore, these readers and tags with known positions can both act as sensors of the unknown position. On the other hand, the object to be localized can be also equipped with a reader and the reader uses its own position-related measurements to find the position of the object. This kind of application appears in many robotic systems.

This chapter is organized as follows. In Section 8.2, we will present some localization algorithms in two major RFID localization classes: geometric class and proximity class. Then the range-estimation and AoA-estimation approaches will be discussed in Sections 8.3 and 8.4, respectively, which will complement the geometric localization algorithms described in Section 8.2. In Section 8.5, one of the most challenging issue in RFID localization, that is, the NLoS issue, will be discussed. Section 8.6 concludes this chapter.

8.2   RFID Localization

In RFID localization, the ToA information is often referred to the signal travelling time (STT) from the reader to tag and back to reader again. Different from radar systems, the STT includes the charging time at the tag for RFID systems and often the charging time is difficult to measure and long enough so that it cannot be neglected in indoor localization.

In this chapter, we confine our focus to the localization in the two-dimensional space. It is not difficult to extend the discussed approaches or algorithms to the case of the three-dimensional space.

8.2.1 Geometric Class

In the geometric class, localization can be considered an application of array processing. If the antenna array is a uniform linear array, combining the AoA and ToA information obtained by the array will uniquely decide the object position. If the antenna elements in the 'array' are distributed irregularly but sufficiently far from each other as in the case of sensor networks (in this case, the word 'array' just means multiple sensors located at different places), the ToA information obtained at several elements in the 'array' can jointly give a unique solution for the object position. In this section, we will present some basic localization algorithms based on the above principle.

8.2.1.1 The Case of Two Readers

The most simple geometric class is the beamforming approach. Its basic idea is to use two readers, each being equipped with a beamformer, to find the position of a tag, as shown in Figure 8.1. It can be clearly seen from this figure that the position of the tag is uniquely decided by the two rays from the two directions, respectively. The positioning accuracy is determined by the main lobe beamwidth of the beamformers at readers.

If the STT can also be obtained at the readers, the information can be integrated together with the AoA information to increase the positioning accuracy. In the following, we discuss some details about how to solve the tag position by using relevant information.

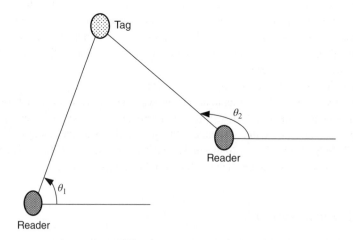

Figure 8.1   RFID Localization via the AoA measurement.

Let the coordinates of the two readers be (x_i, y_i), $i = 1, 2$, respectively, the coordinate of the tag be (x, y), the AoA of the signals received at the two readers be θ_1 and θ_2, respectively, and the STT of the signals received at the two readers be t_1 and t_2, respectively. Then we have

$$\frac{y - y_1}{x - x_1} = \tan(\theta_1), \tag{8.1}$$

$$\frac{y - y_2}{x - x_2} = \tan(\theta_2), \tag{8.2}$$

$$r_1^2 = (x - x_1)^2 + (y - y_1)^2 = \left[c\frac{t_1 - t^0}{2}\right]^2, \tag{8.3}$$

$$r_2^2 = (x - x_2)^2 + (y - y_2)^2 = \left[c\frac{t_2 - t^0}{2}\right]^2, \tag{8.4}$$

where c is the speed of light ($c = 3 \times 10^8$ m/s), t^0 is the tag's charging time plus the processing time[2] and r_i is the distance between the tag and the ith reader.

If there is only AoA information available, the unknown position (x, y) can be solved from equations (8.1) and (8.2). Rewrite both equations in the following matrix form

$$\mathbf{M}_1\mathbf{p} = \mathbf{b}_1, \tag{8.5}$$

where

$$\mathbf{p} = \begin{bmatrix} x \\ y \end{bmatrix}, \quad \mathbf{M}_1 = \begin{bmatrix} \sin(\theta_1) & -\cos(\theta_1) \\ \sin(\theta_2) & -\cos(\theta_2) \end{bmatrix}, \quad \mathbf{b}_1 = \begin{bmatrix} x_1\sin(\theta_1) - y_1\cos(\theta_1) \\ x_2\sin(\theta_2) - y_2\cos(\theta_2) \end{bmatrix}.$$

Then the solution is

$$\mathbf{p} = \mathbf{M}_1^{-1}\mathbf{b}_1.$$

It is clear that the position $\mathbf{p}$ can be uniquely decided if and only if $\theta_1 - \theta_2 \neq k\pi$, where k is an integer.

If the STT information is also available, equations (8.3) and (8.4) can be combined with (8.1) and (8.2) to increase the position estimation accuracy. The first approach is to directly solve a nonlinear least squares (LS) problem based on equations (8.1)–(8.4), which is not easy. Another major drawback of this approach is that the unknown (or only known very inaccurately in most cases) charging time of the tag will make indoor wireless localization nonsense, since the charging time of the tag is typically on the order of a few microseconds, causing a distance error of several hundred metres. The second approach is to use the time-difference-in-STT (TDSTT) information between two readers. In the following, we will present the solution to the problem following the method as outlined in [46]. Define

$$d_{21} := r_2 - r_1 = c\frac{t_2 - t^0}{2} - c\frac{t_1 - t^0}{2} = c\frac{t_2 - t_1}{2}.$$

---

[2] Note that the charging time for different reader to the same tag might be different since the transmitted powers from different readers and the propagation fading from the readers to the tag might be different. Therefore, t^0 should also depend on reader sub-index i. For easy exposition, we omit this kind of dependency.

Note that even though the two t^0 in these equations are different from each other, the difference between them is far less than the two t^0 themselves. Hence it can be neglected in d_{21}.

From equations (8.3) and (8.4), we have

$$
\begin{aligned}
r_2^2 &= (d_{21} + r_1)^2 = (x - x_1 + x_1 - x_2)^2 + (y - y_1 + y_1 - y_2)^2 \\
&= -2(x_2 - x_1)x - 2(y_2 - y_1)y + r_1^2 + x_2^2 - x_1^2 + y_2^2 - y_1^2.
\end{aligned}
$$

Expanding the above equation yields

$$(x_2 - x_1)x + (y_2 - y_1)y = -d_{21}r_1 - \frac{1}{2}\left(d_{21}^2 + x_1^2 - x_2^2 + y_1^2 - y_2^2\right). \tag{8.6}$$

Combining equations (8.5) and (8.6) gives

$$\mathbf{M}_2\mathbf{p} = \mathbf{b}_3 r_1 + \mathbf{b}_2, \tag{8.7}$$

where

$$\mathbf{M}_2 = \begin{bmatrix} \sin(\theta_1) & -\cos(\theta_1) \\ \sin(\theta_2) & -\cos(\theta_2) \\ x_2 - x_1 & y_2 - y_1 \end{bmatrix}, \tag{8.8}$$

$$\mathbf{b}_2 = \begin{bmatrix} x_1\sin(\theta_1) - y_1\cos(\theta_1) \\ x_2\sin(\theta_2) - y_2\cos(\theta_2) \\ \frac{1}{2}\left(x_2^2 - x_1^2 + y_2^2 - y_1^2 - d_{21}^2\right) \end{bmatrix}, \quad \mathbf{b}_3 = \begin{bmatrix} 0 \\ 0 \\ -d_{21} \end{bmatrix}. \tag{8.9}$$

Equation (8.7) is over-determined. The LS intermediate solution for equation (8.7) is

$$\hat{\mathbf{p}} = (\mathbf{M}_2^T\mathbf{M}_2)^{-1}\mathbf{M}_2^T(\mathbf{b}_3 r_1 + \mathbf{b}_2). \tag{8.10}$$

Substituting the solved $\hat{\mathbf{p}}$ into the following equation

$$r_1^2 = (x - x_1)^2 + (y - y_1)^2 \tag{8.11}$$

leads to a quadratic equation in r_1. Solving for r_1 and substituting the positive root back into equation (8.10) gives the final solution for $\mathbf{p}$.

A more accurate estimate $\mathbf{p}$ than that provided by equation (8.10) can be theoretically obtained if the statistical information about the measurement noises contained in θ_i ($i = 1, 2$) and d_{21} is available. In this case, the maximum likelihood estimation (MLE) method can be exploited to estimate $\mathbf{p}$. However, it can be seen from equations (8.8) and (8.9) that the non-linear transforms are performed upon the measurement noises in the final estimation model (8.7), which will make it very difficult to obtain the MLE estimate of $\mathbf{p}$.

8.2.1.2 The Case of More Than Two Readers

The results in the preceding subsection can be easily extended to the case of more than two readers. Suppose that there are now M ($M \geq 3$) readers that can measure the AoA and STT of the relevant signal. Let the coordinates of the M readers be (x_i, y_i), the AoA of the signals received at the ith reader be θ_i and the STT of the signals received at the ith reader be t_i,

$i = 1, 2, \ldots, M$. Similarly, these variables are related to each other according to the following equations

$$\frac{y - y_i}{x - x_i} = \tan(\theta_i),$$

$$r_i^2 = (x - x_i)^2 + (y - y_i)^2 = \left[c\frac{t_i - t^0}{2} \right]^2, \quad i = 1, 2, \ldots, M.$$

In the case where the available information is only the AoA, the measurement model can be written as the following matrix form

$$\mathbf{M}_3 \mathbf{p} = \mathbf{b}_4, \tag{8.12}$$

where

$$\mathbf{M}_3 = \begin{bmatrix} \sin(\theta_1) & -\cos(\theta_1) \\ \sin(\theta_2) & -\cos(\theta_2) \\ \vdots & \vdots \\ \sin(\theta_M) & -\cos(\theta_M) \end{bmatrix}, \quad \mathbf{b}_4 = \begin{bmatrix} x_1 \sin(\theta_1) - y_1 \cos(\theta_1) \\ x_2 \sin(\theta_2) - y_2 \cos(\theta_2) \\ \vdots \\ x_M \sin(\theta_M) - y_M \cos(\theta_M) \end{bmatrix}.$$

Equation (8.12) is over-determined. Its LS solution is given by

$$\hat{\mathbf{p}} = (\mathbf{M}_3^T \mathbf{M}_3)^{-1} \mathbf{M}_3^T \mathbf{b}_4.$$

In the case where the information about the STT, besides the AoA, is also available, we can proceed as follows. Define

$$d_{i1} := r_i - r_1 = c\frac{t_i - t^0}{2} - c\frac{t_1 - t^0}{2} = c\frac{t_i - t_1}{2}, \quad i = 2, \ldots, M.$$

Then following the same procedure as the preceding subsection, we have

$$(x_i - x_1)x + (y_i - y_1)y = -d_{i1}r_1 - \frac{1}{2}\left[d_{i1}^2 + x_1^2 - x_i^2 + y_1^2 - y_i^2 \right]. \tag{8.13}$$

Combining equations (8.12) and (8.13) gives

$$\mathbf{M}_4 \mathbf{p} = \mathbf{b}_5 r_1 + \mathbf{b}_6, \tag{8.14}$$

where

$$\mathbf{M}_4 = \begin{bmatrix} \sin(\theta_1) & -\cos(\theta_1) \\ \sin(\theta_2) & -\cos(\theta_2) \\ \vdots & \vdots \\ \sin(\theta_M) & -\cos(\theta_M) \\ x_2 - x_1 & y_2 - y_1 \\ \vdots & \vdots \\ x_M - x_1 & y_M - y_1 \end{bmatrix},$$

$$\mathbf{b}_5 = \begin{bmatrix} x_1 \sin(\theta_1) - y_1 \cos(\theta_1) \\ x_2 \sin(\theta_2) - y_2 \cos(\theta_2) \\ \vdots \\ x_M \sin(\theta_M) - y_M \cos(\theta_M) \\ \frac{1}{2}\left(x_2^2 - x_1^2 + y_2^2 - y_1^2 - d_{21}^2\right) \\ \vdots \\ \frac{1}{2}\left(x_M^2 - x_1^2 + y_M^2 - y_1^2 - d_{M1}^2\right) \end{bmatrix}, \quad \mathbf{b}_6 = \begin{bmatrix} 0 \\ 0 \\ \vdots \\ 0 \\ -d_{21} \\ \vdots \\ -d_{M1} \end{bmatrix}.$$

The LS intermediate solution for equation (8.14) is

$$\hat{\mathbf{p}} = (\mathbf{M}_4^T \mathbf{M}_4)^{-1} \mathbf{M}_4^T (\mathbf{b}_5 r_1 + \mathbf{b}_6). \tag{8.15}$$

Substituting the solved $\hat{\mathbf{p}}$ into equation (8.11) leads to a quadratic equation in r_1. Solving for r_1 and substituting the positive root back into equation (8.15) gives the final solution for $\mathbf{p}$.

The approach outlined here can be easily extended to the case where only the STT information is available. For the reason of completeness, we repeat the solution as follows. The equivalent measurement equation is

$$\mathbf{M}_5 \mathbf{p} = \mathbf{b}_7 r_1 + \mathbf{b}_8, \tag{8.16}$$

where

$$\mathbf{M}_5 = \begin{bmatrix} x_2 - x_1 & y_2 - y_1 \\ x_3 - x_1 & y_3 - y_1 \\ \vdots & \vdots \\ x_M - x_1 & y_M - y_1 \end{bmatrix},$$

$$\mathbf{b}_7 = \begin{bmatrix} \frac{1}{2}\left(x_2^2 - x_1^2 + y_2^2 - y_1^2 - d_{21}^2\right) \\ \frac{1}{2}\left(x_3^2 - x_1^2 + y_3^2 - y_1^2 - d_{31}^2\right) \\ \vdots \\ \frac{1}{2}\left(x_M^2 - x_1^2 + y_M^2 - y_1^2 - d_{M1}^2\right) \end{bmatrix}, \quad \mathbf{b}_8 = \begin{bmatrix} -d_{21} \\ -d_{31} \\ \vdots \\ -d_{M1} \end{bmatrix}.$$

The LS intermediate solution for equation (8.16) is

$$\hat{\mathbf{p}} = (\mathbf{M}_5^T \mathbf{M}_5)^{-1} \mathbf{M}_5^T (\mathbf{b}_7 r_1 + \mathbf{b}_8). \tag{8.17}$$

Substituting the solved $\hat{\mathbf{p}}$ into equation (8.11) leads to a quadratic equation in r_1. Solving for r_1 and substituting the positive root back into equation (8.17) gives the final solution for $\mathbf{p}$.

8.2.2 Proximity Class

The proximity approach takes full advantage of RFID systems to localize some tagged objects. In this approach, a number of tags are deployed in a specific region. The positions of these tags are known *a priori*. Therefore, these tags are called reference tags.

The basic idea of the proximity approach is to compare the signal strength of the tag to be localized and the signal strengths of reference tags and associate the position of the tag to be localized to the positions of those tags whose signal strengths are near to the signal strength of the tag to be localized.

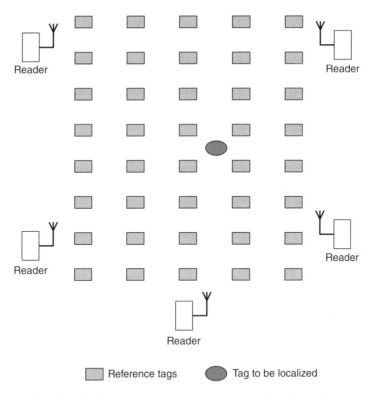

Figure 8.2 An illustration of the deployment of readers and tags for 2D localization problem using the proximity approach.

Let us consider a two-dimensional localization problem. The three-dimensional localization problem can be dealt with similarly. Suppose that there are N reference tags and M readers deployed in the concerned area, as shown in Figure 8.2, where the tag to be localized is denoted by T_{loc}.

Suppose that a general coordinate system has been defined. In this coordinate system, the coordinates of the N reference tags are (x_i, y_i), respectively, $i = 1, 2, \ldots, N$, and the coordinate of tag T_{loc} is (x, y). Let $S_1, S_2, \ldots, S_M$ denote the signal strengths of tag T_{loc} received by the M readers, respectively, and $s_{i1}, s_{i2}, \ldots, s_{iN}$ denote, respectively, the signal strengths of N reference tags received by the ith reader, $i = 1, 2, \ldots, M$. Define

$$\mathbf{S} = \begin{bmatrix} S_1 \\ S_2 \\ \vdots \\ S_M \end{bmatrix}, \quad \mathbf{S}_{\text{ref}} = \begin{bmatrix} s_{11} & s_{12} & \cdots & s_{1N} \\ s_{21} & s_{22} & \cdots & s_{2N} \\ \vdots & \vdots & \ddots & \vdots \\ s_{M1} & s_{M2} & \cdots & s_{MN} \end{bmatrix}, \quad \mathbf{s}_{\text{ref},i} = \begin{bmatrix} s_{1i} \\ s_{2i} \\ \vdots \\ s_{Mi} \end{bmatrix}.$$

Note that matrix $\mathbf{S}_{\text{ref}}$ is organized in such a way that each of its columns is the signal strength vector of all reference tags corresponding to a given reader and each of its rows is the signal strength vector of a specific tag corresponding to all the readers. Two localization algorithms can be developed by exploiting the structure of matrix $\mathbf{S}_{\text{ref}}$ from the aforementioned two views.

The first algorithm takes a column-wise view of S_{ref}. Define the received signal strength error (RSSE) η_i by

$$\eta_i = ||s_{\text{ref},i} - S|| = \sqrt{\sum_{j=1}^{M} (s_{ji} - S_j)^2}, \quad i = 1, 2, \ldots, N.$$

The simplest way for determining the position of tag T_{loc} is to let (x, y) be the coordinate of the reference tag that has smallest η_i, i.e.,

$$\begin{cases} i_1 = \arg \min_i \{\eta_i, \ i = 1, 2, \ldots, N\}, \\ (x, y) = (x_{i_1}, y_{i_1}). \end{cases} \tag{8.18}$$

A drawback of algorithm (8.18) is that its resolution is determined by the density of the reference tags and the signal strength information provided by the reference tags other than the tag with the smallest RSSE is wasted. To overcome this drawback, the so-called k-NN approach is proposed in [34]. In the k-NN approach, we select k reference tags with smallest RSSEs η_i in the sense that the RSSEs of all other reference tags are greater than the RSSEs of these selected reference tags. Denote the indexes of these selected reference tags as $i_1, i_2, \ldots,$ i_k, i.e.,

$$i_1, i_2, \ldots, i_k := \arg \min_{i_1, i_2, \ldots, i_k \in \{1, 2, \ldots, N\}, i_{j_1} \neq i_{j_2} \text{ when } j_1 \neq j_2} \{\eta_i, i = 1, 2, \ldots, N\}. \tag{8.19}$$

Then the estimate of the coordinate of tag T_{loc} is calculated based on the weighted summation of the coordinates of the selected reference tags:

$$(\hat{x}, \hat{y}) = \sum_{j=1}^{k} w_{i_j}(x_{i_j}, y_{i_j}), \tag{8.20}$$

where $\hat{x}$ and $\hat{y}$ are the estimates of x and y, respectively, and w_{i_j}'s are the weighting coefficients. A common way for determining the weighting coefficients is to let w_i be inversely proportional to η_i^2. This leads to [34]:

$$w_{i_j} = \frac{1/\eta_{i_j}^2}{\sum_{j=1}^{k} 1/\eta_{i_j}^2}, \quad j = 1, 2, \ldots, k. \tag{8.21}$$

Equations (8.19)–(8.21) characterize the whole data processing process for the k-NN localization algorithm. It can be easily seen that the accuracy of the approach critically depends on k. As shown in the experiments of [34], $k = 4$ works best in most cases in the investigated scenario.

The second algorithm takes a row-wise view of S_{ref}. Define the RSSE ζ_{ij} by

$$\zeta_{ij} = |s_{ij} - S_i|, \quad i = 1, 2, \ldots, M, \ j = 1, 2, \ldots, N.$$

Let us examine all the ζ_{ij} for any a fixed i, i.e., to compare the signal strength of tag T_{loc} and the received signal strengths of all reference tags at the ith reader. Clearly, when ζ_{ij} is small for some j, tag T_{loc} will be *possibly* near to the jth reference tag. However, some tags whose

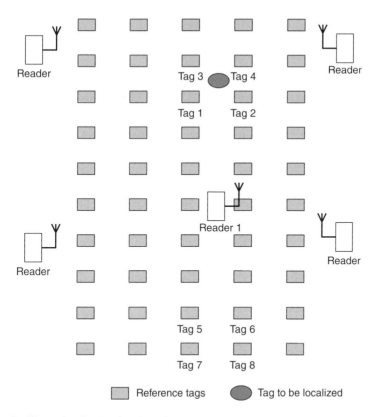

Figure 8.3 An illustration for the situation where some tags are of the same range of RSSEs as that of the tag to be localized, but their positions are far from that of the tag to be localized.

RSSEs are very small might be far from tag T_{loc}, as shown in Figure 8.3. It can be observed from this figure that, for reader 1, in general, the RSSE values of tags 1, 2, 3 and 4 will be in the same range as those of tags 5, 6, 7 and 8, respectively. However, the positions of latter group of tags are far from that of the tag to be localized compared to the former group of tags.

Based on the above consideration, we use a vector T_i to store the indexes of all the tags that are possibly near to tag T_{loc}. To this end, reader i needs to adaptively set up a threshold so that the indexes of relevant reference tags will be stored. Define

$$\xi_i = \sum_{j=1}^{N} |s_{ij} - S_i|^2, \quad i = 1, 2, \ldots, M.$$

Then we can calculate T_i as

$$T_i = \left\{ \text{all } j \in \{1, 2, \ldots, N\} \text{ such that } \frac{\zeta_{ij}^2}{\xi_i} \le \epsilon, \right\}$$

where ϵ is a small constant. For example, we can set $\epsilon = 0.01$.

The indexes of the *true* nearest reference tags can be found in the set

$$T_{\rm c} := \bigcap_{i=1}^{M} T_i.$$

In set $T_{\rm c}$, the indexes of some reference tags that have small RSSEs while being far from tag $T_{\rm loc}$ are removed if the positions of the readers are well arranged.

The estimate of the coordinate of tag $T_{\rm loc}$ can be calculated in two steps. In the first step, we estimate (x, y) based on the information from an individual reader, namely

$$\hat{p}_i = \sum_{j \in T_{\rm c}} \bar{w}_{ij}(x_j, y_j), \quad i = 1, 2, \ldots, M,$$

$$\bar{w}_{ij} = \frac{1/\zeta_{ij}^2}{\sum\limits_{j \in T_{\rm c}} 1/\zeta_{ij}^2}, \quad i = 1, 2, \ldots, M.$$

where $\hat{p}_i$ denotes the estimate of (x, y) based on the information from the ith reader. In the second step, all the $\hat{p}_i$s are fused based on the accuracies of the readers. This gives

$$\hat{p} = \sum_{i=1}^{M} \breve{w}_i \hat{p}_i,$$

$$\breve{w}_i = \frac{1/\alpha_i^2}{\sum\limits_{i=1}^{M} 1/\alpha_i^2}, \quad i = 1, 2, \ldots, M,$$

where $\hat{p}$ stands for the estimate of (x, y) based on all available information and α_i the accuracy of reader i.

8.3   RFID Ranging – Frequency-Domain PDoA Approach

The frequency-domain (FD) phase difference of arrival (PDoA) is a ranging method. It allows coherent signal processing and therefore, at least in theory, can improve range estimation performance of passive RFID tags compared to RSS-based techniques [67].

In this section, the basic design principle of FD PDoA will be addressed. First, we discuss the ranging method for a dual-frequency system. Then the approach is extended to a multi-frequency system.

In a dual-frequency RFID system, the reader transmits a dual-frequency continuous wave to the tag. Suppose that the two carrier frequencies are f_1 and f_2 (assuming $f_2 > f_1$). Then the transmit signal at the reader can be expressed as

$$x(t) = \exp(j2\pi f_1 t + j2\pi f_2 t).$$

Let r be the distance between the reader and tag. Let h represent the channel fading of the combined forward and backward links and let γ denote the backscatter modulation signal at the tag. Note that $\gamma(t)$ is a complex signal, so that both phase and amplitude modulations can be

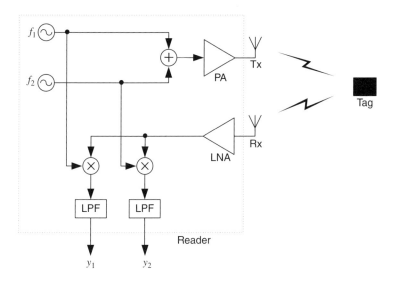

Figure 8.4 A schematic diagram of the transmitter and receiver structure of a dual-frequency RFID system, where Tx, Rx, PA, LNA and LPF stand for transmit antenna, receive antenna, power amplifier, low-noise amplifier and low-pass filter, respectively.

characterized by $\gamma(t)$. The two back tones will be down-converted at two mixers corresponding to the two carrier frequencies and then demodulated at two receiver branches, yielding two separate signals, as shown in Figure 8.4.

The output signals at the reader's receiver can be expressed as:

$$y_i(t) = h\gamma(t)\exp(-\jmath\varphi_i),$$

$$\varphi_i = \frac{4\pi f_i r}{c}, \quad i = 1, 2,$$

where c is the speed of light.

Suppose that the phase offset of the system and the phases contained in the fading h and $\gamma(t)$ do not change with the carrier frequency or can be calibrated out and the tag does not move much (in the sense of much less than the wavelength) during the measurement [35]. Then the distance-related phase information φ_i can be used to calculate the range:

$$\varphi_2 - \varphi_1 = \frac{4\pi(f_2 - f_1)r}{c} \quad \Leftrightarrow \quad r = \frac{c(\varphi_2 - \varphi_1)}{4\pi(f_2 - f_1)}. \tag{8.22}$$

Since the phase difference is bounded by $|\varphi_2 - \varphi_1| \leq 2\pi$, from equation (8.22) we obtain the maximal unambiguous range that the FD-PDoA approach can resolve as follows

$$r_{\max} = \frac{c}{2(f_2 - f_1)}. \tag{8.23}$$

When the actual distance exceeds $r_{\max}$, the distance can be only solved in the sense of $\hat{r} = r \bmod r_{\max}$, where $\hat{r}$ is the distance obtained by the FD-PDoA approach, i.e., range ambiguity happens.

From equation (8.23) it is seen that the frequency separation should be decreased if we want to increase the maximal unambiguous range. However, a small frequency separation would make the phase difference measurement sensitive to noises. Using multi-frequency PDoA approach is an effective way to solve this problem [30].

In a multi-frequency RFID system, the reader transmits to the tag a continuous wave consisting of M frequencies $f_1, f_2, \ldots, f_M$. Without loss of generality, we assume that $f_1 < f_2 < \cdots < f_M$. The transmit signal at the reader can be expressed as

$$x(t) = \exp\left(J2\pi \sum_{i=1}^{M} f_i t\right).$$

The back wave will be down-converted at M mixers corresponding to the M carrier frequencies and then demodulated at M receiver branches, yielding M separate signals. The system architecture is similar to that of the dual-frequency RFID system shown in Figure 8.4.

Using the similar notations as those for the dual-frequency RFID system, we can express the output signals at the reader's receiver as follows

$$y_i(t) = h\gamma(t)\exp(-J\varphi_i),$$

$$\varphi_i = \frac{4\pi f_i r}{c}, \quad i = 1, 2, \ldots, M.$$

Using the phase φ_1 as a reference signal, we obtain $M - 1$ independent phase differences

$$\Delta\varphi_{i1} := \varphi_i - \varphi_1 = \frac{4\pi(f_i - f_1)r}{c}, \quad i = 2, \ldots, M.$$

Suppose that there is a constant D_0 such that all $\frac{c}{f_i - f_1}$ can be expressed as a multiple of D_0, i.e.,

$$\frac{c}{f_i - f_1} = v_i D_0, \quad v_i \text{ is an integer}, \quad i = 2, \ldots, M.$$

Then we have

$$\Delta\varphi_{i1} = \frac{2\pi}{v_i}\frac{2r}{D_0}, \quad i = 2, \ldots, M. \tag{8.24}$$

Considering that r might exceed the maximal unambiguous range, while the measured variable $\Delta\varphi_{i1}$ is always in the range of $[0, 2\pi)$, equation (8.24) should be changed to

$$\Delta\varphi_{i1} = \frac{2\pi}{v_i}\frac{2r}{D_0} - 2\pi\kappa_i = \frac{2\pi}{v_i}\left(\frac{2r}{D_0} - v_i\kappa_i\right), \quad i = 2, \ldots, M. \tag{8.25}$$

to take into account the phase wrapping, where κ_i is an unknown integer.

From equation (8.25) it can be seen that the maximal unambiguous range is given by

$$r_{\max} = \frac{D_0}{2}\mathrm{LCD}(v_2, \ldots, v_M),$$

where LCD denotes the least common denominator of relevant arguments. When $v_2, \ldots, v_M$ are coprime, $r_{\max}$ is maximized (denoted by $\bar{r}_{\max}$):

$$\bar{r}_{\max} = \frac{D_0}{2}\prod_{i=2}^{M} v_i.$$

To get some feeling about the effectiveness of the aforementioned approach, let us consider the following example for an RFID system in UHF range:

Case i: $M = 2, f_1 = 900$ MHz, $f_2 = 930$ MHz.

Case ii (uniform frequency separation): $M = 3$, $f_1 = 900$ MHz, $f_2 = 915$ MHz, $f_3 = 930$ MHz.

Case iii ('coprime' frequency separation): $M = 3$, $f_1 = 900$ MHz, $f_2 = 900 + \frac{100}{7} \approx 914.2857$ MHz, $f_3 = 930$ MHz.

For Case i, from equation (8.23) we can easily find that $r_{max} = 5$ m.

For Case ii, we have

$$\frac{c}{f_2 - f_1} = 20 \text{ m},$$

$$\frac{c}{f_3 - f_1} = 10 \text{ m}.$$

From the above equations we have $D_0 = 10$ m, $v_2 = 2$, and $v_3 = 1$. This gives $r_{max} = 10$ m.

For Case iii, we have

$$\frac{c}{f_2 - f_1} = 21 \text{ m},$$

$$\frac{c}{f_3 - f_1} = 10 \text{ m}.$$

From the above equations we have $D_0 = 1$ m, $v_2 = 10$, and $v_3 = 21$. This gives $r_{max} = 105$ m.

It is seen that the maximal unambiguous range is greatly increased by a proper choice of the carrier frequencies in a multi-frequency RFID system.

When $r < r_{max}$, equation (8.24) holds true. Therefore, there is only one unknown with $M - 1$ equations. Then r can be solved using a simple method of least squares or weighted least squares fit. The latter method makes sense since the phase difference observation with a larger frequency separation can be trusted more and hence weighted more due to the fact that it is not as sensitive to noises as the phase difference observation with a smaller frequency separation.

An important feature of the PDoA approach is that the effect of the uncertain charging time incurred in RFID tags is automatically removed, since the tones at different receiver branches share the same uncertain charging time of a common tag.

In [40], the details for how to measure the PDoA are discussed. It is also shown that the mean absolute error of the range estimation is 0.14 m by using the measurement setup in [40].

8.4 RFID AoA Finding – Spatial-Domain PDoA

The basic principle of the FD-PDoA approach discussed in the preceding section is that it utilizes the different phase delays exhibited by signals with different carrier frequencies when propagating over the same distance between the reader and tag [67]. Another approach is the spatial-domain (SD) PDoA. In this approach, the different phase delays exhibited by signals with the same carrier frequencies but received by different antennas located at different places are exploited to measure the angle information. This is actually equivalent to the traditional beamforming approach. In this section, we will describe the relevant algorithms for calculating the angle of the tags. First the data model, i.e. the measurement equation for the received signal, will be discussed. Then two widely used algorithms will be presented.

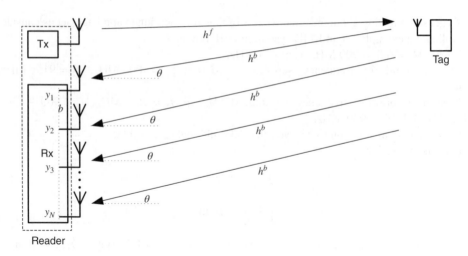

Figure 8.5   An illustration for the reader's transmitter and receiver structure for spatial domain PDoA localization problem: for the case of a single tag.

Let us consider the RFID system as shown in Figure 8.5, where the reader is equipped with a single transmit antenna and N receive antennas. The N receive antennas are located along a line with the distance between two neighbouring antennas being b. The forward channel from the transmit antenna of the reader to the tag is h^f, and backward channel from the tag to the ith receive antenna of the reader is h_i^b. Let $x(t)$, $y_i(t)$ and $\gamma(t)$ denote, respectively, the transmit signal of the reader, the received signal at the ith receive antenna of the reader and the backscattering signal of the tag at time t. In AoA-finding problem, a sinusoidal wave is generally adopted for transmit signal $x(t)$, that is, $x(t) = \exp(j\omega t)$, where ω is the angular frequency of the carrier.

To make further analysis possible, we need the following two assumptions.

Assumption 8.1   The tag is far from the antenna array at the reader so that the incident angles of the signal from the tag to all the receive antennas can be approximated by the same angle θ.

Assumption 8.2   The channel fading of backward links, excluding the excess phase differences caused by different travel distances of the signal, from the tag to all the receive antennas of the reader, can be approximated by the same fading h^b, i.e., $h_1^b = h_2^b = \cdots = h_N^b := h^b$.

Notice that a prerequisite for Assumption 8.2 is that Assumption 8.1 holds true. On the other hand, if Assumption 8.1 holds true, Assumption 8.2 is also a reasonable assumption if the array size $(N - 1)b$ is much less than the distance between the tag and receive antenna array of the reader.

Under Assumptions 8.1 and 8.2, the received baseband signal at the ith receive antenna of the reader can be expressed as

$$y_i(t) = h^f h^b \gamma(t) \exp(-j\varphi_0) \exp(-j\varphi_i) + n_i(t),$$

$$\varphi_i = \omega \frac{(i-1)b \sin\theta}{c}, \quad i = 1, 2, \ldots, N, \tag{8.26}$$

where n_i is the measurement noise, and θ is the AoA to be solved. The phase shift φ_0 is caused by the common travel time of the sinusoidal wave from the reader to the tag and then back to the reader (at the first receive antenna of the antenna array). Note that φ_0 is the same for all the receive antennas. Let

$$h = h^f h^b \exp(-j\varphi_0).$$

Writing equation (8.26) in a vector form gives

$$\mathbf{y}(t) := \begin{bmatrix} y_1(t) \\ y_2(t) \\ \vdots \\ y_N(t) \end{bmatrix} = \mathbf{a}(\theta)h\gamma(t) + \mathbf{n}(t), \tag{8.27}$$

where

$$\mathbf{a}(\theta) = \begin{bmatrix} 1 \\ \exp\left(-j\omega\frac{b\sin\theta}{c}\right) \\ \vdots \\ \exp\left(-j\omega\frac{(N-1)b\sin\theta}{c}\right) \end{bmatrix}, \quad \mathbf{n}(t) := \begin{bmatrix} n_1(t) \\ n_2(t) \\ \vdots \\ n_N(t) \end{bmatrix}.$$

In equation (8.27), there are N equations and two unknowns θ and h. Therefore, the two unknowns can be solved easily from these N equations. An intuitive method is as follows. Suppose that θ is known first and denote it temporarily as $\bar{\theta}$. Pre-multiplying $\mathbf{a}^\dagger(\theta)$ at both sides of (8.27) gives

$$\mathbf{a}^\dagger(\theta)\mathbf{y}(t) = Nh\gamma(t) + \mathbf{a}^\dagger(\theta)\mathbf{n}(t). \tag{8.28}$$

The second term of the right-hand side (RHS) of equation (8.28) is noise, which should be statistically much small compared to the first term of the RHS of (8.28) and hence is neglected. Therefore, we can solve h from equation (8.28). Substituting the solved h into equation (8.27) we can solve θ. Denote the solved θ as $\breve{\theta}$. If $\breve{\theta}$ matches with $\bar{\theta}$ well, then $\breve{\theta}$ is what we need to find. Otherwise take a new θ and repeat the above procedure until θ is solved. This method is actually the well-known beamforming approach, and this recursive procedure is equivalent to sweeping all the possible angle θ.

A drawback of the beamforming approach is that the statistical information of the noise is not exploited and therefore the accuracy of the algorithm is not satisfactory. Another drawback is that it is computationally expensive when multiple tags/sources exist in the concerned area. To overcome these drawbacks, the MUSIC (multiple signal characterization) algorithm was developed during 1970s and 1980s [3, 4, 47, 48, 54]. It fully exploits the geometrical structure and statistical properties of the measurement data. Therefore, this algorithm should be useful in finding the AoA of multiple tags.

In the MUSIC algorithm, several tags can be dealt with simultaneously and statistical properties of relevant signals and measurement noises are needed. Therefore, we extend the data model (8.26) or (8.27) to the case of multiple tags and multiple measurements.

Now suppose that the reader has the same setup as that in Figure 8.5, while there are M tags in the reader's reading zone, as shown in Figure 8.6.

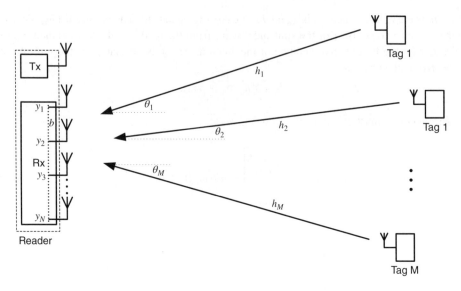

Figure 8.6 An illustration for the reader's transmitter and receiver structure for spatial domain PDoA localization problem: for the case of multiple tags.

The composite channel from the mth tag to the reader is denoted by h_m, which includes the fading of both forward and backward links and the common phase delay from the mth tag to the first antenna of the receive antenna array at the reader. Let $\gamma_m(t)$ denote the backscattering signal of the mth tag at time t. Similar to the single-tag case, the following two assumptions are assumed.

Assumption 8.3 All the tags are far from the antenna array at the reader so that the incident angles of the signals from the mth tag to all the receive antennas can be approximated by the same angle θ_m, $m = 1, \ldots, M$.

Assumption 8.4 The channel fading of backward links from a specific tag to all the receive antennas of the reader can be approximated by the same fading, and furthermore, all the composite fading h_m, $m = 1, \ldots, M$, keep unchanged during the whole measurement phase.

In addition, we need two other assumptions.

Assumption 8.5 The number of antennas in the reader's receive antenna array is greater than the number of tags to be identified, i.e., $N > M$.

Assumption 8.6 The measurement noise satisfies: $\mathbb{E}[\mathbf{n}(t_k)] = 0$ and $\mathbb{E}[\mathbf{n}(t_k)\mathbf{n}^\dagger(t_k)] = \sigma_{\mathbf{n}}^2 \mathbf{I}$, where $\sigma_{\mathbf{n}}^2$ is a constant.

Based on Assumptions 8.3 and 8.4, the received baseband signal at the receive antenna array of the reader can be expressed as

$$\mathbf{y}(t_k) = \sum_{m=1}^{M} \mathbf{a}(\theta_m) h_m \gamma_m(t_k) + \mathbf{n}(t_k) = \mathbf{A}(\theta)\beta(t_k) + \mathbf{n}(t_k), k = 1, 2, \ldots, K, \tag{8.29}$$

where t_k denotes the measurement instant, and

$$\mathbf{A}(\theta) = [\mathbf{a}(\theta_1) \ \mathbf{a}(\theta_2) \ \cdots \ \mathbf{a}(\theta_M)], \ \ \beta(t_k) = \begin{bmatrix} h_1\gamma_1(t_k) \\ h_2\gamma_2(t_k) \\ \vdots \\ h_M\gamma_M(t_k) \end{bmatrix}, \ \ \theta = \begin{bmatrix} \theta_1 \\ \theta_2 \\ \vdots \\ \theta_M \end{bmatrix}.$$

From equation (8.29) and Assumption 8.6 we have

$$\mathbf{R_y} := \mathbb{E}[\mathbf{y}(t_k)\mathbf{y}^\dagger(t_k)] = \mathbf{A}(\theta)\mathbf{R}_\beta\mathbf{A}^\dagger(\theta) + \sigma_\mathbf{n}^2\mathbf{I}, \tag{8.30}$$

where

$$\mathbf{R}_\beta = \mathbb{E}[\beta(t_k)\beta^\dagger(t_k)].$$

To easy exposition, we assume that the following assumption holds true.

Assumption 8.7 The matrix $\mathbf{R}_\beta = \mathbb{E}[\beta(t_k)\beta^\dagger(t_k)]$ is nonsingular and does not depend on measurement time instants.

From equation (8.30) we can see that matrix $\mathbf{R_y}$ holds some very nice properties. Since the angles contained in θ are generically different from each other, matrix $\mathbf{A}(\theta)$ is of full column rank generically. Therefore, matrix $\mathbf{A}(\theta)\mathbf{R}_\beta\mathbf{A}^\dagger(\theta)$ is positive semidefinite with M non-zero eigenvalues. Denote these M non-zero eigenvalues as $\mu_1, \ldots, \mu_M$, respectively. Then there exists a unitary matrix $\mathbf{U}$ such that

$$\mathbf{U}^\dagger\mathbf{A}(\theta)\mathbf{R}_\beta\mathbf{A}^\dagger(\theta)\mathbf{U} = \mathtt{diag}(\mu_1, \mu_2, \cdots, \mu_M, 0, \cdots, 0), \tag{8.31}$$

where $\mathtt{diag}$ stands for a diagonal matrix with its diagonal entries being given by the arguments in that order.

Based on equation (8.31), left- and right-multiplying matrix $\mathbf{R_y}$ by $\mathbf{U}^\dagger$ and $\mathbf{U}$, respectively, yields

$$\mathbf{U}^\dagger\mathbf{R_y}\mathbf{U} = \mathbf{U}^\dagger\mathbf{A}(\theta)\mathbf{R}_\beta\mathbf{A}^\dagger(\theta)\mathbf{U} + \sigma_\mathbf{n}^2\mathbf{U}^\dagger\mathbf{U}$$

$$= \mathtt{diag}(\mu_1, \mu_2, \cdots, \mu_M, 0, \cdots, 0) + \sigma_\mathbf{n}^2\mathbf{I} \tag{8.32}$$

$$= \mathtt{diag}\left(\mu_1 + \sigma_\mathbf{n}^2, \mu_2 + \sigma_\mathbf{n}^2, \cdots, \mu_M + \sigma_\mathbf{n}^2, \sigma_\mathbf{n}^2, \cdots, \sigma_\mathbf{n}^2\right). \tag{8.33}$$

Equation (8.33) says that matrix $\mathbf{R_y}$ has an minimum eigenvalue $\sigma_\mathbf{n}^2$ of algebraic multiplicity $N - M$, and all the other eigenvalues are greater than this minimum eigenvalue.

Let us decompose $\mathbf{U}$ as

$$\mathbf{U} = [\mathbf{U}_\mathrm{s} \ \mathbf{U}_\mathrm{n}], \ \ \mathbf{U}_\mathrm{s} \in C^{N \times M}, \mathbf{U}_\mathrm{n} \in C^{N \times (N-M)},$$

i.e., U_s is the left M columns of U, and U_n is the right $N - M$ columns of U. From equation (8.32) we get

$$U^\dagger R_y U - \sigma_n^2 U^\dagger U = \text{diag}(\mu_1, \mu_2, \cdots, \mu_M, 0, \cdots, 0)$$

$$\Updownarrow$$

$$(R_y - \sigma_n^2 I)U = U \begin{bmatrix} \Upsilon & 0 \\ 0 & 0 \end{bmatrix}$$

$$\Updownarrow$$

$$(R_y - \sigma_n^2 I)[U_s \ \ U_n] = [U_s \ \ U_n] \begin{bmatrix} \Upsilon & 0 \\ 0 & 0 \end{bmatrix}$$

$$\Downarrow$$

$$(R_y - \sigma_n^2 I)U_n = 0$$

$$\Updownarrow \tag{8.34}$$

$$A(\theta)R_\beta A^\dagger(\theta)U_n = 0, \tag{8.35}$$

where $\Upsilon = \text{diag}(\mu_1, \mu_2, \cdots, \mu_M)$.

Equation (8.34) says that the column vectors in U_n belong to the null space of $R_y - \sigma_n^2 I$, while equation (8.35) implies that

$$A^\dagger(\theta)U_n = 0 \quad \Longleftrightarrow \quad a^\dagger(\theta_m)U_n = 0, \ m = 1, \ldots, M \tag{8.36}$$

since $A(\theta)$ and R_β are of full rank.

The MUSIC algorithm is based on equation (8.36). Define

$$F(\theta) := \frac{1}{a^\dagger(\theta)U_n U_n^\dagger a(\theta)}. \tag{8.37}$$

It is seen from equation (8.36) that $F(\theta)$ approaches infinity when θ equals the AoA of one of the tags; otherwise the value of $F(\theta)$ is relatively small. In practical systems, the 'infinity' appears as a peak.

Note that the angle variable θ in equation (8.37) is a scalar. Sweeping all the possible θ in a fine grid of the interval $[0, 2\pi]$ will yield M peaks of $F(\theta)$. For each of these M peaks, the corresponding θ is the AoA of one of the tags to be localized.

Therefore, the remaining issue for the MUSIC algorithm is how to calculate the correlation matrix R_y. This is approximated by

$$R_y = \frac{1}{K} \sum_{k=1}^{K} y(t_k)y^\dagger(t_k). \tag{8.38}$$

The MUSIC algorithm is summarized as follows.

Algorithm 8.1 MUSIC. Initial data: given measured data $\mathbf{y}(t_k)$, $k = 1, \ldots, K$.

Step 1 Calculate $\mathbf{R_y}$ according to (8.38).

Step 2 Perform eigenvalue decomposition for $\mathbf{R_y}$, yielding result (8.33).

Step 3 In terms of the distribution of the eigenvalues of $\mathbf{R_y}$, determine parameters M, σ_n^2, and correspondingly matrix $\mathbf{U_n}$, according to the feature as shown in equation (8.33).

Step 4 Calculate θ_m, $m = 1, \ldots, M$, in terms of (8.37), which is usually performed by sweeping all the points of a fine grid of θ in $[0, 2\pi]$.

The main computational burden is caused by Step 4 in Algorithm 8.1.

The statistical properties of the MUSIC algorithm are well documented in [54, 55]. It is proven in [54] that, when the number of measurements K is sufficiently large, the MUSIC estimator approaches to the maximum likelihood estimator for any $N > M$ if the signals are uncorrelated, and hence in this case, the MUSIC estimator can achieve the Cramer–Rao bound [27, 42], the best achievable performance of any an unbiased estimator. It is also shown in [54] that the error variance of the MUSIC estimator monotonically decreases with increasing N. Therefore, increasing the number of the array elements at the reader's receiver is always beneficial for improving the system performance.

Based on extensive simulations conducted by Lincoln Laboratory of Massachusetts Institute of Technology, it is found that, among currently accepted high-resolution algorithms, MUSIC is the most promising and a leading candidate for further study and actual hardware implementation[3].

There are many modified versions of MUSIC. See, for example, references [55, 56]. The basic idea is to change matrix $\mathbf{U_n}$ in equation (8.37) to $\mathbf{U_n W}$, where $\mathbf{W}$ is a weighting matrix which can be designed based on prior knowledge about the measurement system.

A prerequisite for the SD-PDoA approach is that the directions of impinging waves on the receive array from a tag to be localized are approximately parallel to each other, as shown in Figure 8.5. This requirement becomes critical for RFID systems since the readers' reading range is typically within 10 m or under several tens of metres in some cases. When the distance between the tag and reader is not sufficiently greater than the array size, the phase difference between the received signals at any two array elements depends not only on the directions of the impinging waves but also on the distance between the tag and the reader. This fact will make the corresponding parameter estimation problem difficult to solve. An initial discussion for this problem was reported in [66].

8.5 NLoS Issue

The NLoS issue has been extensively studied in the localization problem for wireless systems. The main studies are focused on two concerns: (i) how to detect the NLoS and (ii) how to correct the error caused by the NLoS. These two concerns also exist for RFID localization problem and can be dealt with using similar approaches.

An early account for the NLoS issue can be found in [61], where the history of the standard deviation of ranging measurements from different receivers is compared. If a receiver

---

[3] See the website http://en.wikipedia.org/wiki/Multiple_signal_classification.

shows consistently a large standard deviation of measurements compared to other receivers, an NLoS for the former receiver can be reliably declared and the ranging bias can be derived from the history of the standard deviation of measurements for that receiver. A prerequisite for this approach is that there are at least some LoS measurements and the accuracies of all the measurements are roughly in the same order. An advantage of this method is that the probability of false alarm of an NLoS is low and the resultant bias estimate is accurate compared to other existing NLoS mitigation approaches.

In [8], the least-median-of-squares technique is used to detect and reject NLoS measurements and a minimum number of range measurements are selected to obtain the estimate of the tag position. In [9], a residual test was proposed to detect NLoS and NLoS-related measurements, if detected, were removed from the localization purpose. In [10], a residual ranking algorithm for NLoS detection was proposed, and this algorithm was modified in [12] by taking into account the asymmetric property of the distribution of the NLoS error.

In [46, 58], the knowledge about the geometrical structure between transmitters and receivers was incorporated into an optimization problem for solving the ToA from measurement data, and the NLoS bias can be correspondingly suppressed. It remains unclear about how much the localization accuracy can be improved, although the NLoS error can be reduced somehow by using this approach.

Basically, all the aforementioned approaches can be extended to RFID localization systems to solve the NLoS issue, even though very few reports have been devoted to this issue. In the following, we will extend the NLoS detection algorithm proposed in [10] and its modified version in [12] to RFID localization systems.

Let us consider an RFID localization system consisting N readers and a tagged object. The readers, whose positions (x_i, y_i) are known, are used to measure the range r_i between the readers and tag, $i = 1, \ldots, N$ ($N \geq 3$). Our purpose is to find the position $\mathbf{p} = (x, y)$ of the object or tag. Generally, the measurement equation can be expressed as

$$r_i = f(\mathbf{p}) + n_i + e_i, \quad i = 1, \ldots, N,$$

where $f(\mathbf{p}) = \sqrt{(x - x_i)^2 + (y - y_i)^2}$, n_i is the ranging noise (which is the noise after the data processing unit, derived from the RF front-end noise) and e_i stands for the bias caused by the NLoS.

In the case of LoS measurement, $e_i = 0$. In the case of NLoS measurement, $e_i > 0$ and it typically holds true that $e_i \gg n_i$ almost certainly. It is the latter property that is used to detect NLoS or mitigate the effect of NLoS.

If the NLoS is not taken into account, the LS estimate of the position $\mathbf{p}$ is given by

$$\hat{\mathbf{p}} = \arg \min_{\mathbf{p}} \sum_{i=1}^{N} (r_i - f(\mathbf{p}))^2. \tag{8.39}$$

Equation (8.39) can be solved by using the approaches as presented in Section 8.2.

For the purpose of removing/mitigating NLoS errors, we can select a subset of the N measurements to formulate a new position estimate $\hat{\mathbf{p}}$. Let S_k, $k = 1, \ldots, K$, are all the possible subsets of N measurements that can be used to decide the position $\mathbf{p}$, where $K = \sum_{i=3}^{N} \binom{N}{i}$.

For a given subset S_k and a given position $\mathbf{p}$, define the related residue squares as

$$R(\mathbf{p}, S_k) = \sum_{i \in S_k} (r_i - f(\mathbf{p}))^2. \qquad (8.40)$$

Now calculate the LS estimate of $\mathbf{p}$ based on equation (8.40) as follows (denoted by $\hat{\mathbf{p}}_k$):

$$\hat{\mathbf{p}}_k = \arg \min_{\mathbf{p}} R(\mathbf{p}, S_k),$$

and the corresponding residue (normalized by the size of the subset used):

$$\tilde{R}(\hat{\mathbf{p}}_k, S_k) = \frac{1}{|S_k|} \sum_{i \in S_k} (r_i - f(\hat{\mathbf{p}}_k))^2, \qquad (8.41)$$

where $|S_k|$ stands for the size of S_k, i.e. the number of measurements contained in the subset S_k.

The final estimate of $\mathbf{p}$ is given by the weighted linear combination of the above intermediate estimate with the weights being inversely proportional to the corresponding residue $\tilde{R}(\hat{\mathbf{p}}_k, S_k)$ [10]:

$$\hat{\mathbf{p}} = \frac{\displaystyle\sum_{k=1}^{K} [\tilde{R}(\hat{\mathbf{p}}_k, S_k)]^{-1} \hat{\mathbf{p}}_k}{\displaystyle\sum_{k=1}^{K} [\tilde{R}(\hat{\mathbf{p}}_k, S_k)]^{-1}}.$$

An immediate modification of the above algorithm is to directly remove those subsets whose residues, as calculated in equation (8.41), are too large compared to the residues of other subsets.

Simulation results in [10] show that substantial performance improvement by using the above algorithm has been achieved compared with the traditional LS estimation approach even when the NLoS measurements are not distinguishable. If the standard deviation of the measurement noise is small, the above algorithm performs so well as if only LoS measurements were used.

A prerequisite for this approach is that there are at least three LoS measurements. The advantage of this approach is that there are no requirements on any *a priori* information about the scenario. The disadvantage is its computational burden. When the number N is large, the computational burden is very heavy since the LS estimation algorithm should be performed for all possible K subsets.

A possible remedy for this disadvantage is first to remove some NLoS measurements by using a modified residual algorithm, as shown in [12].

The residue defined in equation (8.40) has three limitations. First, the information about the variance of measurement noises is not exploited. Second, the property that the residue is not symmetric with respect to the origin is not exploited. Since $e_i \geq 0$, it is less likely that the ith measurement belongs to an NLoS if $r_i - f(\mathbf{p}) < 0$ considering that the measurement noise is relatively small, while the square operation in equation (8.40) removes the sign information of $r_i - f(\mathbf{p})$. Finally, the summation form in equation (8.40) can only indicate how likely a group of measurements, but not which one(s), contains NLoS. Considering the aforementioned limitations, [12] modified the residue as the following:

$$R(r_i, \mathbf{p}) = 1 - Q\left(\frac{r_i - f(\mathbf{p})}{\sigma_i}\right), \tag{8.42}$$

where σ_i^2 is the variance of the ith measurement noise n_i, and $Q(\cdot)$ stands for the Gaussian Q-function defined by equation (4.5) in Chapter 4.

For a fixed $\mathbf{p}$, $R(r_i, \mathbf{p})$ is a monotonously increasing function of r_i. Let $\mathbf{p}$ be the true position of the tag. If $r_i = f(\mathbf{p}) - 3\sigma_i$, which indicates that this measurement most likely belongs to a LoS, then we have $R(r_i, \mathbf{p}) = 0.0013$. On the other hand, if $r_i = f(\mathbf{p}) + 3\sigma_i$, which indicates that this measurement most likely belongs to an NLoS, then we have $R(r_i, \mathbf{p}) = 0.9987$. Therefore, the function $R(r_i, \mathbf{p})$ provides a good measure about if a measurement belongs to an NLoS. The larger the residue $R(r_i, \mathbf{p})$, the more likely the measurement belongs to an NLoS.

To proceed with this modification, it is necessary to find an initial estimate of the position $\mathbf{p}$. This can be done by using all the measurements.

According to this idea, the following localization algorithm was proposed in [12].

Algorithm 8.2   Residue-based localization algorithm. Initial data: given all measurement data $\{r_i, \quad i = 1, \ldots, N\}$ and a threshold λ.

Step 1 Calculate an initial estimate of the position $\mathbf{p}$. Denote it by $\hat{\mathbf{p}}$.
Step 2 Calculate the residues $R(r_i, \hat{\mathbf{p}})$ according to equation (8.42).
Step 3 Remove those measurement data whose residues are larger than the pre-defined threshold λ and use the remained measurement data to calculate the new position estimate.
Step 4 If the sum of the residues based on the new position estimate is smaller than the sum of the residues based on the previously obtained position estimate, stop. Otherwise increase the threshold λ to admit more measurement data and go to Step 2.

How to choose parameter λ is a critical issue for Algorithm 8.2, which affects the false alarm probability about NLoS, the convergence speed of the algorithm and the position estimation accuracy.

The NLoS mitigation method discussed in this section can be only applied to range-based RFID localization systems. For AoA-based RFID localization systems, the NLoS issue is more challenging and has been rarely addressed in the literature.

8.6   Summary

In this chapter, two main localization methods, namely geometric approach and proximity approach, have been addressed for RFID indoor localization. The corresponding algorithms are presented. To use the localization algorithms of the geometric approach, the range between readers and tags or AoA of tags must be reliably measured or estimated from the measured information. Two approaches, namely FD PDoA and SD PDoA for measuring the range and AoA, respectively, are discussed. These two approaches are particularly useful in dealing with the uncertain phase delay caused by charging time of RFID tags, which is often difficult to know exactly. The 'coprime' design of the multiple carrier frequencies in the FD-PDoA approach is an efficient method to deal with the range ambiguity problem. It is showed that, for a UHF RFID system with three-tone FD PDoA, a small shift towards the smaller frequency in the middle tone can increase the maximal unambiguous range to more than 100 m from less than 10 m for the same UHF RFID system with uniform separated tones. The MUSIC algorithm is especially effective in the SD-PDoA approach for estimating the AoA of multiple tags in an RFID reader with multiple receive antennas. However, the number of measurements (snapshots) should be sufficiently large to reliably obtain the correlation matrix of the received baseband signals at the reader's antenna array.

In general radio localization systems, it is commonly admitted that the RSS approach is less favoured than other approaches. The reason is that an exact propagation model for radio signals must be used there. More accurately speaking, the power decay exponent of the concerned wireless channels must be obtained. To get exactly the power decay exponent, a measurement campaign might be even required, which costs a lot. Much worse is that the power decay exponent changes with propagation environments. In the RSS approach for RFID localization systems, all these issues are circumvented by deploying a number of reference tags. Due to the fact that tags are cheap, small and easy to deploy, the RSS approach is much favoured in RFID localization systems.

References

[1] A. Azenha and A. Carvalho. Radio frequency localization for AGV positioning. In *14th Saint Petersburg Int. Conf. Integrated Navigation Systems*, pages 301–302, Saint Petersburg, Russia, 28–30 May 2007.

[2] S. Azzouzi, M. Cremer, U. Dettmar, R. Kronberger, and T. Knie. New measurement results for the localization of UHF RFID transponders using an angle of arrival (AoA) approach. In *2011 IEEE Int. Conf. on RFID*, pages 91–97, Orlando, Florida, USA, 12–14 Apr. 2011.

[3] G. Bienvenu and L. Kopp. Adaptivity to background noise spatial coherence for high resolution passive methods. In *Proc. IEEE Int. Conf. Acoust., Speech, Signal Processing*, pages 89–96, Denver, Colorado, USA, Apr. 1980.

[4] G. Bienvenu and L. Kopp. Optimality of high resolution array processing using the eigensystem approach. *IEEE Trans. Acoust., Speech, Signal Processing*, 31:1235–1248 1983.

[5] S. Boncinelli, P. Citti, E. Del Re, G. Campatelli, L. Pierucci, and L. Bocchi. Real time detection and tracking of gauzes by RFID UWB technique. In *2010 IEEE Int. Conf. on RFID*, pages 97–101, Orlando, Florida, USA, 14–15 Apr. 2010.

[6] M. Bouet and A. L. dos Santos. RFID tags: Positioning principles and localization techniques. In *Proc. IFIP Wireless Days*, Dubai, United Arab Emirates, 24–27 Nov. 2008.

[7] J. L. Brchan, L. Zhao, and J. Wu. A real-time RFID localization experiment using propagation models. In *2012 IEEE Int. Conf. on RFID*, pages 141–148, Orlando, Florida, USA, 3–5 Apr. 2012.

[8] R. Casas, A. Marco, J. J. Guerrero, and J. Falco. Robust estimator for non-line-of-sight error mitigation in indoor localization. *EURASIP J. Applied Signal Processing*, 2006:Article ID 43429, 2006.

[9] Y.-T. Chan, W.-Y. Tsui, H.-C. So, and P.-C. Ching. Time-of-arrival based localization under NLOS conditions. *IEEE Trans. Veh. Technol.*, 55:17–24, 2006.

[10] P.-C. Chen. A nonline-of-sight error mitigation algorithm in location estimation. In *Proc. IEEE Wireless Communications Networking Conf.*, pages 316–320, New Orleans, USA, 21–24 Sept. 1999.

[11] E. Colin, A. Moretto, and M. Hayoz. Improving indoor localization within corridors by UHF active tags placement analysis. In *2014 IEEE RFID Technology and Applications Conf.*, pages 181–186, Tampere, Finland, 8–9 Sept. 2014.

[12] L. Cong and W. Zhuang. Nonline-of-sight error mitigation in mobile location. *IEEE Trans. Wireless Commun.*, 4:560–573, 2005.

[13] M. Cremer, A. Pervez, U. Dettmar, T. Knie, and R. Kronberger. Improved UHF RFID localization accuracy using circularly polarized antennas. In *2014 IEEE RFID Technology and Applications Conf.*, pages 175–180, Tampere, Finland, 8–9 Sept. 2014.

[14] C. C. Cruz, J. R. Costa, and C. A. Fernandes. Hybrid UHF/UWB antenna for passive indoor identification and localization systems. *IEEE Trans. Antennas Propag.*, 61:354–361, 2013.

[15] K. Curran, E. Furey, T. Lunney, J. Santos, D. Woods, and A. Mc Caughey. An evaluation of indoor location determination technologies. *J. Location Based Services*, 5(2):61–78, 2011.

[16] D. Dardari, A. Conti, U. Ferner, A. Giorgetti, and M. Z. Win. Ranging with ultrawide bandwidth signals in multipath environments. *Proceedings of the IEEE*, 97:404–426, 2009.

[17] E. DiGiampaolo and F. Martinelli. A passive UHF-RFID system for the localization of an indoor autonomous vehicle. *IEEE Trans. Industrial Electronics*, 59:3961–3970, 2012.

[18] S. Gezici and H. V. Poor. Position estimation via ultra-wide-band signals. *Proceedings of the IEEE*, 97:386–403, 2009.

[19] S. Gezici, Z. Tian, G. B. Giannakis, H. Kobayashi, A. F. Molisch, H. V. Poor, and Z. Sahinoglu. Localization via ultra-wideband radios. *IEEE Signal Processing Magazine*, 22(4):70–84, 2005.

[20] Y. Gu, A. Lo, and I. Niemegeers. A survey of indoor positioning systems for wireless personal networks. *IEEE Communications Surveys & Tutorials*, 11(1):13–32, 2009.

[21] P. Gulden, S. Roehr, and M. Christmann. An overview of wireless local positioning system configurations. In *IEEE Int. Microwave Workshop on Wireless Sensing, Local Positioning, and RFID*, Cavtat, Croatia, 24–25 Sept. 2009.

[22] S. S. Han, H. S. Lim, and J. M. Lee. An efficient localization scheme for a differential-driving mobile robot based on RFID system. *IEEE Trans. Industrial Electronics*, 54:3362–3369, 2007.

[23] M. Hasani, E.-S. Lohan, L. Sydanheimo, and L. Ukkonen. Path-loss model of embroidered passive RFID tag on human body for indoor positioning applications. In *2014 IEEE RFID Technology and Applications Conf.*, pages 170–174, Tampere, Finland, 8–9 Sept. 2014.

[24] C. Hekimian-Williams, B. Grant, X. Liu, Z. Zhang, and P. Kumar. Accurate localization of RFID tags using phase difference. In *2010 IEEE Int. Conf. on RFID*, pages 89–96, Orlando, Florida, USA, 14–16 Apr. 2010.

[25] J. Hightower and G. Borriello. Location systems for ubiquitous computing. *Computer*, 34(8):57–66, 2001.

[26] T. Kaiser and F. Zheng. *Ultra Wideband Systems with MIMO*. John Wiley & Sons, Ltd, Chichester, 2010.

[27] S. M. Kay. *Fundamentals of Statistical Signal Processing: Vol. 1: Estimation Theory*. Prentice-Hall, Upper Saddle River, New Jersey, 1993.

[28] K. Koski, E. S. Lohan, L. Sydanheimo, L Ukkonen, and Y. Rahmat-Samii. Electro-textile UHF RFID patch antennas for positioning and localization applications. In *2014 IEEE RFID Technology and Applications Conf.*, pages 246–250, Tampere, Finland, 8–9 Sept. 2014.

[29] G. Li, D. Arnitz, R. Ebelt, U. Muehlmann, K. Witrisal, and M. Vossiek. Bandwidth dependence of CW ranging to UHF RFID tags in severe multipath environments. In *2011 IEEE Int. Conf. RFID*, pages 19–25, Orlando, FL, USA, 12–14 Apr. 2011.

[30] X. Li, Y. Zhang, and M. G. Amin. Multifrequency-based range estimation of RFID tags. *In 2009 IEEE Int. Conf. on RFID*, pages 147–154, Orlando, Florida, USA, 27–28 Apr. 2009.

[31] H. Liu, H. Darabi, P. P. Banerjee, and J. Liu. Survey of wireless indoor positioning techniques and systems. *IEEE Trans. Systems, Man, and Cybernetics, Part C*, 37:1067–1080, 2007.

[32] X. Liu, J. Peng, and T. Liu. A novel indoor localization system based on passive RFID technology. In *2011 Int. Conf. Electronic & Mechanical Engineering and Information Technology*, Harbin, China, 12–14 Aug. 2011.

[33] A. Loeffler and H. Gerhaeuser. Localizing with passive UHF RFID tags using wideband signals. In *Radio Frequency Identification from System to Applications*, M. B. I. Reaz (ed.). InTech, Rijeka, pages 85–110, 2013. Available from: www.intechopen.com/books/radio-frequency-identification-from-system-to-applications/localizing-with-passive-uhf-rfid-tags-using-wideband-signals.

[34] L. M. Ni, Y. Liu, Y. C. Lau, and A. P. Patil. LANDMARC: Indoor location sensing using active RFID. *Wireless Networks*, 10:701–710, 2004.

[35] P. V. Nikitin, R. Martinez, S. Ramamurthy, H. Leland, G. Spiess, and K. V. S. Rao. Phase based spatial identification of UHF RFID tags. In *2010 IEEE Int. Conf. on RFID*, pages 102–109, Orlando, Florida, USA, 14–16 Apr. 2010.

[36] M. C. O'Connor. IBSS launches healthcare tracking. In *RFID Journal*, 13 Jan. 2005. Available from www.rfidjournal.com/articles/view?1318.

[37] M. C. O'Connor. Emory healthcare tracks its pumps. In *RFID Journal*, 15 May 2007. Available from www.rfidjournal.com/articles/view?3311.

[38] K. Pahlavan, F. O. Akgul, M. Heidari, A. Hatami, J. M. Elwell, and R. D. Tingley. Indoor geolocation in the absence of direct path. *IEEE Wireless Communications*, 13(6):50–58, 2006.

[39] S. Park and S. Hashimoto. Autonomous mobile robot navigation using passive RFID in indoor environment. *IEEE Trans. Industrial Electronics*, 56:2366–2373, 2009.

[40] A. Povalac and J. Sebesta. Phase difference of arrival distance estimation for RFID tags in frequency domain. In *IEEE Int. Conf. on RFID-Technologies and Applications*, pages 188–193, Sitges, Spain, 15–16 Sept. 2011.

[41] A. Pozzebon. A wireless waterproof RFID reader for marine sediment localization and tracking. In *2014 IEEE RFID Technology and Applications Conf.*, pages 187–192, Tampere, Finland, 8–9 Sept. 2014.

[42] C. R. Rao. *Linear Statistical Inference and Its Applications*. John Wiley & Sons, Inc, New York, 1973.

[43] T. M. Ruff and D. Hession-Kunz. Application of radio-frequency identification systems to collision avoidance in metal/nonmetal mines. *IEEE Trans. Ind. Applicat.*, 37:112–116, 2001.

[44] S. S. Saab and Z. S. Nakad. A standalone RFID indoor positioning system using passive tags. *IEEE Trans. Industrial Electronics*, 58:1961–1970, 2011.

[45] T. Sanpechuda and L. Kovavisaruch. A review of RFID localization: Applications and techniques. In *2008 5th Int. Conf. Electrical Engineering/Electronics, Computer, Telecommunications and Information Technology*, pages 69–772, Krabi, Thailand, 14–17 May 2008.

[46] A. H. Sayed, A. Tarighat, and N. Khajehnouri. Network-based wireless location: challenges faced in developing techniques for accurate wireless location information. *IEEE Signal Processing Magazine*, 22(4):24–40, 2005.

[47] R. O. Schmidt. Multiple emitter location and signal parameter estimation. In *Proc. RADC Spectral Estimation Workshop*, pages 243–258, Griffiss AFB, NY, USA, 3–5 Oct. 1979.

[48] R. O. Schmidt. Multiple emitter location and signal parameter estimation. *IEEE Trans. Antennas Propag.*, 34:276–280, 1986.

[49] B. T. Sieskul, T. Kaiser, and F. Zheng. A hybrid SS-ToA wireless NLoS geolocation based on path attenuation: Cramer-Rao bound. In *Proc. IEEE 69th Vehicular Technology Conf.*, Barcelona, Spain, 26–29 April 2009.

[50] B. T. Sieskul, F. Zheng, and T. Kaiser. A hybrid SS-ToA wireless NLoS geolocation based on path attenuation. *IEEE Trans. Veh. Technol.*, 58:4930–4942, 2009.

[51] B. T. Sieskul, F. Zheng, and T. Kaiser. A hybrid SS-ToA wireless NLoS geolocation based on path attenuation: Mobile position estimation. In *Proc. IEEE Wireless Communications and Networking Conf.*, Budapest, Hungary, 5–8 April 2009.

[52] B. T. Sieskul, F. Zheng, and T. Kaiser. On effects of shadow fading in NLoS geolocation. *IEEE Trans. Signal Process.*, 57:4196–4208, 2009.

[53] B. T. Sieskul, F. Zheng, and T. Kaiser. Time-of-arrival estimation in path attenuation. In *Proc. 10th IEEE Int. Workshop on Signal Processing Advances in Wireless Communications*, Perugia, Italy, 21–24 June 2009.

[54] P. Stoica and A. Nehorai. MUSIC, maximum likelihood, and Cramer–Rao bound. *IEEE Trans. Acoust., Speech, Signal Processing*, 37:720–741, 1989.

[55] P. Stoica and A. Nehorai. MUSIC, maximum likelihood, and Cramer–Rao bound: further results and comparisons. *IEEE Trans. Acoust., Speech, Signal Processing*, 38:2140–2150, 1990.

[56] P. Stoica and K. C. Sharman. Maximum likelihood methods for direction-of-arrival estimation. *IEEE Trans. Acoust., Speech, Signal Processing*, 38:1132–1143, 1990.

[57] S. P. Subramanian, J. Sommer, S. Schmitt, and W. Rosenstiel. RIL–reliable RFID based indoor localization for pedestrians. In *16th Int. Conf. Software, Telecommunications and Computer Networks*, pages 218–222, Split, Dubrovnik, Croatia, 25–27 Sept. 2008.

[58] S. Venkatraman, Jr. J. Caffery, and Y. Heung-Ryeol. A novel ToA location algorithm using LoS range estimation for NLoS environments. *IEEE Trans. Veh. Technol.*, 53:1515–1524, 2004.

[59] J. Vongkulbhisal and Y. Zhao. An RFID-based indoor localization system using antenna beam scanning. In *Proc. 9th Int. Conf. Electrical Engineering/Electronics, Computer, Telecommunications and Information Technology*, Hua Hin, Thailand, 16–18 May 2012.

[60] M. Vossiek, L. Wiebking, P. Gulden, J. Wieghardt, C. Hoffmann, and P. Heide. Wireless local positioning. *IEEE Microwave Magazine*, 4(4):77–86, 2003.

[61] M. P. Wylie and J. Holtzman. The non-line of sight problem in mobile location estimation. In *Proc. 5th IEEE Int. Conf. Universal Personal Communications*, pages 827–831, Cambridge, MA, USA, 1996.

[62] H. Wymeersch, J. Lien, and M. Z. Win. Cooperative localization in wireless networks. *Proceedings of the IEEE*, 97:427–450, 2009.

[63] P. Yang, W. Wu, M. Moniri, and C. C. Chibelushi. Efficient object localization using sparsely distributed passive RFID tags. *IEEE Trans. Industrial Electronics*, 60:5914–5924, 2013.

[64] S.-P. Zhang and F.-L. Yuan. RFID technique and its application in safety management system for people and vehicle in mine shafts. *Southern Metals*, 2006(1):12–14, 2006.

[65] T. Zhang, Z. Chen, Y. Ouyang, J. Hao, and Z. Xiong. An improved RFID-based locating algorithm by eliminating diversity of active tags for indoor environment. *The Computer Journal*, 52:902–909, 2009.

[66] Y. Zhang, M. G. Amin, and S. Kaushik. Localization and tracking of passive RFID tags based on direction estimation. *International Journal of Antennas and Propagation*, doi:10.1155/2007/17426, 2007.

[67] Y. Zhang, X. Li, and M. Amin. Principles and techniques of RFID positioning. In *RFID Systems: Research Trends and Challenges*, M. Bolic, D. Simplot-Ryl, and I. Stojmenovic (eds.) John. Wiley & Sons, Ltd, Chichester, pages 389–415, 2010.

[68] Y. Zhao, Y. Liu, and L. M. Ni. VIRE: Active RFID-based localization using virtual reference elimination. In *Proc. 2007 Int. Conf. Parallel Processing*, pages 357–362, Xi'an, China, 10–14 Sept. 2007.

[69] J. Zhou and J. Shi. RFID localization algorithms and applications–a review. *J. Intelligent Manufacturing*, 20:695–707, 2009.

9

Some Future Perspectives for RFID

9.1 Introduction

The RFID systems discussed in the preceding chapters are in the middle class of RFID in the sense that IC chips are integrated inside the tags, but the power needed for signal transmission in the tags of this kind of RFID should be harvested from the reader's transmitted radio waves. Therefore, the functionality and complexity of these tags are also in the middle class. Obviously, this situation can be extended to two extreme ends. The first is that no IC chips are integrated inside tags. This extension leads to chipless tags. The second is that there is enough power resided in tags that enables more powerful signal processing ability in the tags. This extension leads to active tags.

The traditional tags have been mainly developed for the purpose of inventory. Therefore, a tag can be generally read by any readers in their reading zone. With the developments of requirements, there arises the case that some tags are desired to be read only by some limited readers. This raises the concerns of privacy and security. It is not easy to deal with this issue by using traditional tags. Using active tags provides an efficient way to combat with undesired reading.

The purpose of this chapter is to give a brief view for the developments of RFID systems in the aforementioned two directions. First, we will address the issue of covert radio frequency identification (i.e. covert RFID), to which using active tags together with ultra wideband (UWB) technology is an effective approach. Then we will address the topic of chipless RFID, which often means that the RFID tags are printable and thus the price for a tag can be greatly reduced.

This chapter is organized as follows. In Section 9.2, some basic knowledge about UWB technology will be presented, which lays a foundation for covert RFID and also for some kinds of chipless tags. Section 9.3 is devoted to covert RFID, where UWB time reversal (TR) technique is used to realize the idea of covert communications between readers and tags. In Section 9.4 we discuss two kinds of chipless tags, namely time-domain reflectometry-based chipless tags and frequency-domain spectral-signature-based chipless tags. Section 9.5 concludes this chapter.

Digital Signal Processing for RFID, First Edition. Feng Zheng and Thomas Kaiser.
© 2016 John Wiley & Sons, Ltd. Published 2016 by John Wiley & Sons, Ltd.

9.2   UWB Basics

9.2.1   UWB Definition

A UWB transmission system, by strict definition [8, 35], is a radio system whose 10-dB band-width $(f_H - f_L)$ is at least 500 MHz, and whose fractional bandwidth $(f_H - f_L)/[(f_H + f_L)/2]$ is at least 20%, but many claimed UWB systems do not satisfy this condition strictly. A UWB radio system can co-exist with other kinds of narrow- and wide-band radio systems. Hence its power spectrum density (PSD) is strictly restricted by relevant regulatory authorities. A well known and widely used spectral mask specification is defined by the Federal Communications Commission of the USA [8], where it is required that a UWB device for indoor applications can use the frequency band from 3.1 to 10.6 GHz if its equivalent isotropically radiated power (EIRP) density is below the spectral mask as illustrated in Figure 9.1.

There are two basic approaches to implementing UWB radio systems. The first is the impulse radio (IR) based approach, where a pulse train, in which each pulse is very short in time domain (typically in the order of several tens of pico-seconds), is used to carry out information data. This pulse train will be directly transmitted through the antenna without any carriers. The second is the multiband based approach, where the information data is multiplexed into sub-frequency bands in the entire band from 3.1 GHz to 10.6 GHz or a part of it, each sub-band having a 528 MHz bandwidth.

In this chapter, the IR-based UWB systems will be used to illustrate the concept of covert RFID.

The popularly used waveforms for the monopulse in the IR-based UWB systems are the first and second derivatives of the Gaussian monopulse [40, 42], which are defined, respectively, by

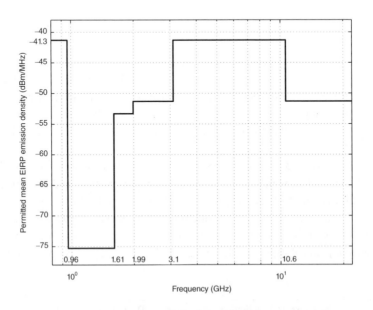

Figure 9.1   FCC spectral mask for UWB indoor applications.

$$w_1(t) = \varsigma_1 t \, \exp\left[-2\pi\left(\frac{t}{\tau_p}\right)^2\right],$$

$$w_2(t) = \varsigma_2\left[1 - 4\pi\left(\frac{t}{\tau_p}\right)^2\right] \, \exp\left[-2\pi\left(\frac{t}{\tau_p}\right)^2\right],$$

where τ_p is a parameter used to adjust the pulse width, and ς_1 and ς_2 are two constants to normalize the peak amplitudes or powers of the pulses $w_1(t)$ and $w_2(t)$, respectively. The monopulses w_1 and w_2 and and their normalized power spectrums are illustrated in Figure 9.2.

9.2.2 UWB Modulation Schemes

The information data can be embedded in either the amplitude or the position of the UWB impulse train, producing pulse amplitude modulation (PAM) and pulse position modulation (PPM), respectively. For multi-user access to the UWB channel, basically there are two kinds of accessing techniques: time hopping (TH) spread spectrum (SS) and direct sequence (DS) SS accessing. Since the transmit power is rather low, one information bit in the IR-based UWB system is generally spread over multiple monocycles to achieve a processing gain in reception.

For the TH-SS accessing, the data modulation can be generally expressed as [34, 41, 43]

$$s_k(t) = \sum_{j=-\infty}^{\infty} a_k(\lfloor j/N_f \rfloor) w(t - jT_f - c_k(j)T_c - m_\delta d_k(\lfloor j/N_f \rfloor)), \tag{9.1}$$

where $\lfloor x \rfloor$ denotes the integer floor of x, s_k is the transmitted signal for the kth user, $w(t)$ the monopulse of duration T_w, N_f the number of frames for one data symbol, T_f the frame duration, T_c the chip duration, m_δ the modulation index, $\{c_k(j)\}$ the TH coding sequence that takes values in $[0, N_c - 1]$ and is assumed to be periodic with period N_f, and a_k and d_k are the transmitted data symbols.

It is assumed that $T_f = N_c T_c$ with N_c being the number of chips in one frame duration, $T_w \ll T_c$, and $m_\delta \ll T_c$.

If $a_k \equiv 1$, equation (9.1) reduces to TH-SS-PPM. If $d_k \equiv 0$, equation (9.1) reduces to TH-SS-PAM.

For the DS-SS accessing, the data modulation is expressed as [43, 45]

$$s_k(t) = \sum_{j=-\infty}^{\infty} a_k(\lfloor j/N_f \rfloor) c_k(j) w(t - jT_f - m_\delta d_k(\lfloor j/N_f \rfloor)).$$

In UWB RFID systems, it is recommended to use TH-SS PPM due to its implementation advantage of not requiring to change, or inverse for binary modulation, the pulse amplitude [45]. Besides, the PSD of a TH-SS PPM signal does not have strong spectral lines since the time hopping information sequence smooths the PSD of the transmitted signal [43, 45].

9.2.3 UWB Channel Models

The most general UWB channel model is described by the Saleh–Valenzuela (S–V) model [33]. In the S–V model, multipath arrivals are grouped into two different categories: a cluster

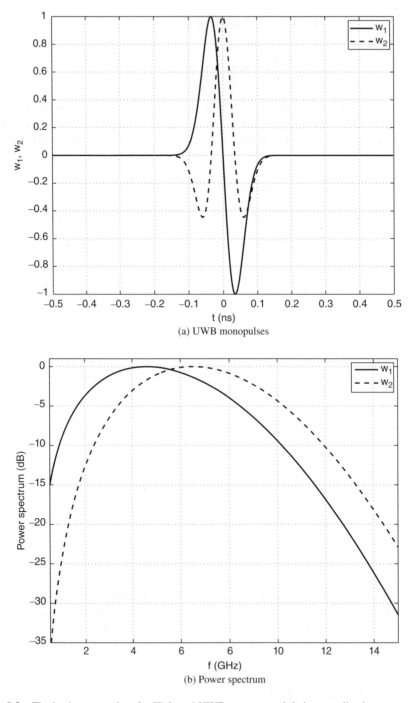

(a) UWB monopulses

(b) Power spectrum

Figure 9.2 The basic monopulses for IR-based UWB systems and their normalized power spectrums ($\tau_p = 0.1225$ ns).

arrival and a ray arrival within a cluster. The CIR of the S–V model is characterized by

$$h(t) = \sum_{k=1}^{K} \sum_{l=1}^{L} \alpha_{k,l} \delta(t - T_k - \tau_{k,l}),$$ (9.2)

where h is the CIR of a UWB channel, δ the Dirac delta function, K the total number of clusters, L the total number of rays within each cluster, $\alpha_{k,l}$ the tap gain (a real number) of the lth multipath component in the kth cluster, T_k the excess delay of the kth cluster, and $\tau_{k,l}$ the excess delay of the lth multipath component in the kth cluster relative to the cluster arrival time T_k. The parameters T_k, $\tau_{k,l}$ and $\alpha_{k,l}$ are random variables. Both the cluster and ray arrivals are modelled by Poisson processes:[1]

$$p_{T_k}(T_k|T_{k-1}) = \lambda_C \exp\ [-\lambda_C(T_k - T_{k-1})], \quad k > 1;$$ (9.3)

$$p_{\tau_{k,l}}(\tau_{k,l}|\tau_{k,l-1}) = \lambda_R \exp\ [-\lambda_R(\tau_{k,l} - \tau_{k,l-1})], \quad k > 1,$$ (9.4)

where λ_C and λ_R are the cluster arrival rate and ray arrival rate, respectively. By definition, $\tau_{k,0} = 0$ for all k. The parameter $1/\lambda_C$ is typically in the range of 10–50 ns, while $1/\lambda_R$ shows a wide variation from 0.5 ns in NLoS situations to more than 5 ns in LoS situations [21]. It is generally assumed that T_k and $\tau_{k,l}$ are mutually independent. For a given T_k and $\tau_{k,l}$, the power delay profile admits the following form:

$$\mathbb{E}[|\alpha_{k,l}|^2|\ T_k, \tau_{k,l}] = \Omega_{1,1} \exp\ \left(-\frac{T_k}{\gamma_C} - \frac{\tau_{k,l}}{\gamma_R}\right),$$ (9.5)

where $\Omega_{1,1}$ is the integrated energy of the first cluster, and γ_C and γ_R are the cluster decay time constant and ray decay time constant, respectively. The cluster decay time constant γ_C is typically around 10–30 ns, while different values (between 1 and 60 ns) have been reported for the ray decay time constant γ_R (see [21, 22] and references therein).

Channel model (9.2) is complete in characterizing the clustering phenomenon of rich mutipaths of UWB channels, but too complicated in theoretical analysis of UWB systems. Therefore, a simple UWB channel model is often adopted in the literature of UWB studies. This is the stochastic tapped delay line model [4]

$$h(t) = \sum_{l=1}^{L} \alpha_l \delta(t - (l-1)\Delta T),$$

where L is the number of multipath components, ΔT the sampling interval, and α_l the amplitude fading in the lth delay bin. The parameter α_l is a random variable. The magnitude of α_l is typically modeled by the Nakagami distribution [4] or lognormal distribution [11, 23], with its variance decaying exponentially [17]. In typical office environments, the number of taps L can be from 50 up to several hundreds, which is much greater than the multipaths of wideband communication channels in typical propagation environments.

---

[1] In the most general case, the parameters λ_C, λ_R, γ_C, γ_R and L for different clusters may have different values and hence it is legitimate to attach a subscript k to all these five parameters. For concision, we choose not to do so.

9.3   Covert RFID

RFID systems possess the functionality of both radar and communications, but in their most simple forms. Only when a tag is triggered by a reader's impinging signal, will the tag send back its signal. In this sense, an RFID system resembles a radar. Since the backscattered wave contains actively modulated information-bearing signals on the other hand, an RFID system looks like a communication system. In this section, we will present a way to integrate some more advanced communication functionalities to a tag. In this way, covert identification can be realized via covert communications between readers and tags. To this end, we need to use UWB active tags and TR technique.

The concept similar to covert RFID has been reported in several studies. The reports [2, 3] introduced the concept of intrapulse radar-embedded communications, where an incident radar waveform is converted into one of K (a given integer) communication waveforms, each of which acts as a communication symbol representing some predetermined information. In [37], a concept of radar-embedded communication combining MIMO and OFDM technology was presented.

9.3.1   Basics of Time Reversal

Time reversal is a technique to use the multipath property inherited by the channels of broadband signals to increase the SNR for the desired signal. In TR technique, the multipath channel with rich scattering is exploited by actively modulating the signal at the transmitter side using the channel information. This technique has been extensively used in acoustic, medical applications and underwater communications [9, 10, 38, 39].

The main advantages of the TR technique are:

- Temporal focusing: The received signal is compressed in time domain, yielding a Dirac-delta-like signal in the time domain.
- Spatial discrimination: The received signal is focused on the intended user at some specific position, while the users at other places, even in a very short distance from the desired user, can only receive noise-like signal if the scatterers in the propagation environments are rich enough.

An illustration for the above properties is shown in Figure 9.3, where a UWB transmitter uses the time reversed channel from the transmitter to the targeted user to modulate its transmitted symbol, while another unintended user also listens to the communications. The CIRs of the equivalent composite channels for a targeted user and for an unintended user are illustrated. The original UWB channels for the two users are generated based on model (9.2)–(9.5), where the parameters used in the model are as follows: $\lambda_C = 0.0667$ ns^{-1}, $\lambda_R = 2.1$ ns^{-1}, $\gamma_C = 14$ ns and $\gamma_R = 7.9$ ns (i.e. the so-called CM3 model in UWB literature [[17], p. 22]). The channels for the two users are independently generated. It is seen that the received energy for the targeted user is compressed into a very narrow time domain, while the received energy for the unintended user is dispersed in a wider time span.

The compressing gain is more prominent when multiple antennas are used. An experimental study is reported in [44], where each of two independent users is equipped with a transmit antenna array with four elements and the receiver has one antenna. The two users are separated

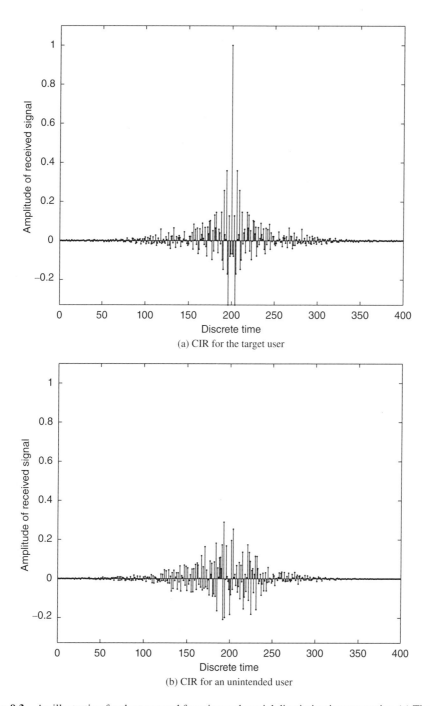

Figure 9.3 An illustration for the temporal focusing and spatial discrimination properties. (a) The CIR of the equivalent composite channel for the targeted user. (b) The CIR of the equivalent composite channel for an unintended user.

by 0.2 m. As shown in Figure 7 of [44], the signal for the targeted user is very strong at the peak instant, while the signal at the instants other than the peak instant for the targeted user and the signal from the undesired user are almost immersed in the background noise. As shown in Figure 8 of [44], the energy drop can be more than 15 dB for the aforementioned scenario. The capabilities of the TR system in temporal focusing and spatial discrimination are clearly witnessed.

It is due to the spatial discrimination property that TR is exploited for covert RFID.

The TR principle can be theoretically derived by considering the following problem: how to design a modulation scheme or a weighting filter at the transmitter to maximize the received SNR at the receiver under the condition that the CIR $\{h_0, h_1, \ldots, h_{L-1}\}$ is available at the transmitter. This problem is shown in Figure 9.4. Mathematically, we need to find an optimal weighting filter $g_0, g_1, \ldots, g_{L-1}$ such that

$$\frac{\mathbb{E}[(y(t))^2|_{t=(L-1)\Delta T}]}{\mathbb{E}(n^2)}$$

is maximized, where ΔT is the sampling interval and the coefficients $g_0, g_1, \ldots, g_{L-1}$ should satisfy

$$\sum_{l=0}^{L-1} g_l^2 = 1. \tag{9.6}$$

Since the weighting filter does not affect the power of the noise, to maximize the output SNR is equivalent to maximize $\mathbb{E}[(y(t))^2|_{t=(L-1)\Delta T}]$ under constraint (9.6). Suppose that the transmitted symbol is S. The transmitted signal after passing through the weighting filter is

$$x(t) = S \sum_{l=0}^{L-1} g_l w(t - l\Delta T), \tag{9.7}$$

where $w(t)$ is the monopulse waveform to carry a basic symbol, e.g., the monopulse waveform of the UWB transmission systems. The received signal is the convolution of $x(t)$ with the CIR $h(t)$ of the wireless channel. After sampling, the received signal at time instant $t = (L - 1)\Delta T$ reads

$$y(L - 1) = S \sum_{l=0}^{L-1} g_l h_{L-1-l} + n.$$

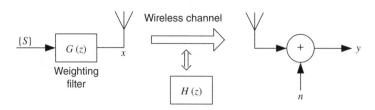

Figure 9.4 Block diagram of a TR filter at the transmitter side, where n is receiver noise and stationary, $\{S\}$ is the information symbol to be transmitted, $G(z) = \sum_{l=0}^{L-1} g_l z^{-l}$, and $H(z) = \sum_{l=0}^{L-1} h_l z^{-l}$. The composite channel $G(z)H(z)$ is called equivalent composite channel of the TR system.

Let us define

$$\mathbf{g} = [g_0 \quad g_1 \quad \cdots \quad g_{L-1}]^T, \quad \mathbf{h} = [h_{L-1} \quad h_{L-2} \quad \cdots \quad h_0]^T.$$

Then we have

$$y(L-1) = S\mathbf{g}^T\mathbf{h} + n.$$

Therefore

$$\mathbb{E}[(y(L-1))^2] = S^2(\mathbf{g}^T\mathbf{h})^2 + \sigma_n^2 \leq S^2(\mathbf{g}^T\mathbf{g})(\mathbf{h}^T\mathbf{h}) + \sigma_n^2, \qquad (9.8)$$

where σ_n^2 is the variance of the noise n. In inequality (9.8), we have used the Cauchy–Schwarz inequality and the equality holds true if and only if

$$\mathbf{g} = c\mathbf{h}, \qquad (9.9)$$

where c is a scalar constant. The scalar form of equation (9.9) reads

$$g_l = ch_{L-1-l}, \quad l = 0, 1, \ldots, L-1. \qquad (9.10)$$

Equation (9.10) shows that *the optimal filter for achieving a maximum output SNR is a time reversal filter.* The transmitted symbol can be decoded by sampling the output signal at the time instant $t = (L-1)\Delta T$.

9.3.2   Covert RFID

The approach for covert RFID that we propose here is to use UWB TR technique. The whole identification process involves three major steps. To ease citation, we formulate it as the following algorithm. The case of single-transmit and single-receive antenna (SISO) is first considered.

Algorithm 9.1 covert RFID

Step 1 The reader sends out a query signal. The query signal includes two parts: a key that associates with the reader's signature and a pilot UWB pulse train, which enables the tag to estimate the channel between the reader and tag $h(l)$, $l = 0, 1, \ldots, L$.

Step 2 Once the tag receives the query signal, the reader's key is first extracted. If the key cannot pass through the reading-enabling check, the tag keeps silent; otherwise, the tag then estimates the channel $h(l)$ and uses the time-reversed channel $h(L-l)$ to modulate the tag ID sequence and transmits it back to the reader.

Step 3 The reader detects the tag ID by detecting the peaking signal at the reader's receiver.

In Step 1 of Algorithm 9.1, the reader's key is a digital sequence that characterizes which class of tags the reader can read. In an extreme situation, each reader can be assigned an ID and the key represents the reader's ID, and in a tag's memory, all the IDs of the readers by which the tag is allowed to read are stored in the tag's memory. In this case, the reading-enabling check is very simple: the tag only needs to check whether or not the received reader's ID coincides with one of its stored IDs.

In Step 2 of Algorithm 9.1, the TR modulation can be carried out based on equations (9.7) and (9.10), and each symbol can be encoded by using equation (9.1).

From the spatial discrimination property of TR it is seen that only the desired reader can most likely detect the peaking signal induced by the backward propagation channel. It can happen that other readers might also intercept the backscattered signal when they are very near to the desired reader, but the chance is low and the situation can be controlled.

From Algorithm 9.1 we can see that the security or privacy of tag identification is implemented through two passwords. The first is the reader's key and the second is the key provided by point-to-point spatial channel, which is effective only for UWB radio.

To further increase the time focusing gain and spatial discrimination effectiveness, we can use multiple antennas at either the reader or the tag or both sides.

Consider an RFID system where the reader and tag are equipped with N_{rd} and N_{tag} antennas, respectively. To make the system expression compact, the system is represented in continuous time. The discrete-time implementation of the system can be easily constructed from its continuous-time representation. Let $h_{i_1 i_2}(t)$ be the CIR of the channel from the i_1th tag antenna to the i_2th reader antenna. The MIMO channel of the forward link from the reader to the tag is characterized by the CIR matrix $\mathbf{H}^f(t)$:

$$\mathbf{H}^f(t) = \begin{bmatrix} h_{11}(t) & h_{12}(t) & \cdots & h_{1N_{rd}}(t) \\ h_{21}(t) & h_{22}(t) & \cdots & h_{2N_{rd}}(t) \\ \vdots & \vdots & \ddots & \vdots \\ h_{N_{tag}1}(t) & h_{N_{tag}2}(t) & \cdots & h_{N_{tag}N_{rd}}(t) \end{bmatrix}.$$

Construct the TR filter matrix, denoted by $\check{\mathbf{G}}(t)$, as follows:

$$\check{\mathbf{G}}(t) = \begin{bmatrix} \frac{h_{11}(T_0-t)}{||h_{11}||} & \frac{h_{21}(T_0-t)}{||h_{21}||} & \cdots & \frac{h_{N_{tag}1}(T_0-t)}{||h_{N_{tag}1}||} \\ \frac{h_{12}(T_0-t)}{||h_{12}||} & \frac{h_{22}(T_0-t)}{||h_{22}||} & \cdots & \frac{h_{N_{tag}2}(T_0-t)}{||h_{N_{tag}2}||} \\ \vdots & \vdots & \ddots & \vdots \\ \frac{h_{1N_{rd}}(T_0-t)}{||h_{1N_{rd}}||} & \frac{h_{2N_{rd}}(T_0-t)}{||h_{2N_{rd}}||} & \cdots & \frac{h_{N_{tag}N_{rd}}(T_0-t)}{||h_{N_{tag}N_{rd}}||} \end{bmatrix}, \qquad (9.11)$$

where T_0 is the maximal length of all the CIRs $h_{i_1 i_2}(t)$, $|| \cdot ||$ denotes the two-norm in the L^2 space defined by

$$||h|| = \sqrt{\int_{-\infty}^{\infty} h^2(t)\mathrm{d}t}.$$

The TR filter matrix $\check{\mathbf{G}}(t)$ in (9.11) is based on the original CIR matrix reversed in time domain and transposed in the spatial domain.

Let $\mathbf{s}(t) := [s_1(t) \quad s_2(t) \quad \cdots \quad s_{N_{tag}}(t)]^T$ be the symbol to be backscattered to the reader. Then the transmit signal, denoted by $\mathbf{x}(t)$, at the tag antennas is

$$\mathbf{x}(t) = \check{\mathbf{G}}(t) * \mathbf{s}(t),$$

where $*$ denotes convolution in both scalar and vector forms.

The system structure is illustrated in Figure 9.5.

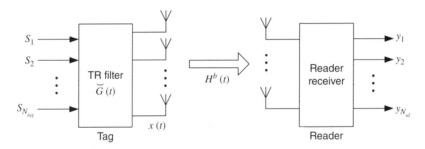

Figure 9.5 Block diagram of the TR system for RFID with multiple antennas at both readers and tags.

The full procedure of covert RFID with multiple antennas at readers and tags is the same as that described in Algorithm 9.1.

In this discussion, we have assumed that the channel reciprocity holds true, i.e., the channel matrix of the backward link from the tag to the reader is the transpose of the channel matrix of the forward link from the reader to the tag. Therefore, the channel estimation can be carried out at the tag. On the other hand, if the channel reciprocity does not hold true, the channel estimation must be made at the reader and then the reader sends the channel matrix back to the tag.

While TR communications for UWB systems have been extensively studied (see, e.g. reference [[17], Chapter 6] for several important issues), the concept of UWB-TR covert RFID is just in its infant stage and experimental validation for this concept has to be performed.

Here, we have briefly presented the application of UWB RFID tags in the covert RFID. Recently, UWB RFID tags have also found application in indoor mapping, localization and guidance [5, 30]. A general introduction about UWB RFID tags can be found in [5, 24], and a detailed design example for UWB RFID tags is provided in [28].

9.4   Chipless RFID

The mainstream type of RFID is passive RFID, in which a tag contains a silicon chip. When an RFID tag does not contain any digital chip, it becomes a chipless tag. The primary potential benefit of chipless RFID is that a chipless tag is often printable and the price for producing and packaging a tag can be down to several cents [6, 27], which would remove the last barrier for RFID technology to replace the barcode technology.

Chipless RFID tags can use either time-domain reflectometry-based or frequency-domain spectral-signature-based techniques to encode a tag's ID. In time-domain reflectometry type RFID, the interrogator sends out a pulse and listens for echoes. The timing of pulse arrivals is controlled by the tag reflectors and thus is used to encode the data. In frequency-signature type RFID, the interrogator sends out radio waves of several frequencies, a broad band pulse, or a chirp signal and monitors the frequency content of the echoes. The presence or absence of certain frequency components in the received waves encodes the data.

The main problem of chipless RFID is that the information that the tag carries cannot be changed once being manufactured. In addition, only a small number of bits can be encoded for both resonant-frequency-based chipless RFID (due to the bandwidth limitation) and time-coded-based chipless RFID (due to finite time resolution in detecting returning pulses and difficulty in integrating long transmission lines in tags).

Recent surveys for chipless RFID can be found in [25, 27]. Detection of chipless tags in some scenarios is especially challenging. Some mathematical tools for the detection are presented in reference [31]. In this section, we will use some simple examples to present the basic principles of two main kinds of chipless tags.

One of the main motivations for developing chipless tags is to reduce the cost of producing and packaging them and an effective way to achieve this goal is make tags printable. Some design methods for printable tags, including antennas and tags' other components, are well documented in [18, 19, 36].

9.4.1   Spectral-Signature-Based Chipless RFID

A well-illustrated example for spectral-signature-based chipless tag is reported by Preradovic *et al.* in [26, 27]. This tag is shown in Figure 9.6 [26].

The tag in Figure 9.6 consists of a receive antenna and transmit antenna separately, a microstrip line connecting the receive antenna and transmit antenna and six cascaded spiral resonators attached to the microstrip line. Both antennas are disc-loaded monopole antennas. In order to minimize the interference between the interrogation signal and retransmitted encoded signal (backscattering signal) at the tag, the receive and transmit antennas are cross-polarized [26]. Each spiral resonator acts as a stopband filter, with the stopping frequency being its resonant frequency.

The layout of the spiral resonator is shown in Figure 9.7 [26]. Its resonant frequency can be calculated in terms of the dimensional parameters and materials of the resonator by using professional software.

This set of cascaded spiral resonators will introduce amplitude attenuation and phase ripple to the interrogation signal at their resonant frequencies. If a reader transmits a frequency-sweeping interrogation signal, the amplitude attenuation or phase ripple can be detected at the reader from the received backscattering signal. Thus the tag ID can be

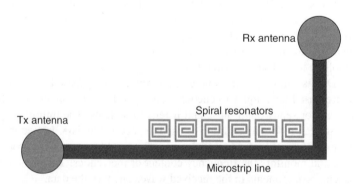

Figure 9.6   A resonator-based chipless tag. (Reproduced from Figure 3, S. Preradovic, I. Balbin, N. C. Karmakar, and G. F. Swiegers. Multiresonator-based chipless RFID system for low-cost item tracking. IEEE Trans. Microwave Theory Tech., 57:1412, 2009.)

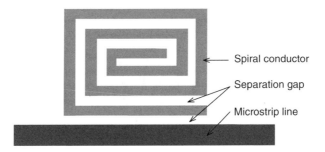

Figure 9.7 Layout of a spiral resonator. (Reproduced from Figure 5, S. Preradovic, I. Balbin, N. C. Karmakar, and G. F. Swiegers. Multiresonator-based chipless RFID system for low-cost item tracking. IEEE Trans. Microwave Theory Tech., 57:1413, 2009.)

identified. An illustration for the amplitude attenuation, characterized by the insertion loss, and the phase ripple is shown in Figure 9.8.

The tag ID in Figure 9.8 is encoded in this way: the presence of a resonator represents logic "0", and the absence of a resonator represents logic "1". The higher digits correspond to the null at lower frequencies, and the lower digits correspond to the null at higher frequencies. It is seen from Figure 9.8 that there is about 7 dB attenuation in the amplitude of tag's backscattered signal and 40° phase jump in the phase of tag's backscattered signal at the resonant frequencies. These amounts of change in amplitude and phase can be easily detected at the receiver of a reader. The tag shown in this example, even though containing only six digits, occupies a frequency band of about 500 MHz. Therefore, it is necessary to use UWB technology to realize this type of chipless RFID. In [26, 27], a 35-bit chipless tag is reported, which is designed based on the same principle as that in the aforementioned example and occupies a frequency band of about 4 GHz (from 3.1 to 7.0 GHz).

A critical problem for spectral-signature-based chipless RFID is that both attenuation in the amplitude of backscattered signal and phase jump in the phase of tag's backscattered signal are severely subjected to the effect of channel fading, and due to the chipless nature, this effect is difficult to calibrate. Therefore, the printable diode technology is proposed to implement printable RFID.

Another problem for spectral-signature-based chipless RFID is that it is difficult to align the notch frequencies with the digits of the tag's ID. This problem becomes more prominent when the resonators corresponding to the higher digits or lower digits are absent. It can be solved by introducing two reference notch frequencies: the lowest reference notch frequency corresponds to the highest digit of a tag's ID and the highest reference notch frequency corresponds to the lowest digit of a tag's ID. Notice that there should be a guard separation frequency between the lowest reference notch frequency and the resonant null frequency corresponding to the highest digit of a tag's ID. Similarly, there should be a guard separation frequency between the highest reference notch frequency and the resonant null frequency corresponding to the lowest digit of a tag's ID. These two guard separation frequencies make it possible to deal with the case of the presence or absence of the resonators with the lowest or highest null frequencies. Between the lowest and highest reference notch frequencies (by adding and subtracting the guard separation frequencies, respectively), the resonant null frequencies can be designed to be uniformly distributed. To mitigate the effect of channel fading or other kind of interference,

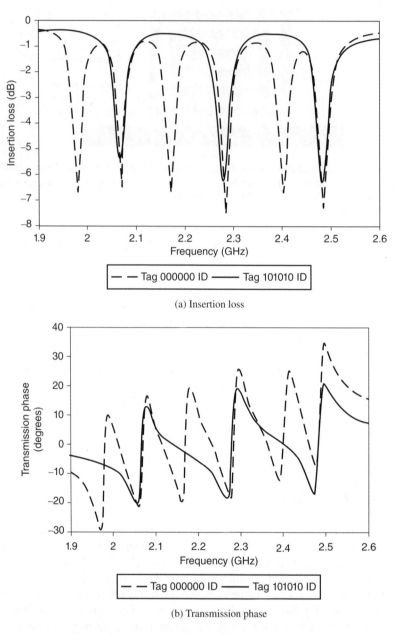

(a) Insertion loss

(b) Transmission phase

Figure 9.8 Amplitude attenuation and phase ripple of the tag's backscattered signal caused by the cascaded resonators. (From [26]. Reproduced with permission from S. Preradovic, I. Balbin, N. C. Karmakar, and G. F. Swiegers. Multiresonator-based chipless RFID system for low-cost item tracking. IEEE Trans. Microwave Theory Tech., 57:1411–1419, 2009.)

the resonators for the two reference notch frequencies can be designed to have stronger losses than other coding resonators.

9.4.2 Time-Domain Reflectometry-Based Chipless RFID

A typical time-domain reflectometry-based chipless tag is the surface acoustic wave (SAW) tag. The fundamental principle of a SAW tag is to exploit the piezoelectricity of some materials, which can transform radio waves into acoustic waves and vice versa. Such materials include lithium niobate ($LiNbO_3$), and quartz and so on. The key components of a SAW tag consist of interdigital transducer (IDT), a piezoelectric substrate and a series of parallel wave reflectors. An IDT is a device that consists of two interlocking comb-shaped arrays of metallic electrodes. These metallic electrodes are deposited on the surface of a piezoelectric substrate in the form of a periodic structure. It is the IDT that converts radio waves to acoustic waves and vice versa.

The operating principle of a SAW-tag RFID system is illustrated in Figure 9.9 [16], where the tag's antenna is directly connected to the IDT, and the wave reflectors are made of aluminum strips. When the tag's antenna receives the interrogation signal (a radio pulse), the IDT transforms it into a very narrow (on the nano-second-scale) acoustic wave. The generated acoustic wave then propagates along the surface of the substrate. For a $LiNbO_3$ substrate, the propagation speed of the acoustic wave is about 4000 m/s [13], far less than the light speed. When the surface acoustic wave meets a reflector, a part of the wave is reflected back towards the IDT, and a part of the wave propagates further to the next reflector. This process repeats till the wave reaches to the last reflector. Therefore, at the IDT, the reflected signal consists of a train of surface acoustic pulses. The time delay between any two consecutive pulses is controlled by the separation of the corresponding two neighbouring reflectors. The IDT will again transform the acoustic pulses into radio pulses and the tag's antenna then transmits them back to the reader.

A typical time response of a SAW-tag RFID system is shown in Figure 9.10 [25], where the tag contains 14 code reflectors, in which 10 are used for encoding the ID of the tag, two are used for error control, and the other two at the very first and very last are used for calibration. In Figure 9.10 we concern only relative delay in time and relative difference in amplitude of the signal. Therefore, the corresponding numerical values are not shown. The simulated numerical values can be seen from Figure 23 of [25]. From Figure 9.10 and Figure 23 of [25] it is seen that spurious pulses, besides the expected coding pulses, also appear in the back-reflected

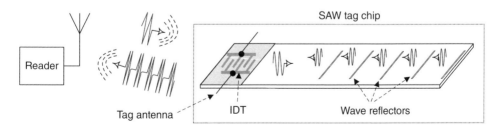

Figure 9.9 Operating principle of a SAW-tag RFID system.

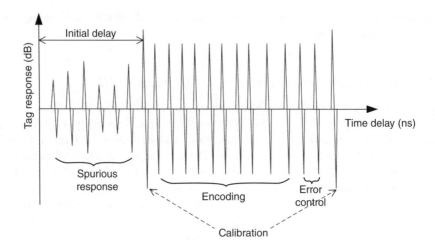

Figure 9.10   A typical time response of a SAW-tag RFID system.

signal. The spurious pulses are caused by the multiple reflections of the SAW pulses among the reflectors. Therefore, it is necessary to have two calibration reflectors in the SAW tag.

From this discussion we can see that a tag's ID can be encoded into the spatial distances of the wave reflectors and it can be identified by calculating the delays among the received pulse train at the reader. In other words, the pulse position modulation is used to encode the ID of a tag. The low propagation speed of the surface acoustic wave on the substrate makes it possible to produce resolvable radio pulses even on a substrate of very small size. For example, on a $LiNbO_3$ substrate, two neighbouring reflectors with a separation of 20 μm will produce two returned pulses with a time-delay separation of 10 ns, which can be easily read out nowadays by a moderate reader.

A key feature of SAW tags is that they do not use any DC power source: they are truly passive devices in the sense that they merely reflect back the impinging signal in an encoded form and they do not contain any power-extraction circuitry (rectifier circuitry). For an IC-based passive tag, the power of the impinging signal received at the tag's antenna must reach a critical threshold of approximately 100 μW and must be maintained at that level throughout the whole process of reading so that the power-extraction circuitry can be switched on and stay in the "on" state [16]. For SAW tags, the impinging signal just needs to be strong enough so that the amplitude of the reflected pulse train is above the noise floor (or even can be below the noise floor if multiple readings are conducted and some joint signal detection algorithms are used). In a typical SAW RFID reading range, SAW tags can operate with RF pulses at a power level of about 10 mW, while IC-based passive tags at the same distance require continuous radiation by a reader on the order of a few watts, i.e., about 100 or 1000 times higher than that for the case of SAW tags [13].

SAW tags were proposed early in the 1970s [7, 25] and have been successfully applied to a few industrial area such as the tracking of railway vehicles [25]. The number of unique codes commercially achievable at present is quite limited (on the order of 10 000 or 14 bits) [16, 25]. Therefore, the final commercial prevalence of SAW tags has not yet come. A significant step for increasing the code capacity is due to Hartmann [14], where the author proposed to combine the pulse phase information with the pulse position information to encode the ID of a tag [12, 15]. This approach can greatly increase the code capacity of a SAW tag with a given size. Now code capacity up to 256 bits is predicted and devices with 128 bits have been demonstrated [15, 25]. A detailed introduction about SAW tags is provided in [13], and a thorough review on SAW tags is given in [25].

A drawback of SAW tags is that they are not yet printable. Therefore, their price is an issue for them to become prevalent in the market.

Another type of time-domain reflectometry-based chipless tag is based on UWB technology. Its operating principle is similar to that of SAW tags. The only difference is that there is no transformation between radio waves and mechanical waves at the tag antennas of time-domain reflectometry-based UWB tags. The UWB pulses propagate directly from the tag's antenna to the reflectors (discontinuous diodes) along the transmission line. A design example for UWB chipless tags is presented in [29], where a delay line of a specifically designed length is connected to the UWB tag antenna, and delay changes appearing in the backscattered UWB pulses are used to encode the ID of the tags. It is reported that a delay change of 150 ps can be detected in the systems developed in [29] and up to 18 states can be coded.

9.5 Concluding Remarks

It is believed that RFID, in the not-too-far future, will replace barcodes in low-end applications and play a crucial role in the Internet of Things (IoT) in high-end applications/services.

The replacement of barcode with RFID has been already heralded by advancement of the technology of printable tags.

In the future IoT, one of the cornerstones is low-cost and ubiquitous sensors and object's ID tracking devices. RFID tags are a natural candidate for this kind of sensors and ID tracking devices [1, 18, 32]. Besides, future intelligent tags can even possess moderate computational ability and thus can create meshes of self-organizing networks and merge them further into pervasive IoT.

As pointed out in [20], "the specific needs of RFID are not restricted to the transfer of data only but most importantly extend to the capture, management, and publication of persistent metadata" in the future. The vision for the new applications of RFID, foreseen in the report [20] from a few years ago, still gives us a vivid image of the future potential of RFID.

The new horizon for the applications of RFID will create new challenges for the signal/data processing of RFID. For example, cloud computing might be an indispensable tool for relevant data management and processing.

References

[1] R. Bhattacharyya, C. Floerkemeier, and S. Sarma. Low-cost, ubiquitous RFID-tag-antenna-based sensing. *Proc. IEEE*, 98:1593–1600, 2010.

[2] S. D. Blunt and P. Yatham. Waveform design for radar-embedded communications. In *Proc. Int. Waveform Diversity and Design Conf.*, pages 214–218, Pisa, Italy, 4–8 Jun. 2007.

[3] S. D. Blunt, P. Yatham, and J. Stiles. Intrapulse radar-embedded communications. *IEEE Trans. Aerosp. Electron. Syst.*, 46:1185–1200, 2010.

[4] D. Cassioli, M. Z. Win, and A. F. Molisch. The ultra-wide bandwidth indoor channel: From statistical model to simulations. *IEEE J. Sel. Areas Commun.*, 20:1247–1257, 2002.

[5] D. Dardari, R. D'Errico, C. Roblin, A. Sibille, and M. Z. Win. Ultrawide bandwidth RFID: The next generation? *Proc. IEEE*, 98:1570–1582, 2010.

[6] R. Das. Chipless RFID — the end game. In *IDTechEx, Cambridge, MA, Internet article*, Feb. 2006. Available: www.idtechex.com/research/articles/chipless_rfid_the_end_game_00000435.asp.

[7] D. E. N. Davies, M. J. Withers, and R. P. Claydon. Passive coded transponder using an acoustic-surface-wave delay line. *Electron. Lett.*, 11(8):163–164, 1975.

[8] FCC docket 02-48. Revision of Part 15 of the Commission's Rules Regarding Ultra-Wideband Transmission Systems. Released on 22 April 2002.

[9] G. F. Edelmann, H. C. Song, S. Kim, W. S. Hodgkiss, W. A. Kuperman, and T. Akal. Underwater acoustic communications using time reversal. *IEEE J. Oceanic Engineering*, 30:852–864, 2005.

[10] M. Fink. Time reversed acoustics. *Physics Today*, 50:34–40, 1997.

[11] J. R. Foerster, M. Pendergrass, and A. F. Molisch. A channel model for ultrawideband indoor communication. Available: www.ieee802.org/15/pub/2003/Mar03/02490r1P802-15_SG3a-Channel-Modeling-Subcommittee-Report-Final.zip, see also www.merl.com/reports/docs/TR2003-73.pdf.

[12] S. Harma, W. G. Arthur, C. S. Hartmann, R. G. Maev, and V. P. Plessky. Inline SAW RFID tag using time position and phase encoding. *IEEE Trans. Ultrason., Ferroelect., Freq. Contr.*, 55:1840–1846, 2008.

[13] S. Harma and V. P. Plessky. Surface acoustic wave RFID tags. In *Development and Implementation of RFID Technology (C. Turcu ed.)*. InTech, Rijeka, pages 145–158, 2009. Available from www.intechopen.com/books/development_and_implementation_of_rfid_technology/surface_acoustic_wave_rfid_tags.

[14] C. S. Hartmann. A global SAW ID tag with large data capacity. In *Proc. IEEE Ultrasonics Symp.*, pages 65–69, Munich, Germany, 8–11 Oct. 2002.

[15] C. S. Hartmann, P. Brown, and J. Bellamy. Design of global SAW RFID tag. In *Proc. 2nd Int. Symp. Acoustic Wave Devices for Future Mobile Communications Systems*, pages 15–19, Chiba, Japan, 3–5 Mar. 2004.

[16] RFSAW Inc. The global SAW tag – a new technical approach to RFID. 2004. Available: www.rfsaw.com/Documents/Paper%20-%20The%20Global%20SAW%20Tag%20-%20a%20New%20Technical%20Approach%20to%20RFID.pdf.

[17] T. Kaiser and F. Zheng. Ultra Wideband Systems with MIMO. John Wiley & Sons, Ltd, Chichester, 2010.

[18] V. Lakafosis, A. Rida, R. Vyas, L. Yang, S. Nikolaou, and M. M. Tentzeris. Progress towards the first wireless sensor networks consisting of inkjet-printed, paper-based RFID-enabled sensor tags. *Proc. IEEE*, 98:1601–1609, 2010.

[19] S. L. Merilampi, T. Bjorninen, A. Vuorimaki, L. Ukkonen, P. Ruuskanen, and L. Sydanheimo. The effect of conductive ink layer thickness on the functioning of printed UHF RFID antennas. *Proc. IEEE*, 98:1610–1619, 2010.

[20] K. Michael, G. Roussos, G. Q. Huang, A. Chattopadhyay, R. Gadh, B. S. Prabhu, and P. Chu. Planetary-scale RFID services in an age of uberveillance. *Proc. IEEE*, 98:1663–1671, 2010.

[21] A. F. Molisch. Ultrawideband propagation channels – theory, measurement, and modeling. *IEEE Trans. Veh. Technol.*, 54(5):1528–1545, 2005.

[22] A. F. Molisch. Ultra-wide-band propagation channels. *Proceedings of the IEEE*, 97:353–371, 2009.

[23] A. F. Molisch, J. R. Foerster, and M. Pendergrass. Channel models for ultrawideband personal area networks. *IEEE Commun. Mag.*, 10(6):14–21, 2003.

[24] F. Nekoogar and F. Dowla. *Ultra-Wideband Radio Frequency Identification Systems*. Springer, New York, 2011.

[25] V. P. Plessky and L. M. Reindl. Review on SAW RFID tags. *IEEE Trans. Ultrason., Ferroelect., Freq. Contr.*, 57:654–668, 2010.

[26] S. Preradovic, I. Balbin, N. C. Karmakar, and G. F. Swiegers. Multiresonator-based chipless RFID system for low-cost item tracking. *IEEE Trans. Microwave Theory Tech.*, 57:1411–1419, 2009.

[27] S. Preradovic and C. Karmakar. Chipless RFID: Bar code of the future. *IEEE Microwave Magazine*, 11(7):87–97, 2010.

[28] A. Ramos, A. Lazaro, and D. Girbau. Semi-passive time-domain UWB RFID system. *IEEE Trans. Microwave Theory Tech.*, 61:1700–1708, 2013.

[29] A. Ramos, A. Lazaro, D. Girbau, and R. Villarino. Time-domain measurement of time-coded UWB chipless RFID tags. *Progress In Electromagnetics Research*, 116:313–331, 2011.

[30] A. Ramos, A. Lazaro, R. Villarino, and D. Girbau. Time-domain UWB RFID tags for smart floor applications. In *2014 IEEE RFID Technology and Applications Conf.*, pages 165–169, Tampere, Finland, 8–9 Sept. 2014.

[31] R. Rezaiesarlak and M. Manteghi. *Chipless RFID: Design Procedure and Detection Techniques*. Springer, Cham, 2015.

[32] S. Roy, V. Jandhyala, J. R. Smith, D. J. Wetherall, B. P. Otis, R. Chakraborty, M. Buettner, D. J. Yeager, Y.-C. Ko, and A. P. Sample. RFID: From supply chains to sensor nets. *Proc. IEEE*, 98:1583–1592, 2010.

[33] A. Saleh and R. A. Valenzuela. A statistical model for indoor multipath propagation. *IEEE J. Sel. Areas Commun.*, 5:128–137, 1987.

[34] R. A. Scholtz. Multiple access with time-hopping impulse modulation. In *Proc. MILCOM Conf.*, pages 447–450, Boston, MA, 1993.

[35] R. A. Scholtz, D. M. Pozar, and W. Namgoong. Ultra wideband radio. *EURASIP J. Applied Signal Processing*, 2005:252–272, 2005.

[36] B. Shao. Fully printed chipless RFID tags towards item-level tracking applications. *Doctoral Thesis*, Royal Institute of Technology, Stockholm, Sweden, 2014.

[37] X. Shi and X. L. Peng. Radar embedded communication technology study. In *2011 IEEE CIE Int. Conf. Radar*, pages 1000–1003, Chengdu, China, 24–27 Oct. 2011.

[38] H. C. Song, W. S. Hodgkiss, W. A. Kuperman, M. Stevenson, and T. Akal. Improvement of time-reversal communications using adaptive channel equalizers. *IEEE J. Oceanic Engineering*, 31:487–495, 2006.

[39] H. C. Song, P. Roux, W. S. Hodgkiss, W. A. Kuperman, T. Akal, and M. Stevenson. Multiple-input-multiple-output coherent time reversal communications in a shallow-water acoustic channel. *IEEE J. Oceanic Engineering*, 31:170–176, 2006.

[40] M. Z. Win and R. A. Scholtz. Impulse radio: How it works. *IEEE Commun. Lett.*, 2(2):36–38, 1998.

[41] M. Z. Win and R. A. Scholtz. Ultra-wide bandwidth time-hopping spread-spectrum impulse radio for wireless multiple-access communications. *IEEE Trans. Commun.*, 48:679–691, 2000.

[42] M. Z. Win and R. A. Scholtz. Characterization of ultra-wide bandwidth wireless indoor channels: A communication-theoretic view. *IEEE J. Sel. Areas Commun.*, 20:1613–1627, 2002.

[43] L. Yang and G. B. Giannakis. Ultra-wideband communications: An idea whose time has come. *IEEE Signal Processing Mag.*, 21(6):27–54, 2004.

[44] C. Zhou, N. Guo, and R. C. Qiu. Time reversed ultra-wideband (UWB) multiple-input multiple-output (MIMO) based on measured spatial channels. *IEEE Trans. Veh. Technol.*, 58:2884–2898, 2009.

[45] W. Zhuang, X. Shen, and Q. Bi. Ultra-wideband wireless communications. *Wireless Communications and Mobile Computing*, 3:663–685, 2003.

Index

Digital Signal Processing for RFID, First Edition. Feng Zheng and Thomas Kaiser.
© 2016 John Wiley & Sons, Ltd. Published 2016 by John Wiley & Sons, Ltd.